£40.00

*PALAEOHYDROLOGY IN*
*PRACTICE*

INTERNATIONAL GEOLOGICAL CORRELATION
PROGRAMME

PROJECT 158

# PALAEOHYDROLOGY IN PRACTICE

## A River Basin Analysis

Edited by

**K. J. Gregory**
*Department of Geography, University of Southampton*

**J. Lewin**
*Department of Geography, University College of Wales, Aberystwyth*

*and*

**J. B. Thornes**
*Department of Geography, University of Bristol*

*A Wiley–Interscience Publication*

**JOHN WILEY & SONS**
Chichester · New York · Brisbane · Toronto · Singapore

Copyright © 1987 by John Wiley & Sons Ltd.

*Library of Congress Cataloging-in-Publication Data:*
Palaeohydrology in practice.
  1. Palaeohydrology—Severn River Watershed (Wales and England)   I. Gregory, K. J. (Kenneth John)
II. Lewin, J., Ph.D.   III. Thornes, John B.
QE39.5.P27P36   1987 / 551.48′09424   87-10460

ISBN 0 471 91618 8

*British Library Cataloguing in Publication Data:*
Palaeohydrology in practice: a river basin
  analysis.—(International Geological
  Correlation Programme; project 158).
  1. Paleohydrology
  I. Gregory, K. J.   II. Lewin, J. (John)
  III. Thornes, John B   IV. Series
  551.48   QE39.5.P27

ISBN 0 471 91618 8

Phototypeset by Input Typesetting Ltd, London SW19
Printed and bound in Great Britain

# List of Contributors

K. E. Barber — Department of Geography, University of Southampton, SO9 5NH

A. G. Brown — Department of Geography, University of Leicester, Leicester, LE1 7RH

G. R. Coope — Department of Geography, University of Birmingham, Birmingham, B15 2TT

M. R. Dawson — Department of Geography, University College of Wales, Llandinam Building, Penglais, Aberystwyth SY23 3DB

V. Gardiner — Department of Geography, University of Leicester, Leicester, LE1 7RH

A. J. Gerrard — Department of Geography, University of Birmingham, Birmingham, B15 2TT

K. J. Gregory — Department of Geography, University of Southampton, Southampton, SO9 5NH

G. Higgs — Department of Geography, University College of Wales, Llandinam Building, Penglais, Aberystwyth SY23 3DB

D. M. Lawler — Department of Geography, University of Birmingham, Birmingham, B15 2TT

J. Lewin — Department of Geography, University College of Wales, Llandinam Building, Penglais, Aberystwyth SY23 3DB

S. Limbrey — Department of Ancient History and Archaeology, University of Birmingham, Birmingham, B15 2TT

D. J. Mitchell — School of Applied Sciences, Wolverhampton Polytechnic, Wolverhampton, WV1 1LY

L. Starkel — Institute of Geography, PAN, ul Jana, 31–512 Krakow, Poland

J. B. Thornes — Department of Geography, University of Bristol, Bristol BS8 1SS

S. N. Twigger          Department    of    Geography,    University    of
                       Southampton, Southampton, SO9 5NH

T. R. Wood             Severn Trent Water Authority, Abelson House, 2297
                       Coventry Road, Sheldon, Birmingham, B26 3PU

# Contents

# Foreword

Sir Malcolm Brown FRS
Chairman of British National Committee for the IGCP

The International Geological Correlation Programme has been both successful and productive. In the IGCP catalogue 1983–85 published by UNESCO in 1986 there are 3425 publications listed as associated with 73 IGCP projects and 117 publications have arisen from this IGCP Project 158. In previous publications relating to the period since 1973 there have been 17,785 publications listed as associated with more than 60 IGCP projects. Although the title, International Geological Correlation Programme, might immediately suggest a research programme rooted in the geological past, one of the principal goals of the programme, as explained in Chapter 1 of this book, is to encourage international research on geological problems related to the identification and assessment of natural resources and the improvement of man's environment. This volume, concerned with IGCP Project 158 on Palaeohydrology, relates to water as a natural resource. It is the interaction between that resource and environment that is the subject of study and the interaction can usefully provide clues as to how the environment may be improved. At a time when it is acknowledged that global environments may be changing as a result of the impact of human activity, it is important that we are able to learn from the past. We should be able to utilize knowledge about environmental evolution and to extend models of environmental change in a way that could colour our expectation of the environmental future.

Through the Royal Society, the British National Committee for the IGCP has published annual reports since 1975 and has encouraged British participation in a range of IGCP projects, and currently this extends to 34 projects. It is always encouraging to see the presentation of the final results of individual projects in a form in which they can be widely disseminated to many countries of the world and to experts in a range of related disciplines.

I am pleased to write the foreword to this volume which contains the

British contribution to a specific IGCP project, because the results of palaeo-hydrological investigations may assist our future understanding of hydrology. The British contribution to this project has involved several university Departments of Geography and has been organized using the expertise of more than 20 British scientists in a range of disciplines. Much of the research has concentrated on the drainage basins of the Severn and the Wye. Although not all of the researchers who have been involved have contributed to this volume, I know that the National Correspondent, Professor K. J. Gregory, who has organized the British contribution, is extremely grateful to all of the researchers who have participated during the past 10 years. In particular it is important to acknowledge the contribution of Professor F. W. Shotton, FRS. He has undertaken research over a number of years in areas of the Severn Basin and his contribution has been extremely important to the progress of research undertaken as part of this IGCP Project. It is upon such research foundations that IGCP projects have been able to develop and prosper.

# *Acknowledgements*

The editors are very grateful to all those scientists who have contributed during the decade 1977–1987 to the research in the Severn Basin and particularly to those who have contributed papers to this volume. Some of the papers were initially presented at a one-day meeting held during the Section E programme of the British Association in Bristol in September 1986. We are also most grateful for secretarial help to June Gandhi in Southampton, Sarah Howell in Bristol and Linda James in Aberystwyth. Mrs Jane Lewin designed the excellent cover for this book and we are most grateful to her and hope that the contents are equal to the quality of the cover. Mrs Rosemary Thornes was of invaluable help during the final stages of the checking of the proofs and we are also most grateful to Mr C. T. Hill in Southampton who has been largely responsible for production of the index. Finally we would like to express our sincere thanks to Sir Malcolm Brown FRS, who as Chairman of the British National Committee for IGCP kindly contributed a foreword to this volume.

Ken J. Gregory
John Lewin
John B. Thornes

Palaeohydrology in Practice
Edited by K. J. Gregory, J. Lewin and J. B. Thornes
© 1987 John Wiley & Sons Ltd

# 1

# *Introduction*

## K. J. GREGORY

*Department of Geography, University of Southampton*

Nothing puzzles me more than time and space;
and yet nothing troubles me less,
as I never think about them.
Letter from Charles Lamb (1775–1834) to T. Manning, 2 January 1810

Studies of environmental change have not always been able to relate to time
and space in the most appropriate ways. Recent progress has been substantial
as a result of new techniques, conceptual advances and more interdisciplinary
research. The International Geological Correlation Programme (IGCP) has
fostered such interdisciplinary research and has begun to further develop
consideration of time and space. This volume is a product of research for
one of the International Geological Correlation Programme research
investigations.

A proposal for establishing the International Correlation Programme
(IGCP) was formulated in October 1967 at a meeting in Prague, Czechoslovakia, by representatives of the International Union of Geological Sciences
(IUGS) and the United Nations Educational Scientific and Cultural Organization (UNESCO). They jointly recognized the need for greater international
effort to solve some of the fundamental geological problems with which the
IUGS is concerned (UNESCO, 1983). The IGCP Programme was formally
initiated in May 1973 as a joint IUGS/UNESCO activity by the first meeting
of the IGCP Board and at that time 20 projects already established by IUGS
were provisionally accepted into the programme. The principal goal of the
programme was to encourage international research on geological problems
related to the identification and assessment of natural resources and the
improvement of man's environment. The scientific achievements of the IGCP
Programme (1978–82) were classified in a 1983 publication (UNESCO 1983),
into three categories concerned with:

1

1. projects related to better knowledge of geological processes, correlations and concepts;
2. projects related to more effective protection and utilization of the environment;
3. projects related to more efficient identification and assessment of resources.

Research projects in these categories include fundamental investigations of the processes that have governed the development of the earth's structure as related to the solution of practical problems concerned with the distribution of natural resources or the impact of geological factors on the human environment (Reinemund and Watson, 1983). An important feature of all IGCP projects is the way in which objective non-governmental scientific participation can utilize results of research on relevant problems so that international research activity can be further developed and strengthened.

IGCP Project 158 was envisaged as a project related to the second category, the effective protection and utilization of the environment. The project utilizes a multidisciplinary research approach to environmental change with special attention to changes in the hydrological regime of the temperate zone (35–70° latitude) which have resulted from climatic fluctuations and human impact. The project has been organized in two parts. Subproject A is devoted to fluvial environments and subproject B to lake and mire environments (Starkel and Berglund, 1983). Although the two subprojects must be intergrated at the conclusion of the 10-year project (1977–86) nevertheless the two parts could proceed somewhat independently. Therefore the fluvial subproject (158A) has focused upon the reconstruction of hydrological and climatic character of fluvial environments in the past 15,000 years and has explored the reasons for changes which have occurred. Subproject 158B concentrated upon interpretation of information from lake and mire sites and from complete sequences of organic deposits, and the intention was that the results should be considered in relation to palaeoecological units or type regions devised by national working groups.

Progress in this international investigation has been achieved by research coordinated nationally for each of the subprojects, and in addition there have been a series of international meetings which have enabled research problems and results to be discussed and evaluated during the 10-year programme. The success of the programme owes much to the originators of the project, Professors L. Starkel and B. Berglund, who have initiated a considerable number of research publications which have described the methods employed and the results achieved during the project.

For subproject 158A a guide to the methods which could be used was published at an early stage (Starkel and Thornes, 1981), and for subproject 158B an extremely thorough review of the techniques available for investigation of lakes and mires has been provided by Berglund (1986). In addition there have been a number of important publications arising from IGCP project 158 and these have included a review of the background to palaeohydrology (Gregory 1983a), and papers produced at international meetings (e.g. Vasari, Saarnisto and Seppala, 1979; Kozarski, 1983; Starkel, Gregory and Gardiner, 1985). IGCP project 158 has therefore already provided opportunities for scientists from several disciplines in a number of contributing countries to elucidate the palaeohydrology of the temperate zone in relation to the controls responsible for the environmental changes which have occurred. In addition there have been other related IGCP projects including project 146 on river flood and lake level changes (1976–84) and project 184 palaeohydrology of low-latitude deserts (1981–85). Together with earlier IGCP projects, these have collectively been able to catalyse international research endeavour.

This interdisciplinary and international effort, under the heading of the IGCP programme, has taken place at a time when there has been considerable progress in understanding environmental change. Such progress has taken place with the recognition of new fields of study. Palaeohydrology has existed as an explicitly defined field of study only since 1954 when it was recognized by Leopold and Miller (1954). Subsequently palaeohydrology has attracted greater interest in the 1970s and 1980s with contributions from several disciplines (Gregory, 1983b). Increased interdisciplinary cooperation has become essential as a range of new techniques has become available including greater facility for chronometric dating, a greater capacity for modelling environmental systems with large elaborate computer models, and a greater knowledge of environmental processes. Such development created the need for cross-discipline and international collaboration to ensure the most effective correlation of research results. Palaeohydrology has been viewed retrospectively, relying upon a greater understanding of contemporary hydrology. Adoption of such a perspective can be the basis for envisaging changes that may occur in the future. Just as Butzer (1980) used late Quaternary world climatic patterns as an analogue for patterns possible in a future higher $CO_2$ world, so a knowledge of the palaeohydrology of the past may be useful in outlining future hydrology. An approach employing retrodiction can be reversed to contribute to predictions.

This volume reflects research undertaken in Britain since 1977 to provide a contribution to IGCP project 158. However, the book is deliberately organized so that it does not simply present results of interest to, and arising from, the Severn Basin which is the focus for British research. The book has been devised to show how palaeohydrologic investigations may be undertaken in a single basin and how they may contribute towards more integrated understanding. This may help towards a more integrated view of environmental change in sympathy with the more holistic views of science, advocated for example by Prigogine and Stengers (1984). To provide a context it is necessary to introduce palaeohydrology in Britain (section 1.1) and then to outline practice in Britain (1.2).

## 1.1  PALAEOHYDROLOGY IN BRITAIN

It is comparatively easy to attempt a review of palaeohydrology in Britain, because prior to the inception of the IGCP project in the 1970s there was little coordinated research on palaeohydrology. This situation arose because although palaeohydrology can be appropriate for geological timescales (e.g. Schumm, 1968), it tends to be most useful in investigations of the postglacial timescale. In a review of progress in geomorphology (Brown and Waters, 1974), there was no reference to palaeohydrology although a section on palaeogeomorphology included papers which emphasized the Cenozoic contribution to landscape morphogenesis. This reflected the emphasis in Britain until the 1970s upon denudation chronology—an approach which paid comparatively little attention to Holocene changes. Although the emphasis in denudation chronology research was often pre-Holocene and in many studies pre-Quaternary, it is only comparatively recently that historical geomorphology in Britain has been reviewed with a retrospective approach (Brown, 1979). It was therefore suggested in a review of fluvial geomorphology in Britain (Gregory, 1978) that palaeohydrology had not been fully explored and offered further opportunities for coordinated research. In an infrequently cited paper D. Kimball (1948) had proposed that historical geology was made up of stratigraphy which deals with what is there and denudation chronology which is concerned with what is not. He argued that denudation chronology was concerned with the remnants of past fluviatile processes. In discussing the approach that could be taken to such processes he employed flow equations including the Manning equation as a basis for showing how changing discharges could lead to alterations in river channel form. However, such approaches were not adopted in the subsequent two decades of research which were dominated instead by concentration on the denudation chronology of different parts of Britain.

Relaxation of the dominance of denudation chronology in British geomorphology came about in at least three main ways. *First* was the contribution made to research overseas because in such investigations where a historical theme was adopted it was often employed in a rather different manner. Thus research undertaken by Vita-Finzi (1969) on Mediterranean valleys involved an historical approach applied to the most recent parts of geological time and included the impact of human activity. A *second* way in which the dominance of denudation chronology was relaxed appeared as investigations of Quaternary environmental change assumed greater significance. Such investigations were greatly enhanced with the development of Quaternary studies in the field of Quaternary geology (e.g. West, 1968). Also important were research developments in other fields of Quaternary research including progress in sedimentology (e.g. Allen, 1977), in palaeoecology (Birks and Birks, 1980), in palaeoclimatology and climatic change (Lamb, 1977) and also in the developing field of geoarchaeology (Davidson and Shackley, 1976). In these fields of study there was enhanced progress in research on environmental change although such research in Britain became integrated under the organization of the Quaternary Research Association (QRA), and somewhat separate from geomorphological research which in Britain was concentrated in the British Geomorphological Research Group (BGRG). Also important in relation to Quaternary research were papers that reported research investigations in particular areas. An excellent example is the paper reporting the detailed analysis of the development of an escarpment dry valley at Brook in Kent (Kerney, Brown and Chandler, 1964). This very successful analysis, although not referring explicitly to palaeohydrology, produced an interpretation of a palaeohydrological nature. It was shown that a chalk dry valley could be dated to a period of 500 years between 10,800 and 10,300 BP and that it originated as a result of niveo-fluvial processes.

A *third* way in which alternatives to denudation chronology emerged in Britain was with the greater investigation of geomorphological processes. Emphasis upon processes has been reflected in a number of publications by the British Geomorphological Research Group including those concerned with hillslopes (Brunsden, 1971) and with fluvial processes (Gregory and Walling, 1974) and derived from an international meeting in 1976 (Embleton, Brunsden and Jones, 1978). One of the major areas to attract process studies was the domain of fluvial processes. In a review of British fluvial geomorphology in the 1980s, Nanson (1986) has argued that 'from a condition of very little significant research in the 1950s and early 1960s, fluvial studies has grown to be probably the most prolific branch of modern British geomorphology.' Nanson went on to argue that unbridled empiricism still prevails in British geomorphology and that greater acknowledgement could be given to the work of sedimentologists and to concern with bedrock channels, major extensive depositional environments, and fluvial glacial environments. His review nonetheless identifies the significance of developments in fluvial geomorphology.

By 1980 denudation chronology and historical geomorphology had become one of several approaches in the spectrum of geomorphological endeavour in Britain. In a volume reviewing progress in research on denudation chronology (Jones, 1980) since the publication of *Structure Surface and Drainage in South East England* (Wooldridge and Linton, 1939), there was a contribution that explored palaeohydrological reconstruction using morphological and sediment information from the Kennet Valley (Cheetham, 1980). The inception of the IGCP project on palaeohydrology was probably very timely because it proffered a means of reconciling the approach of historical geomorphology with the advances achieved in Quaternary environmental change and benefited from the greater knowledge of geomorphological processes. The time was ripe to envisage a closer integration of these three themes and such integration is imperative for palaeohydrological investigations to be effective.

In addition, however, research in a number of specific areas had paved the way for a more palaeohydrological focus. Such research included greater attention accorded to the investigation of terrace deposits and to sequences of river terrace development. Thus late Pleistocene terrace deposits at Beckford were analysed by Briggs, Coope and Gilbertson (1975) and processes and environments in the Upper Thames Valley were investigated by Briggs and Gilbertson (1980) because new sites had become available which provided evidence of the character of the environment during both glacial and interglacial stages of the middle and upper Pleistocene. In the upper Thames Valley it was suggested that interglacial phases were characterized by deposition in meandering channel systems and that the question of when fluvial systems changed from braided systems, associated with terrace aggradation, to meandering forms typical of the interglacial periods was worthy of further research (Briggs and Gilbertson, 1980).

A further strand of research was a growing interest in Holocene alluviation. Although Crampton (1968) had identified aggradation in valleys of southeast Wales which he suggested was associated with Iron Age deforestation, it was not until the late 1970s that such Holocene alluviation was analysed in several areas. The age and significance of alluvium in the Windrush Valley, Oxford (Hazeldine and Jarvis, 1979) and the products of Holocene alluviation in relation to hydrology of the Upper Thames basin (Robinson and Lambrick, 1984) are two examples of the way in which Holocene sediments were identified and dated. In the North York Moors it was shown that a phase of alluvial filling could be dated as post-6270BP (Richards, 1981) in an area where there is other evidence of Mesolithic and Neolithic deforestation. In the Howgill Fells it was possible to identify stages of Holocene landform development and to date these stages and to associate them with soil and vegetation change (Harvey *et al.*, 1981). Recent floodplain sedimentation can be illuminated by analysis of sediment heavy metal chemistry and in the upper Axe Valley, Macklin (1985) was able to show that floodplain sedimentation rates ranged from 8.8 to 15 mm $a^{-1}$ during mining activity compared with 2.4–4.6 mm $a^{-1}$ when mining activity ceased. Archaeological dating also can be of great value in dating Holocene alluviation, and in the Upper Thames basin it was shown that flooding and alluviation were largely restricted to the past 3000 years (Robinson and Lambrick, 1984).

Such investigations have also been supported by research on changes of river channels. It has now been demonstrated that there have been significant changes of river channels in Britain during the period of the Holocene and particularly as a consequence of changes induced by human activity (e.g. Gregory, 1977). Thus the recent history of channel patterns was analysed for the lower part of the River Spey (Lewin and Weir, 1977) and in central Wales a number of important papers have been produced describing changes of river channel patterns (e.g. Lewin, 1983) and demonstrating the relationship of channel plan form change and flood plain sedimentation (Lewin, 1978). In addition there have been some research contributions which have been explicitly palaeohydrological. Notable amongst these has been the application of palaeohydrological methods to late glacial sandur deposits of northeast Scotland and this benefited from comparison of contemporary systems with fossil ones (Maizels, 1983). Arising from studies of river channel patterns it was suggested that palaeohydrology was a potentially fruitful avenue of research (Lewin, 1977) and subsequently palaeohydrological research has been undertaken in specific areas. Thus Burrin (1983) investigated the palaeohydrology of the Cuckmere Valley in south-east England and Cheetham (1980) analysed the ways in which changes in terrace deposits and former channel patterns could be used to interpret former stages of the Kennet river valley during the late Quaternary.

Such developments in specific areas have all tended to encourage a more palaeohydrological and interdisciplinary approach. However, a continuing theme that has been evident in research in Britain for more than 30 years was inspired by Professor G. H. Dury who analysed underfit streams, initially working in the Cotswolds and subsequently in many other parts of the world. This research gave rise to a number of extremely important papers which have advanced our knowledge of stream and valley development and have also provided a context for palaeohydrological endeavour. It is perhaps in the papers by Dury that we have the earliest and most convincing demonstration of the advantages to be gained from retrodiction and from the interpretation of the past in the light of a greater understanding of the present (e.g. Dury, 1977, 1983, 1985).

It would be unfortunate to imply that all investigations which provide the context for palaeohydrological approaches are geomorphological. Particularly important has been the way in which climatology and hydrology can now provide a background for palaeohydrological analysis. Thus the water balance of Britain from 50,000 BP to the present was reviewed by Lockwood (1979) and in the context of the River Severn particularly useful was the attempt to search for trends in flow records over the last 200 years (Rodda, Sheckley and Tan, 1978).

## 1.2   PRACTICE IN BRITAIN

Against this research background it was necessary to select a major drainage basin in the UK for research to contribute to the IGCP project 158A. The choice of such a basin was governed by the need to include a range of relief; to avoid too substantial a human impact by way of river regulation; to have a large basin which could be the basis for comparison with the even larger basins being researched in Europe; and also to employ an area in which some research had already been done. In the British Isles the choice rapidly devolves upon the Thames, the Trent, the Severn and the Shannon. The latter is not easily accessible for frequent interdisciplinary work by scientists from a range of institutions; the Thames is affected by human activity in its lower regions although work has been undertaken there by a number of recent researchers (e.g. Nunn, 1983); and the Trent has comparatively few tributaries from areas of highest relief.

By a process of elimination the Severn and the Wye were selected. The first step was to assemble a list of, and to establish contact with, all scientists currently undertaking research in the Severn Basin. This quickly indicated that there were such scientists in the fields of Archaeology, Ecology, Engineering, Geography, Geology, Hydrology and Soil Science. Having established a list of people who were involved in research in the basin and obtained the agreement of participants it then transpired that at least two meetings of all those involved were required each year. One meeting was essentially a discussion meeting and one was based in a specific field area in the Severn Basin. Each year a report was produced for circulation to members of the Working Group and this report was usually the basis for discussion at the annual meeting. The reports produced were also conveyed to international meetings to indicate the progress made by the Severn Group. It was also necessary each year to provide a progress report to the UK IGCP Board which is convened by the Royal Society. These reports are contained in the publications produced annually by the Royal Society for submission to the IGCP Board (e.g. Royal Society, 1986).

The Severn Basin had the great advantage that there were a number of very active established researchers who were undertaking work on aspects of the area related to palaeohydrology. These scientists were willing to contribute and in some cases to undertake new research. However, in the British research environment the success of a venture of this kind depends very much upon attracting additional researchers who can undertake full-time postgraduate research. Although some of the higher degree research relevant to the project had been inaugurated prior to the IGCP Project, and some was not exclusively concerned with the project, nevertheless the contributions made by postgraduate research at Aberystwyth, University College of Wales (Lewis, 1983), at the University of Leicester (Dawson,

1986), at the University of London (Wiltshire, see Wiltshire and More, 1983), at the University of Reading (Jones, 1983; Hayward, 1982), and the University of Southampton (Brown, 1983) have all been extremely important in contributing to the research results obtained.

Contributions from postgraduate research are a major way in which the practice of palaeohydrology has been effected in Britain and in the Severn Basin in particular. However, additional features of practice relate to the evidence available, the retrospective approach adopted and the perspective which should result.

Evidence available for reconstruction of palaeohydrology of the temperate zone in the last 15,000 years is primarily sedimentological, morphological and historical. Sedimentological evidence includes information derived from physical properties of sediments and from organic deposits and this has enabled the reconstruction of environments in specific parts of the Severn Basin during the last 15,000 years and also facilitated estimation of changing erosion rates. Morphological information has centred particularly upon the analysis of palaeochannels using and developing techniques pioneered by G. H. Dury. An important additional ingredient is the use of historical information which is obtained by the application of range of dating techniques and by using historical sources which facilitate reconstruction of former environments.

It was indicated above that a retrospective approach has been advocated in relation to historical geomorphology (Brown, 1979). In palaeohydrology a retrospective approach is commended first because it is desirable to reconstruct palaeohydrology against the basis of a sure knowledge of the hydrology of contemporary environments. An understanding of contemporary processes should be a prerequisite for interpretation of past hydrological processes. A second reason for a retrospective approach is that it is necessary to attempt to extract a considerable amount of recent human impact before being able to analyse the prehistoric palaeohydrology.

In the basins of the Severn and Wye a notable feature of the contemporary basin is the extent to which the rivers and river courses have been modified as a result of human activity. Figure 1.1 gives an indication of the extent of channelized rivers in the basin according to work by Brookes (1982) who reconstructed the modifications undertaken in the 50 years from 1930 to 1980. Such channelization embraces all engineering practices used to control flooding, to drain wetland areas, to improve navigation or to control streambank erosion. The two major types of channelization shown include land drainage work which itself includes capital engineering works, routine maintenance and pioneer tree clearance, and secondly navigation works which are intended to improve canalized rivers.

Navigation works have been substantial and several rivers were navigable over considerable lengths by the seventeenth century, including the Severn

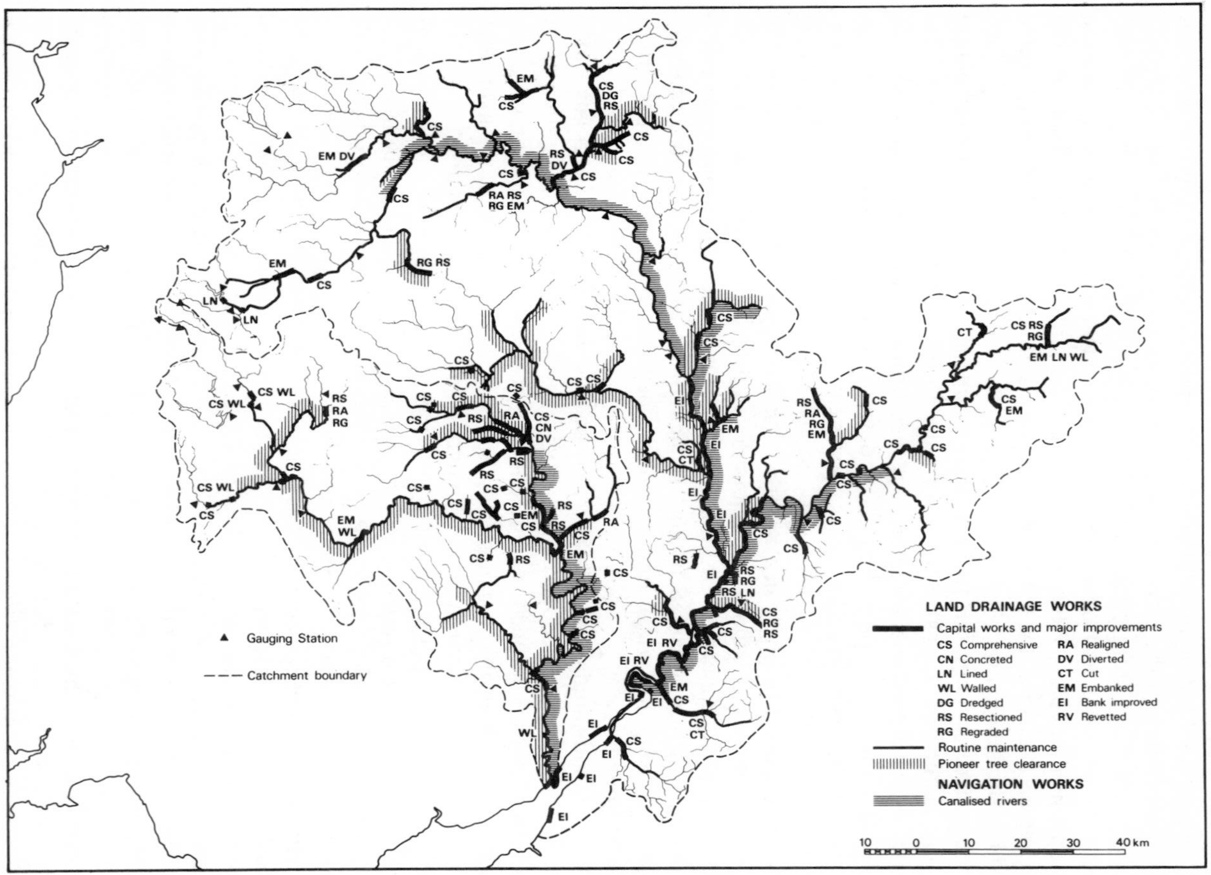

FIGURE 1.1

LAND DRAINAGE WORKS

Capital works and major improvements

| | | | |
|---|---|---|---|
| CS | Comprehensive | RA | Realigned |
| CN | Concreted | DV | Diverted |
| LN | Lined | CT | Cut |
| WL | Walled | EM | Embanked |
| DG | Dredged | EI | Bank improved |
| RS | Resectioned | RV | Revetted |
| RG | Regraded | | |

Routine maintenance

Pioneer tree clearance

NAVIGATION WORKS

Canalised rivers

▲ Gauging Station

--- Catchment boundary

0  10  20  30  40 km

as far as Welshpool, the Wye and Lugg to Leominster, the Warwickshire Avon to within 6 km of Warwick, the Stour and the Salwarpe. Many of these rivers appear to have been suited to navigation and to have required little maintenance so that only in the nineteenth century was significant work carried out on the River Severn between Stourport and Gloucester. Although navigation along the rivers did not continue in the twentieth century, some extensive works in the area were undertaken by the British Transport Commission after the Second World War. Between Stourport and Gloucester dredging has been undertaken to maintain an adequate depth of water and maintenance of the Severn has continued since 1962 under the British Waterways Board; in Warwickshire the Avon is maintained for navigation as far as Stratford as the responsibility of the upper and lower Avon navigation trusts.

Land drainage works for flood alleviation and for agricultural drainage have been undertaken by a number of organizations for more than 500 years. The commissions of sewers set up under the Court of Sewers Act of 1531 became one of the first administrative bodies responsible for drainage. The Land Drainage Act of 1861 saw the establishment of elected drainage boards, later known as internal drainage boards. Such boards had jurisdiction over areas smaller than major drainage basins and it was not until the Land Drainage Act of 1930 that the need for extensive channelization was acknowledged and drainage and flood control works were undertaken by the Severn and Wye catchment boards.

In the 1930s grants towards the cost of major schemes were introduced and land drainage works included capital works and major improvement schemes, routine maintenance and pioneer tree clearance. Major schemes and capital works have included the concrete lining of channels, resectioning of channels by widening and deepening, realignment by artificial cutoffs, and the regrading and embanking of water courses. Some of the earliest capital works included the Monmouth flood alleviation scheme (1936–37) and the widening of the Severn below Gloucester (1936–37). Routine maintenance, like major improvement schemes, is financed by local rates and it includes weed control, dredging of accumulated shelves, removal of debris from urban channels and localized repair of eroding banks. Although the impact of such maintenance schemes is less obvious than capital works, nevertheless they do affect extensive lengths of river. Pioneer tree clearance has often been undertaken, involving the complete removal of root systems, and this may have had a significance influence upon channel stability.

## 1.3 AN APPROACH

The distribution of channelized rivers (Figure 1.1) emphasizes the extent to which the fluvial system of these drainage basins has been modified and such modifications require consideration prior to viewing the changes that have taken place over the last 15,000 years. A retrospective approach is evident in the structure of the book which is organized to proceed from introductory chapters which provide the background context for palaeohydrology and also the background for the Severn/Wye drainage basins, and then proceeds to historical hydrology and subsequently to prehistoric hydrology.

It is hoped that this approach provides a perspective on palaeohydrology by showing how the results from a particular basin can be intergrated. In his review of British fluvial geomorphology in the 1980s Nanson (1986) contended that major extensive depositional environments and fluvial glacial environments were two of the things that had been omitted by British fluvial geomorphology. With the extension of fluvial geomorphology retrospectively by palaeohydrology it is possible to include greater consideration of such environments because, although they do not exist in the Severn Basin today, they did exist in the past. However, when compared with basins being investigated in other parts of the temperate zone it has to be remembered that the Severn/Wye Basin really presents just the upper and possibly middle reaches of a basin.

## REFERENCES

Allen, J. R. L. (1977). Changeable rivers: some aspects of their mechanics and sedimentation, in Gregory, K. J. (ed.), *River Channel Changes*, Wiley, Chichester, pp. 15–45.

Berglund, B. (ed.) (1986). *The Handbook of Holocene Palaeoecology and Palaeohydrology*, Wiley, Chichester.

Birks, H. J. B. and Birks, H. H. (1980). *Quaternary Palaeoecology*, Arnold, London.

Briggs, D., Coope, G. R. and Gilbertson, D. (1975). Late Pleistocene terrace deposits at Beckford, *Geological Journal*, **10**, 1–16.

Briggs, D. and Gilbertson, D. (1980). Quaternary processes and environments in the upper Thames Valley, *Transactions Institute of British Geographers*, **5**, 53–65.

Brookes, A. (1982). River channelization in England and Wales: downstream consequences for the channel morphology and aquatic vegetation, unpublished Ph.D thesis, University of Southampton.

Brown, A. G. (1983). Late Quaternary palaeohydrology, palaeoecology and floodplain development of the Lower River Severn, unpublished Ph.D thesis, University of Southampton.

Brown, E. H. (1979). The shape of Britain, *Transactions Institute of British Geographers*, ns 4, 449–62.

Brown, E. H. and Waters, R. S. (1974). Progress in geomorphology, *Institute of British Geographers Special Publication No. 7*.

Brunsden, D. (ed.), 1971. Slopes form and process, *Institute of British Geographers Special Publication No. 3*.

Burrin, P. (1983). The character and evolution of floodplains with specific reference to the Rivers Ouse and Cuckmere, unpublished Ph.D thesis, University of London.

Butzer, K. W. (1980). Adaptation to global environmental change, *Professional Geographer*, **32**, 269–78.

Cheetham, G. H. (1980). Late Quaternary palaeohydrology: the Kennet valley case study, in Jones, D. K. C. (ed.), *The Shaping of Southern England*, Academic Press, London, pp. 203–23.

Crampton, C. B. (1969). The chronology of certain terraced river deposits in the south east Wales area, *Zeitschrift für Geomorphologie*, **13**, 245–59.

Davidson, D. L. and Shackley, M. L. (eds) (1976). *Geoarchaeology*, Duckworth, London, 408 pp.

Dawson, M. R. (1986). Late Devensian fluvial environments of the lower Severn Basin, U.K., unpublished Ph.D thesis, University of Leicester.

Dury, G. H. (1977). Underfit streams; retrospect, perspect and prospect, in Gregory, K. J. (ed.), *River Channel Changes*, Wiley, Chichester, pp. 280–93.

Dury, G. H. (1983). Osage-type underfitness on the River Severn near Shrewsbury, Shropshire, England, in Gregory, K. J. (ed.), *Background to Palaeohydrology*, Wiley, Chichester, pp. 399–412.

Dury, G. H. (1985). Attainable standards of accuracy in the retrodiction of palaeodischarge from channel dimensions, *Earth Surface Processes and Landforms*, **10**, 205–14.

Embleton, C., Brunsden, D. and Jones, D. K. C. (eds) (1978). *Geomorphology: Present Problems and Future Prospects*, Oxford University Press, Oxford.

Gregory, K. J. (ed.) (1977). *River Channel Changes*, Wiley, Chichester.

Gregory, K. J. (1978). Fluvial processes in British basins, in Embleton, C., Brunsden, D. and Jones, D. K. C. (eds), *Geomorphology Present Problems and Future Prospects*, Oxford, Oxford University Press, pp. 40–72.

Gregory, K. J. (ed.) (1983a). *Background to Palaeohydrology*, Wiley, Chichester, 486 pp.

Gregory, K. J. (1983b). Introduction, in Gregory, K. J. (ed.), *Background to Palaeohydrology*, Wiley, Chichester, pp. 3–23.

Gregory, K. J. and Walling, D. E. (eds) (1974). Fluvial processes in instrumented watersheds, *Institute of British Geographers, Special Publication No 1*, 196 pp.

Harvey, A. M., Oldfield, F., Baron and Pearson (1981). Dating of post-glacial landforms of the central Howgills, *Earth Surface Processes and Landforms*, **6**, 401–12.

Hayward, M. (1982). Floodplain landforms, sediments and soil formation: the River Severn, Shropshire, unpublished Ph.D thesis, University of Reading.

Hazeldine, J. and Jarvis, N. G. (1979). Age and significance of alluvium in the Windrush Valley, Oxfordshire, *Nature*, **282**, 291–2.

Jones, D. K. C. (ed.) (1980). *The Shaping of Southern England*, Academic Press, London.

Jones, M. D. (1983). The palaeogeography and palaeohydrology of the River Severn, Shropshire, during the late Devensian glacial stage and the early Holocene, unpublished M. Phil thesis, University of Reading.

Kerney, M. P., Brown, E. H. and Chandler, T. J. (1964). The late-glacial and post-glacial history of the Chalk escarpment near Brook, Kent, *Philosophical Transactions of the Royal Society*, Series B **248**, 135–204.

Kimball, D. (1948). Denudation chronology. The dynamics of river action, *Occasional Paper Number 8, University of London, Institute of Archaeology*, 21 pp.

Kozarski, S. (ed.) (1983). Palaeohydrology of the temperate zone, *Quaternary Studies in Poland*, **4**.

Lamb, H. H. (1977). *Climate: Past, Present and Future*, 2 vols, London, Methuen.

Leopold, L. B. and Miller, J. P. (1954). Postglacial chronology for alluvial valleys in Wyoming, *U.S. Geological Survey Water Supply Paper 1261*, 61–85.

Lewin, J. (1977). Channel pattern changes, in Gregory, K. J. (ed.), *River Channel Changes*, Wiley, Chichester, pp. 167–84.

Lewin, J. (1978). Meander development and floodplain sedimentation: a case study from mid-Wales, *Geological Journal*, **13**, 25–36.

Lewin, J. (1983). Changes of channel patterns and floodplains, in Gregory, K. J. (ed.), *Background to Palaeohydrology*, Wiley, Chichester, pp. 303–19.

Lewin J. and Weir, M. J. C. (1977). Morphology and recent history of the lower Spey, *Scottish Geographical Magazine*, **93**, 45–51.

Lewis, G. W. (1983). The geomorphology of alluvial cutoffs: geometry, processes, development rates and sedimentation, unpublished Ph.D thesis, University of Wales.

Lockwood, J. G. (1979). Water balance of Britain 50,000 year BP to the present day, *Quaternary Research*, **12**, 297–310.

Macklin, M. G. (1985). Floodplain sedimentation in the upper Axe Valley, Mendip, England, *Transactions Institute of British Geographers*, **10**, 235–44.

Maizels, J. (1983). Palaeovelocity and palaeodischarge determination for coarse gravel deposits, in Gregory, K. J. (ed.), *Background to Palaeohydrology*, Wiley, Chichester, pp. 101–39.

Nanson, G. A. (1986). British fluvial geomorphology in the 1980s: a review of recent reviews, *Australian Geographer*, **17**, 87–91.

Nunn, P. D. (1983). The development of the River Thames in central London during the Flandrian, *Transactions Institute of British Geographers*, **8**, 187–213.

Prigogine, I. and Stengers, I. (1984). *Order out of Chaos: Man's new dialogue with Nature*, London, Heinemann.

Reinemund, J. A. and Watson, J. V. (1983). Updated excerpts from 'Achievements of the International Geological Correlation Programme as related to human needs', *Geologic Correlation*, UNESCO, Paris, pp. 9–11.

Richards, K. S. (1981). Evidence of Flandrian Valley alluviation in Staindale, North York Moors, *Earth Surface Processes and Landforms*, **6**, 183–6.

Robinson, M. A. and Lambrick, G. H. (1984). Holocene alluviation and hydrology in the upper Thames basin, *Nature*, **306**, 809–14.

Rodda, J. C., Sheckley, A. V. and Tan, P. (1978). Water resources and climatic change, *Journal of the Institute of Water Engineers and Scientists*, **32**, 76–83.

Royal Society (1986). *United Kingdom Contribution to the International Geological Correlation Programme*. 1986 Report, prepared on behalf of the British Natural Committee for the IGCP, 87 pp.

Schumm, S. A. (1968). Speculations concerning palaeohydrologic controls of terrestrial sedimentation, *Bulletin Geological Society of America*, **79**, 1573–88.

Starkel, L. and Berglund, B. J. (1983). Foreword, in Gregory, K. J. (ed.), *Background to Palaeohydrology*, Wiley, Chichester, pp. xv–xvi.

Starkel, L. and Thornes, J. B. (1981). Palaeohydrology of river basins, *British Geomorphological Research Group, Technical Bulletin*, No 28, 107 pp.

Starkel, L., Gregory, K. J. and Gardiner, V. (eds) (1985). Progress in Palaeohydrology, *Special Issue of Earth Surface Processes and Landforms*, **10**.

UNESCO, 1983. *Geological Correlation, Science Resources and Developing Nations. A review and a look into the future*, Unesco, Paris.

Vasari, Y., Saarnisto, M. and Seppala, M. (eds) (1979). Palaeohydrology of the temperate zone: Proceedings of the working session on Holocene-Inqua (Eurosiberian Subcommission) Hailuoto-Oulanka-Kevo 1978. *Acta Universitatis Ouluensis Series A, Scientiae Rerum Naturalium No. 82, Geologica No. 3*, University of Oulu, Oulu, 176 pp.

Vita-Finzi, C. (1979). *The Mediterranean Valleys*, Cambridge University Press.
West, R. G. (1968). *Pleistocene Geology and Biology: with special reference to the British Isles*, Longman, London.
Wiltshire, P. E. J. and Moore, P. D. (1983). Palaeovegetation and palaeohydrology in Upland Britain, in Gregory, K. J. (ed.), *Background to Palaeohydrology*, Wiley, Chichester, pp. 433–51.
Wooldridge, S. W. and Linton, D. L. (1939). *Structure, Surface and Drainage in south-eastern England*, Institute of British Geographers, London.

Palaeohydrology in Practice
Edited by K. J. Gregory, J. Lewin and J. B. Thornes
© 1987 John Wiley & Sons Ltd.

# 2

# Models for Palaeohydrology in Practice

## J. B. THORNES

*Department of Geography, Bristol University*

## INTRODUCTION

The most productive and widespread approach to establishing palaeohydrological responses to various changes in the external forcing agents has, of necessity, been inductive. From primary field observations of sediments, channel characteristics and valley morphologies it has been the practice to derive expected changes in discharge and sediment availability and to infer from these, in turn, something of the controls on the input by climate and by man. The limitations of this approach have been recognized over a long period of time. These include the paucity of the primary data, the spatial and temporal variability of natural systems and the existence of several possible causes to a simple morphological or sedimentological response, to mention but a few. Moreover, the direction of investigation is the reverse of the direction of causation as established by sister sciences, such as geomorphology, hydrology, geology and climatology, as shown in Figure 2.1.

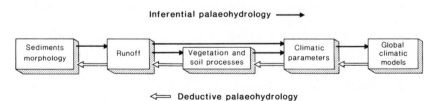

FIGURE 2.1  Relationship between elements in conventional and modelling approaches to palaeohydrology

The generally accepted reasons for this state of affairs are twofold. On the one hand the related sciences have yet to provide an adequate understanding of the relationships within and between the compartments in Figure 2.1. On the other hand, the researchers of past geographies have been unable to establish anything other than crude initial and boundary conditions which could be mobilized in the models. In the last decade there have been very significant improvements in all these respects, which lead us to ask again (cf. Thornes and Brunsden, 1977) whether the problem can, in fact, be approached deductively and in the direction of perceived causality. The purpose of this chapter, then, is to speculate on the extent to which the improvements of the last decade enable us to generate expected changes resulting from climate and other inputs and to trace these through, ultimately to fluviatile responses in specific instances, and most notably in the catchment of the Severn. In the author's view it will never be possible to provide a deterministic response to climatic and/or land use changes in a large river catchment for reasons which are discussed in the section below. However, it may be possible to proceed more rationally towards expected hydrological responses, with all the attendant errors, through the construction of even elementary models. This should be the ultimate goal for two reasons. First, it is easier to refute inadequate models on the basis of very sparse historical data than to build them from such data. Second, interpretations of field data, however abundant, are only as good as the explanatory models available. In a more general context, we may identify three categories of models which throw light on the plethora of problems posed in palaeohydrology which arise from the almost infinite number of combinations of changing climatic and land-use parameters.

1. Category 1—These are conceptual models which substantially improve our knowledge of the theory of the basic processes and the interrelationships between them but may not be directed to the specific problems in hand, or be capable of immediate application. Usually they are necessary but not sufficient. The models of Schumm (1965) and Thornes (1987) fall into this class.
2. Category 2—Models based on physical reasoning, parameterized from field data and applied in an experimental fashion to evaluate assumptions about vegetation or climatic changes and sometimes tested against actual changes. The International Biological Programme Ecosystems models, e.g. ECOWAT are of this type. They often require extensive parameterization with complex variables which can only be obtained for a few localities. They are experimental models in that they are capable of reasonably simulating the expected effects of changes to produce a more objective and less speculative reasoning when compared with that normally deployed in palaeohydrological work.
3. Category 3—These are coarser scale models which, though incomplete and comprising grey and black box components, are parameterized at a level which may be matched by palaeohydrological data. Of the

models discussed below, those of Kirkby (1975, 1987) and Eagleson (1978) appear to fall into this class. Inevitably, with these as with the experimental models, their flexibility in parameters is furnished at the expense of incorporating many known specific effects. Models of this type have the great advantage that their tolerance and range are high, so that as more palaeohydrological information becomes available, or as better understanding of the relationships is developed, they may be incorporated. These models are then essentially pragmatic, in that some detail is sacrificed to provide a way forward.

There have been three highly significant developments since the mid 1970s which have an important bearing on the development of palaeohydrological models. These are the investigation of non-linear model structures and their implications; the related initiation of integrated models of climate–biosphere–hydrosphere interactions and of climate–glacier–ocean interactions and, thirdly, the revolution in the knowledge of the oscillatory variations in the volumes of the ice sheets arising from oxygen isotope studies. The last has been well documented and will not be further discussed here, even though the deeper implications are only poorly recognized by many.

E. Lorenz (1963, 1980) observed that simple equations for describing climatic behaviour could give rise to extremely complex outcomes in time, three of which are of particular interest in this context. First he demonstrated that they could give rise to several unique stable solutions. Seen from the 'field-end' of the problem, apparently different equilibrium forms can arise from the same basic cause by simple variations in parameter values in the models. Second, he showed that a shift to one or other of these alternative configurations of the system can arise from random perturbations to the system if it is in a 'critical' state which can be defined in terms of specific parameter values. Third, he derived the conditions for oscillatory behaviour of the system without recourse to periodic external forcing functions such as Milankovitch mechanics. These ideas have subsequently been developed and applied in ecosystems (May, 1974), glacial systems (Oerlermans, 1981), river systems (Thornes, 1981) and vegetation-erosion systems (Thornes, 1987). Given that complex outcomes may arise from relatively simple causes coupled with random perturbations, they introduce staggering problems for the induction or inference of causes from field observations. At the same time they suggest that however good our causal model building, it may not be capable of defining a unique outcome, especially as more non-linear elements are identified and incorporated into the models.

The second major development has been research into integrated models of the interrelationships between climate, vegetation cover and hydrology. Although agricultural climatology has vigorously pursued the relations between crop productivity and climate for many years the new models, exemplified by the work of Eagleson (1978 *et seq.*) are distinguished by their concern for (a) natural vegetation and (b) feedback from this natural cover to both atmospheric processes and soil and catchment hydrology. This reflects the earlier work of Thornthwaite (1948) and Walther (1963) but differs from them in its attempts to come to grips with soil characteristics as an input to the system. Although the work has evident shortcomings, most notably in the failure to consider nutrients in plant growth (the model is governed by the soil moisture control on actual evapotranspiration), it seems destined to become one of the major innovations of the decade. In a later section we consider how this and similar approaches might provide a key feature of palaeohydrological models.

## THE FAMILY OF MODELS

As the various subjects contributing to palaeohydrology have developed, the earlier black boxes have been augmented, though not replaced by studies on a more detailed scale. One of the earliest quantitative studies was that of S. A. Schumm, who attempted to combine the relationship between precipitation, temperature and runoff of Langbein *et al.* (1949) with the relationship between runoff and sediment yield described by Langbein and Schumm (1958). This was used to produce the well-known synthetic curves for assessing the effects of shifts in weighted mean annual temperature and mean annual precipitation on runoff, sediment yield and concentration and the latter's effect of valley scour and fill. In this work vegetation is subsumed in the empirical relationship between precipitation and sediment yield data which was collected from the United States. It also reflects vegetation as affected by human activity and is based on a very small number of data points. The work mainly of interest here is that which fills out the assumptions made by Schumm and provides a basis for further understanding and development of the causal chain implied in Figure 2.1.

River channel behaviour is dominantly controlled by the resistance of bed and bank materials to the stress applied by flowing water along in mobile boundary. For alluvial rivers these are a function of channel slope, basin sediment yield and river regime. The slope can be locally adjusted through increase in channel length but basin sediment yield and river regime can be regarded as externally governed over the timescales of interest. The principal controls of these are snowmelt in sparsely vegetated periglacial environments, vegetation cover in temperate acultural landscapes and cultivation and grazing in cultural landscapes.

### Snowmelt

Prior to 10,000 BP snowmelt would have been a major control of runoff from British rivers. According to Male and Gray (1981) there are three main considerations in all attempts to measure runoff from snowmelt. These are the estimation of the extent and variability of the snow cover and its water equivalents, the estimation of the energy available to melt the snow cover over a given area, and the estimation of the effects of storage on the movement of melt quantities during transition from the snow surface to the stream channel.

There is little to help us with the first of these. There has been a debate on the extent of snow cover in the Devensian based on stratigraphic evidence and assumptions about permafrost. Obviously changes in the prevailing temperature and the frequency of westerly winds in middle latitudes have big effects on snow cover, especially where it is a seasonal event. The factors relevant to runoff are the duration of the snow pack, its water equivalent at the end of the winter season and its spatial variability. Under the present UK conditions duration is a non-linear function of altitude below 400 m and linear above that. Jackson (1978) provides the following equations:

$$N_H = N_0 \exp (H/300) \text{ for heights less than 400 m}$$
and
$$N_H = N_{400} (1 + (H\text{-}400)/310) \text{ for heights greater than 400 m}$$

where $N_H$ is the duration (in number of days) at height $H$ (metres) for a 'sea level' duration $N_0$ days. Similar results are found in other areas with larger snow volumes. Snow depths and water equivalents are correlated (though not contemporaneous) and also increased with altitude in the UK (Jackson, 1978). From these observations we might expect large variations within the Severn catchment in lying snow, if $N_0$ were to increase.

With respect to snowmelt the models are better physically based and more generally applicable. They comprise simple regressions and more complex energy budget models (for a recent review see Morris, 1985) but most incorporate the basic idea that

$$M = M_f (T_i - T_b)$$

wherein $M$ is snowmelt (mm day$^{-1}$), $M_f$ is a snowmelt index and $T_i$ and $T_b$ are the air temperature and a reference temperature respectively. The latter is usually taken to be zero. $M_f$ ranges from 2–6 and under unforested conditions is complex, reflecting a vegetation transmission coefficient for radiation, which is an exponential function of canopy density, and albedo. This general form applies to 'ripe' snow, i.e. that snow which is at or very close to 0°C so that any energy added can translate the water directly to surface runoff. This general formulation applies more readily to areas covered by forest, because then the long wave radiation transfer between the canopy and the snow is the most important element of the energy flux. In open areas the heat flux and the short wave radiation are not linearly related to air temperature and rely strongly on other parameters such as albedo and wind speed.

Certainly the forest has significant effects on both accumulation and runoff rates. Under forest the water equivalent accumulation is reduced by about half and the accumulation under conifers is less than under deciduous forest. Under rain, even for heavily forested areas, there is a significant and linear increase in snowmelt. Zuzel and Cox (1975) found that if only one model is to be used, daily air temperature provides the best single parameter for runoff.

The third factor of Male and Gray (1981), is the translation of snowmelt to the river channels. So far there has been little real progress in this field that would be of help in palaeohydrology and it is generally assumed that snowmelt provides a simple point source input to a more general hydrological routing procedure.

The only attempt to model Late Glacial melt characteristics in Lowland Britain is Lockwood (1983) for 11,000 BP. His reconstruction of ablation rates is based on the model of Pollard (1980) in which mean monthly ablation rates are a function of mean monthly solar radiation and air temperature. At 11,000 BP he accepts a July sea level mean monthly temperature of 11.3 °C and precipitation values similar to those of today. From these he is able to produce potential ablation rates which show, first, a steep rise in melt rates in May at sea level and, second, a principal melt period in June and July in the upper part of the basin. Since the zero accumulation line by this time was at about 1450 m, there would not have been permanent snow lying in the catchment.

**Vegetation**

The great advances in knowledge of the composition and evolution of post-glacial vegetation cover have enabled reconstruction of the cover composition, as described in other chapters of the book. In hydrological and sedimentological terms more information is necessary on cover, biomass, leaf area index and litter and soil organic matter. First because they are relevant quantities for runoff and erosion modelling and second because, to some extent, clearance can be expressed in these terms. Hitherto palaeohydrological assumptions have been based on speculations as to both the relationship between climate and vegetation and that between vegetation cover and runoff. However, neither of these sets of relationships are quite so poorly understood as the literature suggests.

That vegetation cover is related to mean annual temperature and precipitation underlies the classical classification of climatic types. Only recently, however, have sufficient data become available for them to be quantified. The simplest models are based on empirical relationships between vegetation and broad climatic parameters reinforcing the results obtained from agronomy—that plant productivity is related to the ratio of actual to potential evapotranspiration. Leith (1975) produced curves of net primary productivity as a function of mean annual temperature, mean annual precipitation and annual actual evapotranspiration. The first of these, perhaps most significant for the palaeohydrology of the Severn, has the form

$$y = 3000/(1 + \exp (1.135 - 0.119x))$$

where $y$ = mean annual net primary productivity (g m$^{-2}$ yr$^{-1}$), and $x$ = mean annual temperature in °C. For mean annual temperatures of around $-5$–$0$ °C this produces values in the range 400–600 g m$^{-2}$ yr$^{-1}$ and for 5 °C about 1200 gm$^{-2}$ yr$^{-1}$ and at 10 °C, 1400 g m$^{-2}$ yr$^{-1}$. Since these data are not reflecting mineral or moisture deficiencies, they are only useful at a scale which integrates soil variability. They may be compared with values given by Rosenzweig (1968) whose compilation indicates that annual productivity for arctic tundra is about 160; for heath 400; for coniferous forest about 1000 and for deciduous forest about 1600 g m$^{-2}$ yr$^{-1}$.

Although net productivity is of interest mainly for runoff, since it is an indicator of the water consumed in transpiration, it correlates well with other properties that are of greater interest in terms of erosion. In particular Whittaker and Marks (1975), in discussing methods of assessing territorial productivity, suggest that biomass is strongly correlated with above ground net primary productivity for climax forest and woodland, as are leaf area indices for shrub and forest communities. Leaf area indices for temperate forests are found to be typically in the range 5–14 (m$^2$ m$^{-2}$) whereas values for tundra are typically of the order 0.5–1.5. The function of leaf area index against net primary productivity for broadleaf species rises steeply to about 1000 g m$^{-2}$ yr$^{-1}$ and then remains relatively constant while values for ever-green needle leaf species are generally much higher.

In one of the first attempts to model the relationship between erosion, climate and vegetation from a deductive point of view Kirkby (1975) expressed soil water storage capacity (and hence lack of water for runoff) as proportional to the square of the ratio of the actual to potential evapotran-spiration. This reflects the fact that plant productivity is a function of the ratio, as noted in the previous section, and that soil moisture storage is related to the organic matter content of the soil (but see Francis *et al.*, 1986, for semi-arid environments). In a more recent model (Kirkby and Neale, 1987), the same basic principles are used by include seasonal effects by using a monthly accounting procedure. The model provides for litter fall at a total rate which increases with biomass, linearly at first but at a slower rate for a larger standing crop. Allowance can also be made for cropping of plant cover by its influence on both biomass and organic matter, the latter being controlled by a decomposition formula. In order to express the ability of

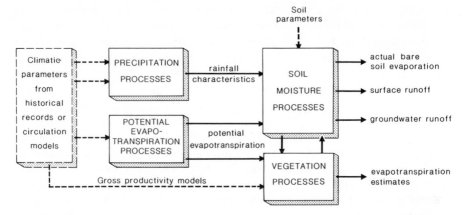

FIGURE 2.2   Components of an interactive soil–vegetation–hydrology model

plants to accommodate short-run fluctuations a persistence factor, in fact a first order Markov process, is added to biomass. This type of model has become very widespread in recent years though generally speaking the parameters required outstrip the general state of knowledge of palaeoclimate. If they are reasonable approximations to reality and if we may assume that the plant cover is in equilibrium with the climatic parameters driving the model, then they extend the understanding of post-glacial vegetation when expressed in terms of usable parameters.

The Kirkby and Neale (1987) model indicates that under equilibrium conditions and above about 800 mm, the living plant biomass is mainly controlled by temperature on the large scale (i.e. excluding soil effects) and thus reflects the empirical data used to build the model. At 1000 mm yr$^{-1}$ precipitation they obtain 1 kg m$^{-2}$ yr$^{-1}$ (cf. Leith, 1975, 0.1–3 for tundra) for a temperature of 0°C, at 6°C a value of 6 kg m$^2$ yr$^{-1}$ (cf. heath woodland, 2–20) and at 12°C they obtain 20 kg m$^2$ yr$^{-1}$, a figure close to those for boreal and temperate forests. While these values understate the significant variability found in nature, they probably provide good approximations to circumstances prevailing in, say, the Severn catchment until man arrived on the scene.

The Kirkby and Neale model uses a general monthly accounting procedure for the various elements of the process whereas Eagleson (1978) has provided analytical solutions for a mean annual moisture budget and its various components. The general structure of these approaches is shown in Figure 2.2. The core of Eagleson's model is the soil moisture budget, and this is operated through expectations of losses through drainage, bare soil evaporation and evapotranspiration from the vegetal cover. The model is quite complex, but like other models assumes that interaction between vegetation cover and soil moisture operates through water-limiting constraints. The

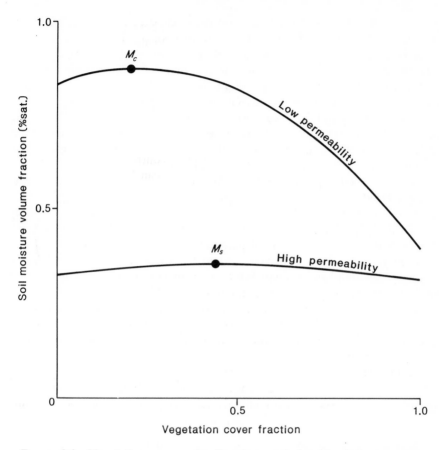

FIGURE 2.3   Vegetation cover and soil moisture relationships for a temperate environment and two distinct soil types. $M_c$ and $M_s$ represent the optimum vegetation covers for the two soil types. Based on Eagleson (1978)

rates of actual to potential transpiration are again expressed through a plant coefficient ($K_v$) which is specific to the cover species. For a given cover and potential evapotranspiration the loss from the soil moisture store can be determined. Other factors also control soil moisture but by holding them constant, the sensitivity of moisture to plant cover, as expressed by the conditions of the model, can be observed. Some typical relationships are shown in Figure 2.3. In a humid zone, for a clay soil, soil moisture depletes at high vegetation cover, whereas for sandy soil the curve is only slightly convex, indicating a relative insensitivity of moisture to cover. Eagleson argues that in this trajectory of possible equilibria the most probable cover is that which leaves the plant under least moisture stress, i.e. at $M_c$ and $M_s$ for clay and sand respectively. These loci of maximum vegetation cover also depend on $K_v$. It is assumed that plant communities will seek to maximize biomass productivity and this is the state in which $M_o$ and $K_v$ are jointly maximized. It is also the condition of maximum evapotranspiration. By differentiating the basic equations, having inverted the analytical model, Eagleson is able to solve for the conditions under which biomass will be at a maximum. Although the two hypotheses relating to vegetation cover are rather restrictive, and Eagleson provides solutions for only two cases and four soil types, this is the first major analytical attempt to provide runoff models which accommodate not only soil characteristics, climate and vegetation cover, but also some of the feedbacks between them.

Both the Eagleson and the Kirkby and Neale models are heavily parameterized. However, they do involve a clear strategy which can be adapted and adopted for deductive palaeohydrology. This involves providing appropriate distribution functions which effectively 'distribute' rainfall characteristics from more general parameters, e.g. by estimation of rainfall per rainday from changes in volume and length of season. These provide inputs to soil moisture budgets which in turn control vegetation cover at equilibrium, provided separate estimates of $K_v$ can be obtained. Alternatively, cover can be provided from the type of equations provided by Leith (1975). The soil moisture storage then controls the infiltration and percolation processes and through them the ground surface and groundwater runoff regimes respectively.

One important result from most models, and especially from Eagleson's, is that in humid areas climate and vegetation are much more important controls on runoff than are soil characteristics.

### Changes in vegetation cover

Although changes in vegetation cover resulted in a relative response in soil moisture and related characteristics for humid areas in the Eagleson model, changes in the $K_v$ value, enhancing the ability to use water, were fairly critical. This is consistent with results obtained by other modellers. For example Lockwood (1983) using models developed by Sellers and Lockwood (1981a,b) used a multilayered crop model to simulate interception and evaporation of rainfall as well as transpiration by grass, wheat, pine and oak canopies. He obtained monthly runoff for present-day conditions for the different covers in central lowland England, showing that for grass, wheat and oak the percentage of rainfall as runoff were 41.7, 57.6 and 38 per cent respectively, whereas for pine there was only a 21 per cent runoff. Lockwood argues that areas which were formerly under deciduous forest would have had more concentrated late winter runoff than under grass but that annual runoff would have been significantly reduced when compared with the present. This modelling is consistent with what is known experimentally from a wide range of catchments (Huff and Swank, 1985), that forest clearance results generally in increased water yield. Experimental observations on catchment behaviour (Hewlett, Fortson and Cunningham, 1977; Hewlett and Bosch, 1985) suggest, however, that this does not result mainly from overland flow. It is hardly surprising that the assessment of forest treatment on hydrology continues to be confused in the literature. The possible variations in clearance practices are enormous and as yet poorly classified. They vary from uniform reduction by thinning, non-uniform reduction in clumps, strips and patches, and all possible combinations of these two extremes.

Some attempts to model these effects have been made in the Eastern Deciduous Forest Biome and in the Coniferous Forest Biome (Waring, Rogers and Swank, 1981). In the latter this was done using ECOWAT, a forest-hydrology simulation model, the essential components of which are shown in Figure 2.4. This was applied to three areas on undisturbed forest, forest reduced by thinning and clearing and to changes in the type of vegetation cover. Results from the Coweeta catchment were actually tested against the observed effects of the same clearance. At Coweeta (South Carolina) the soils are deep and permeable on steep slopes, originally of oak–hickory forest. Three clear cutting treatments produced substantial increases in streamflow and reductions in evaporation (of 58 per cent) and transpiration (33–41 per cent), the latter being smaller than the former due to increased exposure of the remaining uncut vegetation. Here and at the H. J. Andrews Experimental Forest the results were most sensitive to leaf area index, which is the single most important variable. There was clear evidence that aspect and topographic position of the areas cleared were also significant controls of the response to actual as well as simulated clearing. The work emphasizes the need to stratify watersheds into response units in areas where heterogeneity of soils and topography have significant effects on the vegetation and the hydrological response of these areas.

In Coweeta, significantly, changes in vegetation type have more significant effects than thinning due to changes in the amount and seasonal distribution of leaf area, the density and depth of rooting, the season and duration of active growth and the albedo and aerodynamics of the vegetation cover. With conversion from oak–hickory to white pine, the simulations showed a decrease in streamflow and an increase in evapotranspiration of 25–27 per cent due to increased foliar storage capacity during the dormant season, when hardwoods are normally bare. Taken as a whole, these simulations, coupled with other evidence cited in the previous section, suggest that perhaps the most relevant parameter around which to model the impact of climate on hydrology, at least for the now-temperate middle latitudes, is leaf area index, in view of its close relationship to biomass and productivity as well as litter and soil organic matter development.

We have more or less assumed that under the forest conditions prevailing

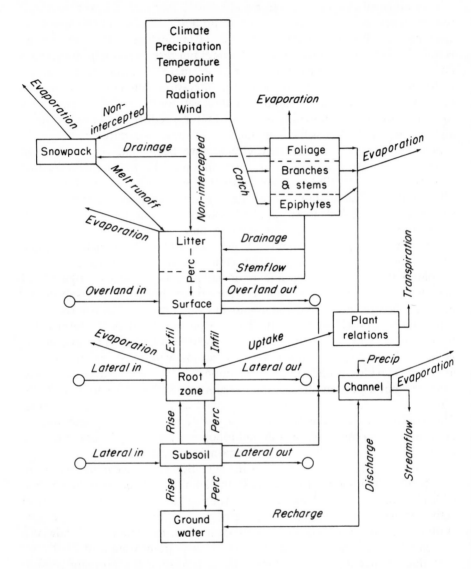

○ = Connects to other land units

Figure 2.4 Components of ECOWAT (Reproduced by permission of Cambridge University Press after Waring, Rogers and Swank, 1981)

during most of the middle Holocene, soil erosion would have been negligible and that the density, composition and elevation of the forest cover would mainly control runoff. After 3000 BP clearance for grazing and later for cultivation occurred and soil erosion models became more relevant for their effects on both runoff and sediment yield.

## Soil erosion

As with assertions about forest clearance, the knowledge of the effects of both climatic change and clearance on soil erosion and sediment supply are inductive or speculative.

The role of palaeoecology on erosion was clearly identified by Schumm (1965) but it is only recently that attempts have been made to establish the quantitative effects of plant cover and animal activity and model them.

The generally accepted belief that soil erosion becomes significant below about 30 per cent cover has received some support in the published literature. For example, Elwell and Stocking (1976) showed a steep exponential decline in the amount of erosion (as a fraction of the bare soil value) as seasonal cover increased. This result shows a strong levelling off at between 25 and 35 per cent cover. A similar general result was obtained by Noble (1963) for subalpine rangeland, though the curves is less steep and only levels off at about 60 per cent. Such vegetation cover effects are compounded with crusting and other multiplicative factors once vegetation has been removed and it might be that the relationship for vegetation *alone* should be more or less linear.

An attempt has been made by Thornes (1987) to produce a category 1 type model of erosion in relation to vegetation cover using the general exponential relationship defined by Elwell and Stocking (1976) in conjunction with vegetation growth and a power-law relation to excess overland flow. In this model the capacity vegetation cover is determined by the temperature and precipitation and the capacity erosion by the total quantity of overland flow. The system is then solved for equilibria in the competition between vegetation and erosion for both wash and gully-type erosion. To these cases we can add the impact of grazing, either as a simple predation term on the vegetation cover, reducing its efficacy to compete with erosion (Figure 2.5), or as a further differential equation governing the growth of the grazing population and interacting competitively with the vegetation. The latter type of model has also been developed for grazing–vegetation interactions in African savannahs by Noy-Meir (1978). Although the effects of grazing on erosion are well documented (see Jansson, 1982), there are no generally con-ceptual models which specifically make the role of grazing central to the issue.

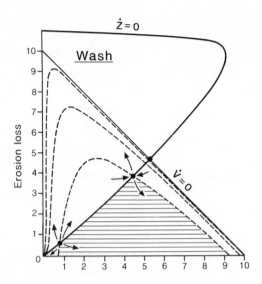

FIGURE 2.5 Interaction between vegetation cover soil erosion cover with controlled grazing. The lines $\dot{Z} = 0$ and $\dot{V} = 0$ are the lines of no change in soil erosion rate and vegetation growth rate respectively. The packed lines show the effects of grazing when the herd population size is not a function of vegetation cover and the bold points show equilibrium states. Horizontal axis is vegetation cover

There are several major attempts to model erosion by closed form equations (Foster and Meyer, 1975, for example) and these have been discussed at length elsewhere, while several have been developed separately or as part of more extensive digital simulation models such as CREAMS and SPUR. As with the category 2 models for vegetation, of which the erosion component is commonly a subset, these require heavy parameterization, tend to be at field rather than basin scale, and are likely to be of greater value to palaeohydrology in an experimental context than in application. These models are more or less valuable to the extent that they include vegetation cover.

Jansson (1982) has looked at the great number of models for predicting erosion on a statistical basis. The global model of Langbein and Schumm (1958) expressed as:

$$S = \frac{K_1 + 10^{-4}P^{2.3}}{K_2 + 10^{-8}P^{3.33}}$$

where $P$ = effective precipitation, $S$ = the sediment yield in m tonnes km$^{-2}$

$yr^{-1}$. The upper term is thought to represent bare soil, the lower the effect of vegetation, with $K_1$ and $K_2$ as constants which differ for sediment station data and for reservoirs. Like the Fournier (1960) equations, these are developed from suspended sediment yield data. Kirkby (1975) simulated erosion for both creep and wash and obtained a stronger temperature dependence on erosion with a deep trough at about 1000 mm and a further rise in ground lowering at higher rainfalls. The latter appears consistent with recorded data from elsewhere and with the more complex patterns obtained from empirical data by Walling and Webb (1983).

For agricultural land, the Universal Soil Loss Equation is the most extensively developed and parameterized empirical model in existence. It has been criticized on its weak theoretical basis and its failure to accommodate channelized flow. It rests dominantly on the $K$, or soil erodibility, factor which in turn involves the rainfall intensity as its key parameter. This has proved valuable in predicting soil loss from unvegetated slopes in many environments. Its use in palaeohydrological studies depends on obtaining a reliable distribution of rainfall intensities from more general statistics as mean rainfall/rain day or from mean rainfall coupled with information on past rainfall types.

With any point production model, whether the source be soil erosion or glacial sediment available to streams, the material still has to be routed downstream. Williams (1977) expresses the problem by the equation

$$S = \sum_{i=1}^{n} S_i e^{-B_i T_i \, d}$$

in which the sediment yield at $S$ is derived from a number of subbasins $i = 1$ to $n$ according to the routing coefficient $B_i$ the travel capacity $T_i$ (which is the ratio of distance to relief between the location and its upstream source) and the particle size $d$. Smaller particles travel further in steeper channels. Even assuming a constant routing coefficient, this model goes a long way to accounting for the differences in the sediments found in the various terraces of the palaeo-Severn found by Dawson and Gardiner (Chapter 13), given what we know about the nature and location of sediment sources in the Severn now and at the time of origin of the terraces, even if we assume nearly constant rates of sediment production. Much more could and should be made of general routing models of this type in hydrology.

**Channel response**

The response of channels to water and sediment discharge is complex and indeterminable (Schumm, 1977). Despite numerous attempts to constrain the degrees of freedom in channel response by recourse to general laws, such as maximum sediment transport efficiency, uniform rate of energy expenditure and others, there is so far no adequate deductive model of channel behaviour, that could be evolved in response to sediment and water discharge. Essentially, aggradation must occur when the rate of sediment supplied to a reach is greater than that which can be transported. Conversely, incision must be reflected by net output being greater than net input of sediment to a reach over a series of discharge events. So far there is no general proposition relating to sediment transport which enables us to deal with this case. However, to a first approximation we may obtain the likelihood of wide shallow channels versus narrow deep channels using either the theory developed by Lane (1957) and Henderson (1966) or the empirical observations of Leopold and Wolman (1957) and Osterkamp (1978). These papers distinguish between braided and meandering streams on the basis of a discriminant function of slope and discharge, and in the case of Henderson (1966), sediment size. In general terms the discrimination is provided by the line

$$S = KQ^{-m}$$

where $S$ = slope, $Q$ = discharge (usually mean annual) and $m$ = an exponent. Lane suggested that for sandbed streams $m \simeq 0.25$, Begin (1981) suggested $m = 0.33$ and Williams (1983) incorporates Henderson's (1966) criterion for bedload movement of a particle of size $d$ with hydraulic geometry exponents to obtain:

$$S = K\,(d/c)Q^{-f}$$

with $d$ in mm, $Q$ as discharge and $c$ and $f$ hydraulic geometry exponents. Begin and Schumm (1984) have attempted to interpret these relationships in terms of relative shear stress from a large body of empirical data:

$$Tm = Q_m^{0.33}S$$

$$Tm = 0.005$$

If $Q_m$ = the mean annual discharge and $S$ = channel slope and these can be estimated from palaeoflows with $Tm$ the mean for all the observed values, then $Tm$ can be determined and used to provide the expected channel behaviour.

In a recent review of the problem Carson (1984) argues that much of the discrimination is illustory and instead of a specific discriminatory function there is a family of thresholds increasing with bed particle diameter and reflecting the threshold $Q$-$S$ relationship for bed material itself. In fact, he shows that the Leopold–Wolman function corresponds to a power of approximately 30 Wm$^{-2}$ (cf. Ferguson, 1981; ~50 Wm$^{-2}$) which is the power required to move gravel of approximately 15–25 mm. He suggests that a more likely criterion for the onset of braiding is the power available *in excess* of that required to move the individual particles. This is then available to cause bank scour at even small depths, so that power and width are not independent. This to some extent suggests an intrinsic structural instability as the condition for braiding and for alternating wide and narrow channel reaches, as suggested by Thornes (1981). Despite these objections, which relate to the mechanism whereby different dominant channel types come about, there seems to be good evidence that given slope discharge and particle size (as a function of the sediment routing algorith and identified source materials) the essential quality of the channel can be modelled, as can gross changes in its behaviour caused by changes in any of these three. This seems a firmer basis for post-dicting channel behaviour than the more general models of channel metamorphosis provided by Schumm (1969).

## CONCLUSIONS

The detailed contents of other chapters in this book show that the Severn palaeohydrology can be defined in terms of three major environments:

1. Snow- and ice-dominated, to about 10,000 BP.
2. Forest-dominated, to about 5000 BP.
3. Anthropogenically-dominated to an increasing extent since 5000 BP.

Some of the major available models for considering hydrological response to climatic controls in these three environments have been considered in this chapter. Although we will probably never reach a general model of river basin response which can be driven by climatic forcing functions and modified by human activity, palaeohydrologists can and should be making more use of even approximate and deductive models. Schumm, in his many papers on geomorphology and palaeohydrology, showed the way. What has changed is the greater understanding of the ecological controls of erosion and their importance. This will eventually shift palaeohydrology away from a preoccupation with pollen diagrams and fluvial sediments to dynamic modelling of the catchment ecological systems in a way that provides a better understanding of water and sediment yield. If this chapter has done no more than to draw attention to these ecological priorities, then it will have served its purpose.

## REFERENCES

Begin, Z. B. (1981). The relationship between flow-shear stress and stream pattern, *J. Hydrol.*, **52**, 307–319.
Begin, Z. B. and Schumm, S. A. (1984). Gradational thresholds and landform singularity: significance for Quaternary studies, *Quarternary Res.*, **21**, 267–75.
Carson, M. A. (1984), The meandering-braided river threshold: a reappraisal, *J. Hydrology*, **73**, 315–34.

Eagleson, P. E. (1978). Climate, vegetation and soil, *Water Resources Research*, **14**(5), 705–76.

Elwell, H. A. and Stocking, M. A. (1976). Vegetal cover to estimate soil erosion hazard in Rhodesia, *Geoderma*, **15**, 61–70.

Ferguson, R. I. (1981). Channel form and channel changes, in Lewin, J. (ed.), *British Rivers*, Allen and Unwin, London, pp. 90–125.

Foster, G. R. and Meyer, L. D. (1975). Mathematical simulation of upland erosion by fundamental erosion mechanics, *U.S. Dep. Agric., Agric. Res. Service (ARS-S-40)*, 190–207.

Fournier, M. F. (1960). *Climat et érosion*, Presses Universitaires de France, Paris.

Francis, C. F., Thornes, J. B., Romero-Diaz, A., Lopez-Bermudez, F. and Fisher, G. C. (1986). Topographic control of soil moisture, vegetation cover and land degradation in a moisture stressed Mediterranean environment, *Catena*, **13**(2), 211–225.

Henderson, F. M. (1966). *Open Channel Flow*, Macmillan, New York.

Hewlett, J. D. and Bosch, J. M. (1985). The dependence of storm flows on rainfall intensity and vegetal cover in South Africa, *J. Hydrology*, **75**, 365–81.

Hewlett, J. D., Fortson, J. C. and Cunningham, G. B. (1977). The effect of rainfall intensity on storm flow and peak discharge from forest land, *Water Resources Research*, **13**(2), 259–66.

Huff, D. D. and Swank, W. T. (1985). Modelling changes in forest evaporation, in Anderson, M. G. and Burt, T., *Hydrological Forecasting*, Wiley, Chichester, pp. 125–51.

Jackson, M. C. (1978). Snow cover in Great Britain, *Weather*, **33**, 298–308.

Jansson, M. B. (1982). Land erosion by water in different climates, *UNGI Rapport No. 57*, Uppsala University, 151 pp.

Kirkby, M. J. (1975). Hydrological slope models: The influence of climate, in *Climate and Geomorphology*, ed. Derbyshire, E., Wiley, Chichester, pp. 247–67.

Kirkby, M. J. and Neale, R. H. (1987). A soil erosion model incorporating seasonal factors, *Proc. 1st International Geomorphology Conference*, Wiley, Chichester, in press.

Lane, E. W. (1957). A study of the shape of channel formed by natural streams flowing in erodible material. *U.S. Army Engineers Division, Sediment Series No. 7*, Omaha, Nebraska.

Langbein, W. B. *et al.* (1949). Annual runoff in the United States, *U.S. Geological Survey Circular*, **52**, 14 pp.

Langbein, W. B. and Schumm, S. A. (1958). Yield of sediment in relation to mean annual precipitation, *Trans. American Geophysical Union*, **39**, 1076–84.

Leith, H. (1975). Modelling the primary productivity of the world, in Leith, H. and Whittaker, R. H. (eds), *Primary Productivity of the Biosphere*, Springer Verlag, New York.

Leopold, L. B. and Wolman, M. G. (1957). River channel patterns: braided, meandering and straight, *U.S. Geological Survey. Prof. Paper*, 282–8, 39–85.

Lockwood, J. (1979). Water balance of Britain 50,000 yr BP to present day, *Quarternary Res.*, **12**, 297–310.

Lockwood, J. G. (1982). Snow and ice balance in Britain at the present time and during the last glacial maximum and the Late Glacial periods, *J. Climatology*, **2**, 209–32.

Lockwood, J. G. (1983). Modelling climatic change, in Gregory, K. J. (ed.), *Background to Palaeohydrology*, Wiley, Chichester, pp. 25–50.

Lorenz, E. N. (1963). Deterministic non-periodic flows, *J. Atmos. Sci.*, **36**, 1685–99.

Lorenz, E. N. (1980). Attractor sets and quasi-geostrophic equilibrium, *J. Atmos. Sci.*, **36**, 1685–99.

Male, D. H. and Gray, D. M. (1981). *Handbook of Snow. Principles, processes and use*, Pergamon, Toronto, 765 pp.

May, R. M. (1974). *Stability and complexity in modern ecosystems*, Princeton.

Morris, E. M. (1985). Snow and ice, in Anderson, M. G. and Burt, T. P. *Hydrological Forecasting*, Wiley, Chichester, pp. 153–82.

Noble, E. L. (1963). Sediment reduction through watershed rehabilitation, U.S. Dep. Agric. *Misc. Publ.* 970, Pap. No. 18, 114–23.

Noy-Meir, I. (1978). Stability of simple grazing models: effects of explicit functions, *Journal of Theoretical Biology*, **71**, 347–80.

Oerlermans, J. (1981). Some basic experiments with vertically integrated ice-sheet models, *Tellus*, **33**, 1–11.

Osterkamp, W. R. (1978). Gradient, discharge and particle size relations of alluvial channels in Kansas, with observations on braiding, *American Journal of Science*, **278**, 1253–68.

Pollard, D. (1980). A simple parameterization for ice sheet ablation rate, *Tellus*, **32**, 384–8.

Rosenzweig, N. (1968). Net primary productivity of terrestrial communities: prediction from climatological data, *American Naturalist*, **102**, 67–74.

Schumm, S. A. (1965). Quaternary palaeohydrology, in *The Quaternary of the United States*, ed. Wright, H. E. and Frey, D. G. Princeton University Press, pp. 783–95.

Schumm, S. A. (1969). River metamorphosis, *Proceedings American Society Civil Engineers, J. Hydraulics Div.*, **HY1**, 255–73.

Schumm, S. A. (1977). *The fluvial system*, Wiley, New York.

Sellers, P. J. and Lockwood, J. G. (1981a). A computer simulation of the effects of differing crop types on the water balance of small catchments over long periods of time, *Quarterly J. Royal Met. Soc.*, **107**, 395–414.

Sellers, P. J. and Lockwood, J. G. (1981b). A numerical simulation of the effects of changing vegetation type on surface hydroclimatology, *Climatic Change*, **3**, 121–36.

Thornes, J. B. (1981). Structural instability and ephemeral stream channel behaviour, *Zeitschrift für Geomorphologie*, Supplementband, **36**, 233–44.

Thornes, J. B. (1987). Erosional equilibria under grazing, in Bintliff, J. and Shackley, M., *Environmental Archeology*, Edinburgh University Press.

Thornes, J. B. and Brunsden, D. (1977). *Geomorphology and Time*, Methuen, London, 208 pp.

Thornthwaite, C. W. (1948). An approach to the rational classification of climate, *Geographical Review*, **38**, 55–94.

Walling, D. E. and Webb, B. (1983). Patterns of sediment yield, in *Background to Palaeohydrology*, ed. K. J. Gregory, Wiley, Chichester.

Walther, H. (1963). Zur klärung des spezifischen Wasserzustande in Plasma und in der Zellwand bei der höheren Pflanze und seine Bestimmung, *Ber. Botanische Gesellschaft*, **76**, 40–70.

Waring, R. H., Rogers, J. J. and Swank, W. T. (1981). Water relations and hydrological cycles, in *Dynamic Properties of Forest Ecosystems*, ed. Reichle, D. E., Cambridge University Press, pp. 205–54.

Whittaker, R. H. and Marks, P. L. (1975). Methods of assessing terrestrial productivity, in Leith, H. L. and Whittaker, R. H. (eds), *Primary Productivity of the Biosphere*, New York, Springer-Verlag, pp. 55–118.

Williams, G. P. (1983). Improper use of regression in earth science, *Geology*, **11**, 195–7.

Williams, J. R. (1977). Sediment delivery ratios determined with sediment and runoff models, *IAHS Publ.* 122.
Zuzel, J. F. and Cox, L. M. (1975). Relative importance of meteorological variables in snow melt, *Water Resources Research*, **11**, 174–6.

Palaeohydrology in Practice
Edited by K. J. Gregory, J. Lewin and J. B. Thornes
© 1987 John Wiley & Sons Ltd.

# 3

# *The Contemporary River: The Basin*

## J. LEWIN

*University College of Wales, Aberystwyth*

### INTRODUCTION

In round figures the Severn and Wye catchments approach 15,000 km² in area when put together: both rivers have their nominal sources on Plynlimon (752 m) in Mid-Wales and less than 4 km apart. Although in world terms the catchments thus appear small with low overall relief, they do contain some remarkable contrasts, and indeed such a diversity of drainage history as to provide a highly varied basis for palaeohydrological developments in the last 15,000 years.

These contrasts lie particularly in relief, as between the higher Welsh plateaux of the West and the lower East; in lithology, with Palaeozoic rocks again to the West and in contrast to the Mesozoic cover in the Avon system and underlying the Shropshire plain; and in terms of a Quaternary history during which the area has been much affected by having straddled ice-terminal positions on two, and quite possibly more, occasions. These dichotomies in form, lithology and Quaternary history form the themes of this chapter.

### MORPHOLOGY

Figure 3.1 shows the catchments of the Severn and Wye, land over 1000 ft (350 m), and the major areas of Quaternary sediments. The Wye (4040 km² to the lowermost gauging station near the tidal limit at Cadora) flows dominantly within the upland plateaux of Wales and the borderland. The lowest section passes through a classic series of incised meanders, examined over 50 years ago by A. A. Miller (Miller, 1935) but rather little since. Just below Hereford, the Wye is joined by the Lugg (886 km²) which, together with the Teme and its tributaries (1580 km²) flowing directly to the Severn,

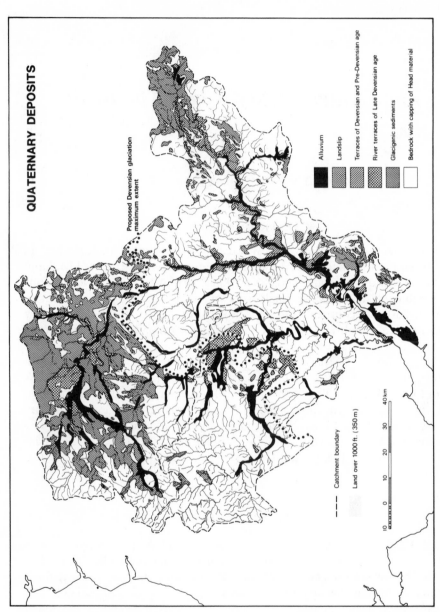

## QUATERNARY DEPOSITS

Proposed Devensian glaciation
maximum extent

Alluvium

Landslip

Terraces of Devensian and Pre-Devensian age

River terraces of Late Devensian age

Glacigenic sediments

Bedrock with capping of Head material

Catchment boundary

Land over 1000 ft. (350 m)

10    0    10    20    30    40 km

FIGURE 3.1  Relief, drainage and Quaternary deposits in the Severn–Wye basin, compiled by members of the UK Project 158A Group

drains the major part of the Welsh borderland. These rivers are characterized by broader alluvial basins alternating with gorge sections and they have been considerably modified by glaciation. Above Hereford, the Wye valley is broad, on Lower Devonian rocks, while beyond Hay-on-Wye its valley is again entrenched but now rather straight in plan. Major tributaries to the upper Wye are the Irfon, Ithon and Clywedog, the latter now dominated by a series of water storage reservoirs built in 1891–1904, with a further one (the Claerwen) completed in 1952.

The 10,000 km² of the Severn is strikingly diverse. Its major tributary, the Avon (2790 m²) is a lowland river at the foot of the Cotswold Hills, and there are only two other left-bank tributaries of any great size. The Severn itself can be divided into four morphologically rather distinct zones: the broad valley with its alluvial terraces below Bridgnorth; the Ironbridge gorge near the limits of Devensian ice (Figure 3.1); the southern Shropshire plain centred on Shrewsbury and including the right-bank Rea Brook and the larger left-bank River Tern (852 km²); and finally the Welsh mountain areas drained by the Severn itself and the Vyrnwy (878 km²) which is nearly as large in area as the Severn itself where they join. The Vyrnwy reservoir (completed in 1891) regulated the flow of the Vyrnwy, while the Severn is regulated by the Clywedog (completed in 1968).

It is this fourth 'Welsh mountain' zone which most closely resembled the Wye, though without the latter's lower incised meander section. Major morphological contrasts within the Severn basin may further be appreciated by considering Table 3.1. This gives data for channel slope and for stream frequency per unit area both down the River Severn itself with increasing catchment area, and for selected subcatchments of intermediate size. While in the headwaters gradients and drainage densities are high, both decrease with increasing catchment area. For catchments of around the same size, the uplands are naturally steeper in gradient, with higher drainage densities. In part this relates to lithologies and soils, but also reflects much greater effective precipitation in the West. For example, the almost adjacent Vyrnwy (878 km²) and Tern (852 km²), of comparable size, have mean discharges of 20 and 7 m³s⁻¹ respectively, the former draining the uplands with high precipitation and the latter the Shropshire plain with lower precipitation.

Morphologically, therefore, there is an East–West split between uplands with high unit discharge and gradient and the less dramatic English valley systems to the East. Such morphological contrasts are important in producing persistent hydrological contrasts in the basin and for providing a contrasted legacy by way of palaeohydrological evidence. Thus the narrower western valley floors and their discontinuous alluvial basins (Figure 3.1) have commonly been actively reworked by rivers in the last 15,000 years; terraces of Devensian and pre-Devensian age are preserved in the East, not only

because they are beyond Devensian ice limits, but also because river energy
has not caused such alluvial reworking.

## GEOLOGICAL EVOLUTION

The Severn–Wye Basin contains a considerable range of Palaeozoic and
Mesozoic rocks (Table 3.2). These are of contrasted origins reflecting a rather
remarkable history of crustal evolution. This can only be briefly outlined
here; much more detailed discussion is available in Anderton *et al.* (1979).

TABLE 3.1   Catchment areas, stream frequency and channel slope

| River | Station | Area (km²) | Stream frequency (km.km⁻²) | S10/85 (m.km⁻¹) |
|-------|---------|------------|----------------------------|-----------------|
| Severn | Plynlimon weir | 8 | 3.6 | 63.5 |
| | Abermule | 580 | 1.8 | 3.6 |
| | Montford | 2030 | 1.5 | 1.2 |
| | Bewdley | 4330 | 0.9 | 0.6 |
| | Upton | 6990 | 1.1 | 0.5 |
| Avon | Evesham | 2210 | 0.7 | 0.9 |
| Teme | Tenbury | 1130 | 0.9 | 3.0 |
| Lugg | Lugwardine | 886 | 0.6 | 2.3 |
| Tern | Walcot | 852 | 0.1 | 1.5 |

Source: NERC Flood Studies Report (1975).

TABLE 3.2   Pre-Tertiary geology of the Severn–Wye Basin

| Era | System | Surface outcrop (%) |
|-----|--------|---------------------|
| Mesozoic | Jurassic ⎱ Triassic ⎰ | 19 |
| | | 17 |
| Palaeozoic | Permian | |
| | Carboniferous | 5 |
| | Devonian | 24 |
| | Silurian ⎱ Ordovician ⎬ Cambrian ⎰ | 34 |
| Precambrian | | <1 |

The earliest rocks in the area are Precambrian in age; their outcrop areas
are small, and these are to be found near the Church Stretton fault (the
Uriconian volcanics and Longmyndian sedimentaries of Shropshire) and the
East Malvern fault (gneisses and volcanics of the Malvern Hills) in Figure

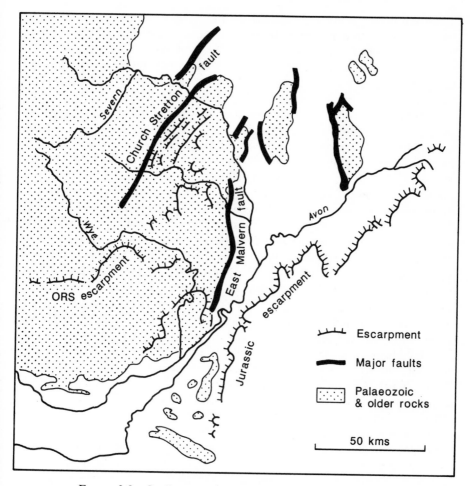

FIGURE 3.2    Outline geology, escarpments and major faults

3.2. The succeeding Lower Palaeozoic rocks (Cambrian, Ordovician, Silurian) cover about one-third of the study area and consist of marine turbidites, with some manganese shales deposited in shallow enclosed basins, and also some volcanics. These sediments dominate the northwest of the region in Wales; their thinly-bedded gritstone and sandstone members provide the gravel-size bed material of contemporary upland Welsh rivers, and the grey-coloured 'Welsh' till of later times.

In the Devonian, marine conditions ended with the closing of a proto-Atlantic Ocean and the region became part of a major continental land mass welded together and including what is today Scandinavia, Greenland and north-west Canada. What is called the 'Anglo-Welsh cuvette' was one of the major embayments on the southern border of this continent which largely became filled with continental fluvial sediments. These now dominate in the middle Wye basin. J. R. L. Allen (1974) in particular has reconstructed former Devonian fluvial environments in great detail and sophistication, and his studies are an example to the palaeohydrologists of this volume, working with sediments some 400 my younger.

Carboniferous times were marked by renewed marine sedimentation and by oscillating conditions which provided both marine limestones, sandstones and shales, and coal deposits. These are found in South Wales and on the Palaeozoic margins of the Severn Basin. Starting in the late Carboniferous, Variscan (Armorican) earth movements were associated with uplift, in particular imposing a SW/NE alignment of structures in Wales, veering to more nearly N/S in the Lower Severn area (see fault alignments in Figure 3.2). Subsequent late Palaeozoic and early Mesozoic sedimentation, largely in continental and generally arid conditions, was involved with the erosion of mountain areas and the infilling of marginal basins. This period saw also the initial opening of what is now the North Atlantic, but it included rifting in the North Sea, and block faulting in the present study area. Continental conditions (Triassic) were succeeded by deltaic and marine ones of considerable spatial complexity in Jurassic times, and it is under these fluctuating conditions that the Mesozoic rocks underlying the Avon basin developed.

Later Mesozoic and Cenozoic sediments are not present in the study area, but more recent tectonic history has been associated with continued widening of the North Atlantic, subsidence of the North Sea basin and a probable differential W/E uplift across our area. It is on this W/E slope, generally tilted away from the rifted margins of the North Atlantic, that early drainage lines were initiated, possibly on an emergent cover of Cretaceous Chalk sediments. In the Tertiary, western Britain was again affected by igneous activity and rifting, linked with a phase of Atlantic sea floor spreading, but also to 'Alpine' effects associated with southern European mountain-building.

Earlier models of British landscape evolution, developed in south-east England (Wooldridge and Linton, 1955), put strong emphasis on such 'European' influences, and on the disruption of an early Tertiary peneplain by the 'Alpine storm'. However, with out present knowledge of plate tectonic ideas, it is probably more satisfactory to consider the 'Atlantic' influence associated with the complexities of trailing plate margin development as being more significant for the development of British landforms as a whole. Such development (like the generation of hard rock assemblages just outlined) has not, however, followed a simple pattern. Inherited Variscan trends may be found picked-out by Severn–Wye drainage alignments. Differential erosion and fault reactivation in the Palaeozoic rocks has led to the persistence of features related initially to the opening and closure of oceans more ancient than the present Atlantic. Ideas on drainage evolution, initiated by J. B. Jukes in souther Ireland, systematized by W. M. Davis, and applied to the gently warped Mesozoic terrains of south-east England, are not so readily put to use in the tilted and faulted blocks and basins of the Midlands and West. Thus in recent years doubts have been expressed as to the former extent of a blanketing chalk cover on which the present drainage lines were assumed to have developed (George, 1974a,b). Evidence from a coastal borehole at Mochras, Merioneth (Woodland, 1971) suggested a local absence of chalk by Palaeogene (earlier Tertiary) times and that this period saw the production of a low relief landscape, but also with considerable contemporary earth movements and fault basin sedimentation (see also Walsh and Brown, 1971, for a discussion of other Welsh Palaeogene sediments, and Curry *et al.*, 1978, for an appraisal of the British Tertiary more generally). Neogene (later Tertiary) faulting and warping were also considerable, eventually leading to the production of upland plateaux whose precise origin, however, remains debatable. E. H. Brown (1960), after carefully mapping these surfaces, interpreted the higher ones as subaerial and the lower marine; George (1974a,b) favoured a marine origin in general with a late development of the drainage system on emergent marine surfaces; Battiau-Queney (1984) regarded the surfaces as being relatable to features outlined as far back as Devonian (post-Caledonian folding) times, with an important and prolonged period of subaerial tropical weathering since the Cretaceous, with the fundamental surfaces so produced being deformed by Neogene tectonic activity into separate blocks and then eroded by a newly entrenched drainage system.

## THE QUATERNARY

The 'pre-glacial' landscape of the Severn Basin was thus one of contrasts, as between 'highland' and Palaeozoic Wales in the west, and the 'lowland' Mesozoic English Midlands. However, the valley system of the Severn was then still considerably different from that seen today. Early erratic material in the Thames Basin gravels suggests at least a connection between the Thames and North Wales (possibly by the introduction of glacial materials into the Midlands and thence to the London basin by an ancestral Thames; see Green, Hey and McGregor, 1980), while later glaciations appear not only to have diverted the Lower Thames (see Gibbard, 1986), but also a formerly NE-flowing proto-Avon (Shotton, 1953) and a northerly-flowing Upper Severn (Hamblin, 1986) (see Figure 3.3). The present Severn, therefore, is a surprisingly 'modern' river system, parts of which have only become connected, or reconnected, in the last 20,000 years.

Table 3.3 gives in major outline a post-Hoxnian Quaternary stratigraphy for the Severn Basin. In fact only the most recent Quaternary stages are as yet locally and unequivocally in evidence, covering only about a quarter of

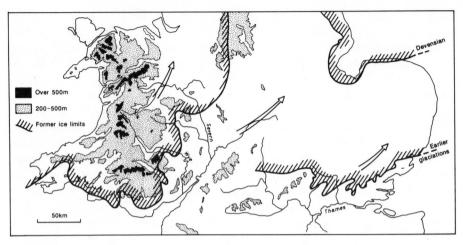

FIGURE 3.3   Central England and Wales, showing relief, ice limits and former drainage

the whole Quaternary. The complexity and equivalence elsewhere of some of the stages listed in Table 3.3 are also the subject of continued vigorous discussion; the interested reader is referred to the listed references, and reviews by Mitchell *et al.* (1973), Shotton (1977, 1985) and Worsley (1985). The last, or main Devensian, glaciation reached its greatest extent comparatively late in that stage (between 25,000 and 18,000 BP). This was associated with the breaching of a divide between Upper and Lower Severn running NE/SW through Buildwas (Hamblin, 1986). A series of subglacial channels just north of the Severn is associated with ice advance over this watershed, but Ironbridge gorge was believed by Hamblin to have formed as an ice-marginal channel along the edge of a stagnant ice sheet. Ice limits have also been determined in the Wye valley and Herefordshire (Lewis, 1970; Luckman, 1970), while there are several (later) valley moraines of uncertain age in both the Severn and Wye systems. The Devensian ice limit is plotted in Figure 3.1, showing that nearly half of the basin was glaciated (66 per cent of the present Wye and 42 per cent of the present Severn valleys). Thin till deposits (which are not mapped) do commonly blanket the Welsh uplands where they may be part of a heterogeneous mixture of slope deposits reworked by periglacial processes (see Potts, 1971). Devensian sediments are most extensive in Shropshire where they include complex till, outwash and lacustrine deposits (see Worsley, 1985). Here the effects of an interaction between ice of Welsh and Irish Sea origin, and of deglaciation, have been worked out in some detail.

TABLE 3.3 Quaternary stratigraphy in the Severn Basin

| Stage | Type locality | Notes on Severn Basin | References |
|---|---|---|---|
| Flandrian (Holocene) | | Post 10,000 BP; present floodplain | Shotton (1978); A. G. Brown (1983) |
| Devensian | Four Ashes (SJ 914082) | Late: 26,000 BP–10,000 BP. Major glaciation and diversion of Severn through Ironbridge gorge. Main, Worcester and Power House Terraces on Severn | Tomlinson (1925); Wills (1938); Shotton (1977); Hamblin (1986) |
| | | Middle: 50,000–26,000 BP. Includes Upton Warren interstadial complex. Avon no. 2 terrace | Coope et al. (1961); Morgan (1973) |
| | | Early: pre-50,000 BP. Includes Chelford interstadial | Worsley et al. (1983); Worsley (1985) |
| Ipswichian | Bobbitshole, Ipswich | Interglacial sediments also present at Four Ashes. Probably more than one stage involved. Avon terrace no. 3 | |
| Wolstonian | Wolston, Warwicks (SP 411748) | Major glaciation, reversal of Avon drainage | Shotton (1953, 1976); Sumbler (1983) |
| Hoxnian | Hoxne, Suffolk | Interglacial lake sediments at Nechells and Quinton | Kelly (1964); Shotton and Osborne (1965); Shotton (1985) |

Source: After Mitchell et al. (1973), with additions.

The pre-glacial Devensian contains interstadial complexes (Chelford, Upton Warren) but in general appears to have been rather cold. The Devensian type-locality at Four Ashes has been particularly important in determining the sequence of events, and Lockwood (1979) has attempted some palaeohydrological reconstructions for this period. Important also from the present point of view has been the analysis of terraces and their sediments in the Severn and Avon valleys, beyond the Devensian glacial limits and dating both to Devensian and earlier stages (Tomlinson, 1925; Wills, 1938). Thus five Avon terrace stages (numbered from lowest to highest) have been distinguished; in the Severn, the Main, Worcester and Power House terraces are believed to be Late Devensian. These terraces are discussed in some detail in Chapter 13.

The pre-Devensian Quaternary cannot be discussed here in detail; in part it is still controversial. Both tills and fluvial sediments of earlier date are important in the Avon basin (Figure 3.1).

## DISCUSSION

This brief review of the morphology, geology and Quaternary development of the Severn Basin has shown that the basin as a whole provides a highly heterogeneous basis for the subsequent hydrological developments discussed in this volume. The changes of the last 15,000 years are in no way as dramatic as those involving drainage diversion and meltwater drainage in the earlier Devensian, while the earlier stages of the Quaternary and the origins of the drainage system itself remain obscure and controversial. However, it is important to appreciate that such a history has provided so varied a heritage: relief contrasts and a complex cover of superficial deposits (Figure 3.1) must mean that the hydrological response to environmental change in more recent times is one of variety, with a spatial diversity to the evidence available for palaeohydrological interpretation. The Severn Basin, though small, does contain a range of temperate zone environments appropriate for the detailed analysis of later chapters.

## REFERENCES

Allen, J. R. L. (1974). Studies in fluviatile sedimentation: implications of pedogenic carbonate units, Lower Old Red Sandstone, Anglo-Welsh outcrop, *Geological Journal*, **9**, 181–208.

Anderton, R., Bridges, P. H., Leeder, M. R. and Sellward, B. W. (1979). *A Dynamic Stratigraphy of the British Isles*, Allen & Unwin, London, 301 pp.

Battiau-Queney, Y. (1984). The pre-glacial evolution of Wales, *Earth Surface Processes and Landforms*, **9**, 229–52.

Brown, A. G. (1983). Floodplain deposits and accelerated sedimentation in the lower Severn basin, in Gregory, K. J. (ed.), *Background to Palaeohydrology*, Wiley, Chichester, pp. 375–97.

Brown, E. H. (1960). *The Relief and Drainage of Wales*, University of Wales Press, Cardiff, 187 pp.

Coope, G. R. (1977). Fossil Coleopteran assemblages as sensitive indicators of climatic changes during the Devensian (Last) cold stage, *Philosophical Transactions of the Royal Society*, Series B, **280**, 313–40.

Coope, G. R., Shotton, F. W. and Strachan, I. (1961). A Late Pleistocene fauna and flora from Upton Warren, Worcestershire, *Philosophical Transactions of the Royal Society*, Series B, **244**, 279–421.

Curry, D., Adams, C. G., Boulter, M. C., Dilley, F. C., Eames, F. E., Funnell, B. M. and Wells, M. K. (1978). *A Correlation of Tertiary Rocks in the British Isles*, Geological Society of London Special Report no. 12, 72 pp.

George, T. N. (1974a). The Cenozoic evolution of Wales, in Owen, T. R. (ed.), *The Upper Paelaeozoic and post-Palaeozoic Rocks of Wales*, University of Wales Press, Cardiff, pp. 341–71.

George, T. N. (1974b). Prologue to a geomorphology of Britain, *Institute of British Geographers special publication no. 7*, 113–25.

Gibbard, P. L. (1985). *The Pleistocene History of the Middle Thames Valley*, Cambridge University Press, Cambridge.

Green, C. P., Hey, R. W. and McGregor, D. F. M. (1980). Volcanic pebbles in

Pleistocene gravels of the Thames in Buckinghamshire and Hertfordshire, *Geological Magazine*, **117**, 59–64.

Hamblin, R. J. O. (1986). The Pleistocene sequence of the Telford district, *Proceedings of the Geologists' Association*, **97**, 365–77.

Kelley, M. R. (1964). The Middle Pleistocene of North Birmingham, *Philosophical Transactions of the Royal Society*, Series B, **247**, 533–92.

Lewis, C. A. (1970). The Upper Wye and Usk Regions, in Lewis, C. A. (ed.), *The Glaciations of Wales and Adjoining Regions*, Longman, London, pp. 147–73.

Lockwood, J. G. (1979). Water balance in Britain, 50,000 yr BP to the Present Day, *Quaternary Research*, **12**, 297–310.

Luckman, B. B. (1970). The Hereford Basin, in Lewis, C. A. (ed.), *The Glaciations of Wales and Adjoining Regions*, Longman, London, pp. 175–96.

Miller, A. A. (1935). The entrenched meanders of the Herefordshire Wye, *Geographical Journal*, **85**, 160–72.

Mitchell, G. F., Penny, L. F., Shotton, F. W. and West, R. G. (1973). *A Correlation of Quaternary Deposits in the British Isles*, Geological Society of London Special Report, no. 4, 99 pp.

Morgan, A. V. (1973). The Pleistocene geology of the area north and west of Wolverhampton, Staffordshire, England, *Philosophical Transactions of the Royal Society*, Series B, **265**, 233–97.

Natural Environmental Research Council (1975). *Flood studies report* (5 vols), Wallingford.

Potts, A. S. (1971). Fossil cryonival features in central Wales, *Geografiska Annaler*, **53A**, 39–51.

Shotton, F. W. (1953). The Pleistocene deposits of the area between Coventry, Rugby and Leamington and their bearing upon the topographic development of the Midlands, *Philosophical Transactions of the Royal Society*, Series B, **237**, 209–60.

Shotton, F. W. (1976). Amplification of the Wolstonian stage of the British Pleistocene, *Geological Magazine*, **113**, 241–50.

Shotton, F. W. (1977). Chronology, climate and marine record, the Devensian stage, *Philosophical Transactions of the Royal Society*, Series B, **280**, 107–18.

Shotton, F. W. (1978). Archaeological inferences from the study of alluvium in the lower Severn-Avon valleys, in Limbrey, S. and Evans, I. G. (eds), *The Effect of Man on the Landscape: the Lowland Zone*, C.B.A. research reported 21, 27–32.

Shotton, F. W. (1985). Quaternary glaciations in the Northern Hemisphere. Final Report by the National Correspondent, *Quaternary Newsletter*, **45**, 28–36.

Shotton, F. W. and Osborne, P. J. (1965). The fauna of the Hoxnian interglacial deposits of Nechells, Birmingham, *Philosophical Transactions of the Royal Society*, Series B, **248**, 353–78.

Simpson, I. M. and West, R. G. (1958). On the stratigraphy and palaeobotany of a late-Pleistocene organic deposit at Chelford, Cheshire, *New Phytologist*, **57**, 239–50.

Sumbler, M. (1983). A new look at the type Wolstonian glacial deposits of Central England, *Proceedings of the Geologists' Association*, **94**, 23–31.

Tomlinson, M. E. (1925). River-terraces of the lower valley of the Warwickshire Avon, *Quarterly Journal of the Geological Society of London*, **81**, 137–63.

Walsh, P. T. and Brown, E. H. (1971). Solution subsidence outliers containing probable Tertiary sediments in North East Wales, *Geological Journal*, **7**, 299–320.

Wills, L. J. (1938). The Pleistocene development of the Severn from Bridgenorth to the sea, *Quarterly Journal of the Geological Society of London*, **94**, 161–242.

Woodland, A. W. (ed.) (1971). The Llanbedr (Mochras Farm) borehole, *Institute of Geological Sciences*, report 71/18, 115 pp.
Wooldridge, S. W. and Linton, D. L. (1955). *Structure, Surface and Drainage and South-east England*, Philip, London 176 pp.
Worsley, P. (1985). Pleistocene history of the Cheshire-Shropshire Plain, in Johnson, R. H. (ed.), *The Geomorphology of North-west England*, Manchester University Press, Manchester, pp. 201–21.
Worsley, P., Coope, G. R., Goode, T. R., Holyoak, D. T. and Robinson, J. E. (1983). A Pleistocene succession from beneath Chelford sands at Oakwood Quarry, Chelford, Cheshire, *Geological Journal*, **8**, 1–16.

Palaeohydrology in Practice
Edited by K. J. Gregory, J. Lewin and J. B. Thornes
© 1987 John Wiley & Sons Ltd.

# 4

# Spatial Variability in the Climate of the Severn Basin: a Palaeohydrological Perspective

D. M. LAWLER

*Department of Geography, University of Birmingham*

## INTRODUCTION

It is often overlooked that the degree of spatial variation in the contemporary climatic character of a region may be much greater than inferred temporal changes of climate at individual locations. Also, 'spatial variability is as important as temporal change in palaeohydrology' (Thornes, 1983, p. 52). The aim of this short chapter is not to attempt to review the 'climate of the Severn Basin' but to examine briefly through a series of maps the nature of spatial variability within the catchment of a few key meteorological elements most pertinent to palaeohydrological investigations. First, some comments are made on the need for, the utility of, and the limitations imposed by, climatic data in relation to palaeohydrological enquiries. The material should complement the hydrological issues raised in Chapter 3 and 5 and act as a background to a consideration of historical climatic change within the project area that is reserved for Chapter 6. Unless specified otherwise, the 'Severn Basin' here is taken to include the Wye and Avon catchments.

## THE NEED FOR CLIMATIC INFORMATION

If the objectives of palaeohydrology are taken to include 'reconstruction of the components of the hydrological cycle, of the water balance, and of sediment budgets for the time before continuous records, and necessarily embracing an understanding of the way in which changes in the hydrological cycle occurred and establishing how they differed from the contemporary hydrological picture' (Gregory, 1983, p. 10), then the specific contribution that climatology may make towards the achieving of these objectives should first be defined. Two main areas can be identified as follows: the relationships between contemporary climatic characteristics and the operation of hydrological systems (the outputs of which may also help to control geomorphological phenomena such as sediment yield and channel form) and the direct impact of climate upon geomorphological processes. At least three problems, however, need to be specified at this stage. First, 'the relationship between climatic inputs and geomorphic processes remains for the most part poorly documented' (Derbyshire, Gregory and Hails, 1979, p. 40) and is often couched in simple bivariate terms. Although significant advances in this respect have been made over the last 10 years, great uncertainty still surrounds the operation of thresholds, system inertia and sensitivity. Second, many relationships in climatic geomorphology have, often necessarily, been obtained through a spatial approach (e.g. the relationship between annual rainfall and drainage density: Gregory and Gardiner, 1975; Walsh, 1980). It is still unclear, however, whether these spatially-derived relationships can be used with confidence in a temporal framework to infer likely process or form change consequent upon a specified climatic change—the so-called ergodic, or space-time substitution, problem. Third, many relationships that have been defined are expressed through surrogate (independent) variables: for example, mean annual rainfall substitutes as an index of flow character in ungauged catchments and Stevenson-Screen air temperatures are often used to infer conditions at a distant soil surface. Little work has been achieved in refining the relationships between these surrogate variables and the real controlling factors involved, particularly with regard to the presence of non-linearities and the temporal stability of such relationships.

Notwithstanding these problems, a number of climatically-influenced processes and forms, of interest to palaeohydrologists, have been defined at various levels of resolution and with various degrees of success (Table 4.1). Many of these associations are fully reviewed in Derbyshire (1976), but a few points may be made here. Mean annual rainfall has often been used as a simple predictor of runoff volumes, sediment yields and channel geometry (Table 4.1), partly because the relevant precipitation data are reasonably easy to obtain on a global scale at a sufficiently fine spatial resolution, with usable records often extending over the last 100 years. However, at least one worker has suggested that, in terms of predicting sediment yield, 'the role and importance of climatic variables often undergo a seasonal change and [therefore] mean annual values are useless' (Douglas, 1976, p. 281). Hence many workers have tried to look at the relationships between geomorphological phenomena and more meaningful rainfall indices (Table 4.1) such as effective precipitation (which accounts for evapotranspiration losses) (Langbein and Schumm, 1958), rainfall intensity (e.g. Wischmeier, 1959) and rainfall seasonality (e.g. Fournier, 1960). Similarly, although mean annual temperature has been used as an index of, for example, weathering rates (Peltier, 1950) or to weight precipitation values for evaporative or vegetative effects (Langbein and Schumm, 1958), it is now more usual to look at more finely resolved indices. Thus, the nature of freeze–thaw cycles has been used in explanations of rock breakdown (e.g. McGreevy, 1982) and slope erosion (Soons and Greenland, 1970), while Lawler (1986, 1987) has demonstrated the strong control of frost variables over bank erosion in a South Wales catchment. Furthermore, there has been a distinct tendency recently to embrace surface and subsurface microclimatology within process investigations generally, rather than simply using standard data recorded at meteorological stations under conditions of questionable geomorphological relevance. For example, Thorn (1980) and Lawler (1987) measured, respectively, rock and river-bank temperatures instead of relying exclusively on data drawn from Stevenson screens.

## SPATIAL PATTERNS OF CLIMATE

### Introduction and climatic background

The purpose of this section is to examine the relative magnitudes and spatial distribution of those climatic parameters more appropriate to a palaeohydrological investigation (many of which feature in Table 4.1). Thus discussion concentrates on rainfall or temperature variables rather than, for example, fog incidence or wind speed. Wind erosion has been shown to be important in some exposed localities of the UK such as north-east Scotland (Pidgeon and Ragg, 1979) but in Wales, for example, with a relatively low frequency of gale-force winds (>33 knots or 17.0 m s$^{-1}$) (Sumner, 1978a) and lack of exposed, bare soil areas (cultivated or otherwise), it is probably a generally insignificant process (cf. Reed, 1979).

As a precursor to more detailed consideration of individual meteorological variables, it would be helpful to outline some of the main features of the climate of the basin. Few regional climatologies of the project area have been written, however, with the western Midlands, Welsh borders and Mid-Wales largely unrepresented in national climatological series (see Gregory, 1976, his Table 15.1 and Figure 15.1). Some useful accounts which include parts of the Severn Basin (or areas peripheral to it) have been produced, though, for the following areas: Wales (Meteorological Office, 1983; Sumner, 1978a); Birmingham area (Saward, 1950) the Midlands (Meteorological Office, 1982); Warwickshire (Warwick, 1971); Keele area (Beaver and Shaw, 1970); South Wales (Faulkner and Perry, 1974; Oliver, 1971; Perry, 1979). More specific localities are dealt with by, *inter alia*, Beckinsale (1934), Burt (1975), Garrett (1913), Harrison (1977) and Newson (1979). A reasonably extensive bibliography of climatic literature relevant to the Severn Basin can

TABLE 4.1  Some studies of the relationships between simple climatic and hydrological and geomorphological variables

| 'DEPENDENT' VARIABLE | CLIMATIC VARIABLE | | | | | | |
|---|---|---|---|---|---|---|---|
| | Mean annual rainfall | Rainfall seasonality | Rainfall intensity | Effective precipitation[a] | Antecedent precipitation | Mean monthly/ annual temp. | Frost indices |
| Mean annual runoff/flood hydrology | Langbein et al. (1949)[c] Marsh & Littlewood (1978)* NERC (1975) Nash & Shaw (1966)[c] | Howe et al. (1967)* | Howe et al. (1967)*[b] NERC (1975)[b] Osborn & Lane (1969)[c] | Hamlin (1971)* | Hamlin (1971)* | Hamlin (1971)* Langbein et al. (1949)[c] | Dreibelbis (1949) |
| Overland flow | | | Anderson et al. (1984) Horton (1933, 1945)[c] Walsh (1980) | | | | |
| Throughflow/ pipeflow | Hewlett & Hibbert (1967) | | Dunne (1978) Jones (1978)* Walsh (1980) | | Arnett (1974) Jones (1978)* | | |
| Soil erosion | | | Ellison (1945)[c] Fullen & Reed (1986)* Reed (1979)[c] Wischmeier (1959) | | Dragoun (1962)[c] Wischmeier & Smith (1958)[c] | | Gradwell (1955) Soons & Greenland (1970) |
| River bank erosion | | | Gardiner (1983) Hooke (1979) Kesel & Baumann (1981) Lawler (1986)* | | Hooke (1979) Lawler (1986)* | | Blacknell (1981)* Gardiner (1983) Hill (1973) Lawler (1986, 1987)* Thorne (1978)* Wolman (1959) |

| Process | | | | | | |
|---|---|---|---|---|---|---|
| Gully erosion | Wilson (1973) | | Harvey (1974)<br>Thompson (1964)[e] | | Bridges & Harding (1971)* | Bridges & Harding (1971)*<br>Gardiner & Dackombe (1980)*<br>Harvey (1974) |
| Mass movement | | | Henkel & Skempton (1955)* | | Prior et al. (1971) | Rapp (1960)[e]<br>Wilson (1973)  Clauzon & Vaudour (1971)[c] |
| Sediment yield/ concentration | Fournier (1960)[c]<br>Guy (1964)[e]<br>Langbein & Schumm (1958)[e]<br>Walling & Webb (1983)<br>Wilson (1973) | Douglas (1969b)[c]<br>Fournier (1960)[c] | Dragoun (1962)[e]<br>Guy (1964)[e] | Langbein & Schumm (1958)[c] | Douglas (1970)[c]<br>Dragoun (1962)[e]<br>Finlayson (1978) | Corbel (1964)[e]<br>Guy (1984)[e]<br>Schumm (1965) |
| Weathering | Peltier (1950)<br>Weinert (1971)[d]<br>Wilson (1973) | | | | | Peltier (1950)<br>Wilson (1973)  McGreevy (1982)<br>Thorn (1980) |
| Drainage density/ texture/ network volume | Abrahams (1972)[f]<br>Chorley (1957)[e]<br>Gregory & Gardiner (1975)<br>Gregory & Ovenden (1979)<br>Morgan (1970)<br>Sumner (1978b)*<br>Walsh (1980) | Morgan (1970) | Chorley (1957)[e]<br>Chorley & Morgan (1962)[e]<br>Gregory & Ovenden (1979)<br>Melton (1957)[f] | Chorley (1957)[e,a] | | |

* Studies from the Severn basin (or near to it e.g. Wales, Midlands).

[a] Representing, too, 'precipitation-effectiveness'—a slightly different concept.

[b] Strictly the magnitude of a 2-day rainfall likely to be exceeded every two years.

[c] Cited in Douglas (1976).

[d] Cited in Stoddart (1969).

[e] Cited in Gregory and Walling (1973).

[f] Cited in Richards (1982).

be found in Lawler (1979). Many chapters in Chandler and Gregory (1976) contain useful analyses of weather patterns and individual meteorological phenomena for the British Isles as a whole, and the contributions by Gregory (1976), Perry (1976) and Taylor (1976) are especially relevant to the Severn Basin.

The Severn Basin is relatively large (its east–west length at the widest point is 190 km, compared to a total width across England and Wales at the same latitude of 400 km) and also includes a large range of altitude (~0–900 m O.D.): hence a number of different climate-types are embraced. Essentially, the climate, particularly in the west, is dominated by north-westerly and south-westerly airstreams, with both air-masses being moist because of long oceanic fetches, and associated with considerable precipitation especially when fronts and depressions move in from the Atlantic. Rain falls on average on almost two out of every three days in the west. With such a strong maritime influence, temperatures are moderate, relative to the more continental parts of Europe, with mild winters and cool summers. Gregory (1976) has produced a useful regional classification of the British Isles climate, based on the length of the growing season, annual rainfall conditions and the seasonal incidence of rainfall, and Figure 4.1 shows the relevant portion of his map. Six of the fifteen categories used by Gregory are represented in the Severn Basin, but the main trend is one of increasing wetness, a growing prominence of a winter rainfall maximum, and a shortening of the growing season (from 7–8 to 5–6 months) as one moves west across the catchment (Figure 4.1). This is because of an increase in both oceanicity and altitude in this direction. North–south trends are minimally important.

**Nature of the climate data**

A crucial factor in any examination of spatial variation in climatic characteristics is the nature of the data available, in terms of areal coverage and density of meteorological stations, the length of record obtained, and the quality of the resultant information. Although serious and organized collection and publication of *rainfall* data in England and Wales began in the 1860s with the appearance of *English Rainfall*, followed by *British Rainfall*, meteorological stations measuring a *range* of climatic variables really only started to become established in the Severn Basin from the 1930s (although a number of individuals had kept earlier quantitative or semi-quantitative weather diaries at various sites—see Lawler (1979) for a selected list). A time of relative stagnation in network development followed around the 1958–67 period, succeeded by rapid growth through the 1970s (Figure 4.2). Subsequently, the 1980s have seen little development in the network, partly because resources available for scientific public bodies in general have declined, but also because some rationalization was thought to be necessary on efficiency grounds. In 1976 there were 45 'official' (Meteorological Office

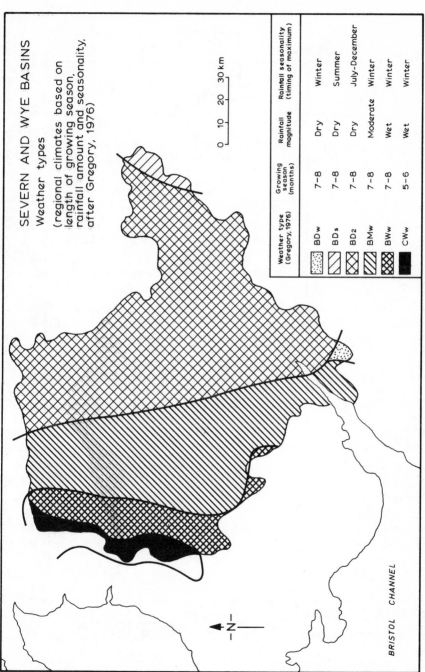

FIGURE 4.1   Weather types over the basin (reproduced by permission of Longman from Chandler and Gregory, 1976)

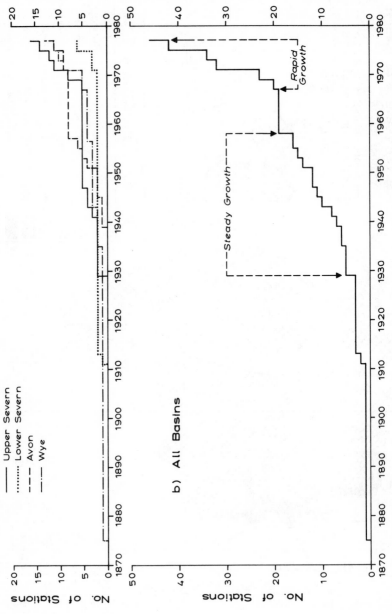

a) Individual Basins

Upper Severn
Lower Severn
Avon
Wye

b) All Basins

FIGURE 4.2   Growth of the meteorological station network in (a) the subcatchments and (b) the whole basin of the Severn

recognized) stations in operation: 22 in the Severn Basin, 13 in the Avon Basin and 10 in the Wye catchment, although only 16 of these had been established before 1957, making the calculation of long-term averages and hence an assessment of spatial variability moderately difficult (Lawler, 1980). Figure 4.3 shows the distribution of these climatological stations and, while the average density of stations is quite representative of the rest of the UK (Severn Basin has one station per 330 km$^2$), it is clear that the coverage is not at all uniform. In particular, the whole of the Teme catchment, the Severn Basin below Gloucester and the southern third of the Wye catchment are very poorly represented, compounding the difficulties noted above of a meaningful analysis of spatial patterns. The distribution becomes yet more uneven when only the longer-period stations are considered. Figure 4.3 also attempts to identify the potentially more useful stations on the basis of record length (as suggested by Starkel and Thornes, 1981, p. 11) and the comprehensiveness of the range of meteorological variables sensed. A tabulated breakdown of meteorological station details appears in Lawler (1980).

In terms of precipitation stations, the situation is much more encouraging. In line with a generally high density of raingauges in Great Britain—with around 7000 in total, and second in the density of coverage only to Israel (Smith, 1972, p. 42)—the Severn Basin boasted on a November 1979 printout of *Rainmaster* (the computerized register of raingauges maintained and regularly updated by the Meteorological Office) a total of 547 rainfall stations (see Figure 4.4A). This gives a raingauge density slightly better than the Great Britain average of one gauge for every 39 km$^2$ (Smith, 1972, p. 42). Some of the gauges, operating on the tipping-bucket principle, also provide data on precipitation intensity, duration and timing, and are interrogable from a remote centre (Figure 4.4B). Such information can prove invaluable in flood-forecasting, rainfall-runoff modelling and the explanation of geomorphological processes. The network, despite some non-uniformity in distribution, appears to present few problems for subsequent spatial analysis.

The standard period adopted by the World Meteorological Organization and the Meteorological Office, for which averages of sites are calculated and compared, runs from 1941–70. Early climatological literature usually adopted either the period 1931–60 or 1916–50, although it was the opinion of at least one reviewer that 'the period 1931–60 . . . was probably one of the most abnormal 30-year periods in the last thousand years' (Mason, 1976, p. 478). These periods, along with non-standard alternatives where necessary, are adopted in the discussion below.

## Annual rainfall

The main pattern of average annual precipitation is a clear and pronounced increase (from less than 600 mm to over 2400 mm per annum) towards the western edge of the basin (Figure 4.5). Indeed, this gradient is listed as one reason for choosing the Severn Basin for palaeohydrological study (Gregory and Thornes, 1980). Most commentators agree that altitude is the dominant control over this distribution, through both orographic uplift and an encouragement of local convective showers (Atkinson and Smithson, 1976)—but cf. Atkinson (1983). In addition, Newson (1979) reports that, in the Welsh uplands, higher rainfalls are found to the lee of prominent watersheds with lower values to the windward. Oceanicity probably exerts a subsidiary influence. Analysis of annual average precipitation data 1941–70 for 541 rainfall stations in the Severn and Wye Basins revealed a very strong, albeit heteroscedastic, positive relationship between annual catch and station altitude, with a product-moment correlation coefficient of 0.866 ($r^2 = 74.9$ per cent) (Figure 4.6). Furthermore, the strength of this relationship is largely preserved for each individual month considered, although it weakens slightly for the summer months of June, July and August and for November (Table 4.2). This suggests that the orographic effect is more pronounced with the 'frontal' types of weather system more commonly encountered in the winter (as one might intuitively expect, with depressions moving in from the west) rather than with convective activity more characteristic of summer—a suspicion that is supported by a perusal of maps of monthly rainfall averages for the Severn Basin produced by Rowsell (1963, his Figure 1). Bivariate regression analysis yielded the following expression linking average annual precipitation in millimetres ($P$) to altitude in metres ($A$)

$$P = 480 + 3.06 A \tag{4.1}$$

implying a 3.1 mm increase in annual rainfall for every metre of elevational increase. This altitudinal gradient is shown by Table 4.3 to be quite close to the values found by other workers elsewhere in Britain—e.g. 2.42 mm m$^{-1}$ for the whole of the British Isles (Bleasdale and Chan, 1972); 2.7 mm m$^{-1}$ for Wales (Sumner, 1978a)—therefore, in this sense at least, the Severn basin may be considered fairly representative. However, it will be noted from Figure 4.6 that, as Ballantyne (1983) and Harrison (1973) (cited in Taylor, 1976) found for north-west Scotland and coastal west Wales respectively, the relationship shows some evidence of curvilinearity, whereby the *rate* of increase of rainfall with altitude itself increases towards the higher elevations. Furthermore, equation (4.1) tends to underestimate rainfall for the upland stations in the west of the Severn Basin, and overpredict slightly for the

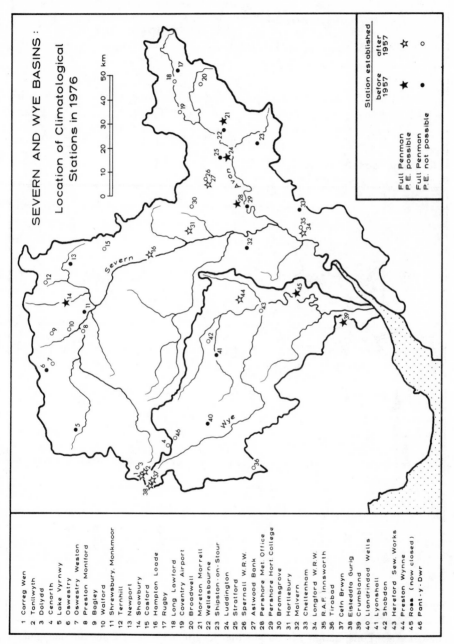

FIGURE 4.3   Climatological stations in the Severn Basin, classified according to length of record available and suitability for the calculation of potential evapotranspiration by the Penman (1963) method (based on data in Severn-Trent Water Authority, Directorate of Operations (1977) and information supplied by the Welsh Water Authority)

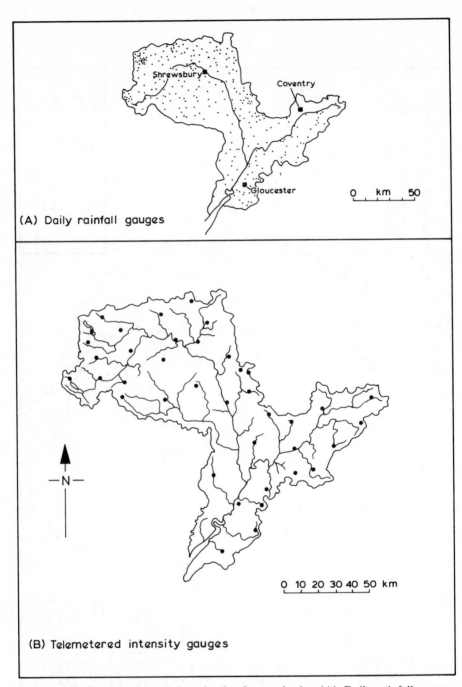

(A) Daily rainfall gauges

(B) Telemetered intensity gauges

FIGURE 4.4 Precipitation stations in the Severn basin: (A) Daily rainfall gauges (Severn-Trent Water Authority, Directorate of Operations, 1977); (B) Telemetered intensity gauges, as in June 1986 (from information kindly supplied by Dr T. R. Wood of the Severn-Trent Water Authority)

eastern localities, suggesting an exhaustion effect (or increasing 'continentality' towards the centre of England) which should be accounted for in any statistical model of the spatial pattern. A simple allowance for this is made by introducing a second independent variable (as Ferguson, 1977, did for some exemplary Scottish data in which similar effects emerged)—the 'easting' (of the O.S. grid) of each of the 541 stations analysed—which produces the multiple regression equation

$$P = 1758 + 2.22 A - 3.22 E \qquad (4.2)$$

in which $E$ = distance east on the Ordnance Survey grid (km). The $r^2$ value significantly improves (using an $F$ test at $p = 0.001$) from 74.9 to 81.2 per cent. The negative partial regression coefficient for eastings, of course, implies that, *ceteris paribus*, a site at a given altitude receives less rainfall in the east of the basin than the west. The multivariate method allows estimation of the relative effect of each variable on rainfall magnitude with the other held constant. For example, *at a given distance east*, it would be expected from equation (4.2) that annual rainfall increases at a rate of 2.22 mm m$^{-1}$ altitude (rather than at 3.06 mm m$^{-1}$ without the attempt to correct for oceanicity): similarly, *at a given altitude*, annual rainfall receipt on average declines 3.22 mm for every kilometre one moves eastwards. With so few upland stations in Wales (Harrison, 1977), and in Britain generally (Taylor, 1976), reliance must often be placed on techniques such as these for the tentative estimation of values in areas of no data.

TABLE 4.2   Correlation coefficients ($r$) between average monthly precipitation and altitude for Severn Basin raingauges ($n = 541$)

| Month | Jan | Feb | Mar | Apr | May | Jun | Jul | Aug | Sep | Oct | Nov | Dec |
|-------|-----|-----|-----|-----|-----|-----|-----|-----|-----|-----|-----|-----|
| $r$ | 0.860 | 0.869 | 0.869 | 0.861 | 0.864 | 0.848 | 0.830 | 0.843 | 0.862 | 0.861 | 0.729 | 0.860 |

## Rainfall seasonality

Some index of the seasonal distribution of rainfall has often been used to help explain hydrological and geomorphological phenomena (Table 4.1). The index used here is a simple one based on (mean) monthly totals (Walsh and Lawler, 1981):

$$\bar{S}I = \frac{1}{P} \sum_{n=1}^{n=12} \bar{X}_n - \frac{P}{12} \qquad (4.3)$$

where $\bar{S}I$ = Seasonality Index, $P$ = mean annual rainfall and $\bar{X}_n$ = mean rainfall of month $n$. The index would assume a value of 0.0 if all months receive equal rainfall, and 1.83 if all rain falls in a single month. Seasonality

TABLE 4.3  Selected studies of altitudinal gradients of mean annual precipitation in upland areas of Great Britain

| Area | Number of raingauges examined | Range of altitude considered (m O.D.) | Precipitation gradient (mm m⁻¹) General | West slope | East slope | Source |
|---|---|---|---|---|---|---|
| *Wales* | | | | | | |
| Western Mid-Wales | 6 | 3–450 | — | 2.28 | — | Harrison (1973*, 1977) |
| Glamorgan Hills | 23 | 10–430[a] | 3.25 | — | — | Hill et al. (1981) |
| Upper Severn and Wye Basins | 39 | 320–740 | 1.71 | — | — | Newson (1976) |
| Ystwyth Basin, Mid-Wales | 19 | 8–457 | — | 1.67 | — | Rodda (1962) |
| Wales (whole?) | ? | ? | 2.67 | — | — | Sumner (1978a) |
| Snowdonia | 47 | 4–713 | 4.58 | — | — | Unwin (1969) |
| *Scotland* | | | | | | |
| North-west Scotland | 4 | 23–671 | — | — | 2.67 | Ballantyne (1983) |
| North-west Scotland | 5 | 15–465 | — | 4.65 | — | Ballantyne (1983) |
| Southern Scotland | 20 | 140–540 | 2.38 | — | — | Ferguson (1977) |
| Scottish Highland | ? | 0–540 | — | 2.81 | ~0.83 | Gloyne (1968) |
| *England* | | | | | | |
| Southern Pennines | 137 | 35–546 | 1.81 | 1.32 | 2.02 | Burt (1980)* |
| Central Penines | ~5 | ~-10–510 | — | 1.88 | 0.98 | Pearsall (1968)* |
| *Other* | | | | | | |
| UK | >6500 | ~0–1093 | 2.42 | — | — | Bleasdale and Chan (1972) |
| Severn, Avon and Wye Basins | 541 | 6–550 | 3.06 | — | — | Author |

* Cited in Ballantyne (1983).

[a] Altitude is computed as mean height over 4 km square, centred 1.5 km to the southwest of each gauge.

in the British Isles is generally low by world standards but interesting differences still emerge when the spatial pattern is examined, even at a regional
scale. Figure 4.7 is a map of $\bar{S}I$ for the Severn Basin and a clear trend of
increasing rainfall seasonality in a westerly direction can be seen. The winter
maximum becomes much more pronounced as one moves towards the Mid-
Wales uplands, and this is in accord with the national pattern (Walsh and
Lawler, 1981) in which seasonality is generally positively related to annual
rainfall amounts in humid temperate regions (but negatively so in some
tropical areas).

### Rainfall event magnitudes

An index of flood-producing rainfall has been mapped for the Severn Basin
in Figure 4.8. This is the '2-day M5' rainfall, or 'the rainfall exceeded on
average once in five years' (Sutcliffe, 1978, p. 43), calculated from an annual
maximum series and mapped for the whole country by the Institute of
Hydrology as an aid to the design engineer needing information on probable
mean and maximum floods in given catchments (NERC, 1975). Again, a
marked east–west (altitudinal) gradient emerges and, as with annual rainfall,
amounts received in the Welsh uplands are approximately three times those
received in the lowlands of the Avon Basin, and the Severn valley downstream of Shrewsbury (Figure 4.8). Turning to daily amounts, the Welsh
uplands (e.g. Lake Vyrnwy) may expect, on average, a 24-hour rainfall of
at least 50 mm every year, whereas eastern areas (e.g. Edgbaston, Birmingham) can only expect a daily total of 29 mm to be exceeded once every
year. Perry (1979) also makes the point that there is very little variation in
the number of rain days (and in the duration of rain falling) between the
coastal lowlands of South Wales and the uplands (about 200–225 days per
annum), implying that the higher precipitation totals recorded in the latter
areas occur because rainfall *intensities* are higher there.

### 'Winter Rain Acceptance Potential'

Although spatial variations in meteorological inputs can help to explain much
of the hydrological response of a basin, information on the character of the
catchment itself is needed to complete the picture. This covers the many
aspects of, for example, land use, soil type, topographic characteristics,
geology and nature of human interference and many of these facets are
explored in other chapters of this volume, but it is useful at this point to
consider briefly an ordinal index of flood runoff potential used by NERC
(1975) and fully described by Farquharson *et al.* (1978) called the 'Winter
Rain Acceptance Potential' (WRAP). The Soil Survey of Great Britain,
using existing information, produced the first WRAP map of England and

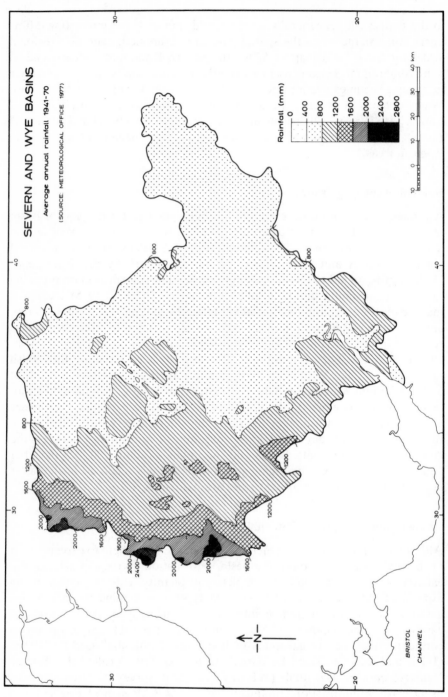

FIGURE 4.5  Average annual rainfall (1941–70) in the Severn Basin from data supplied by the Meterological Office

Wales (at a scale of 1:625,000), which was published by NERC (1975). Essentially, each soil profile is assigned a runoff potential class (from 1, the best drained, to 5, the least permeable) on the basis of soil-water regime (a subjective assessment), depth to an impermeable horizon, permeability above impermeable horizons and slope angle (Farquharson *et al.*, 1978, p. 7). The relevant (redrawn) portion of the original 1975 map (Figure 4.9) shows that (a) only very small parts of the Severn Basin (Mid-Wales mountains) qualify for the highest class of runoff potential; (b) much of the 'High' category (Class 4) land is in the north of the Severn Basin and in the Avon catchment; and (c) at least half of the basin, including almost all of the Wye catchment, is classified as having 'Very Low' or 'Low' runoff potential. An improved map is currently being developed by J. Hollis and A. Gustard of the Soil Survey of England and Wales and Institute of Hydrology respectively.

**Mean annual temperature**

Although 45 climatological stations existed in 1976 (having been recently augumented by the Institute of Hydrology instrumentation at Plynlimon— Figure 4.3), only 21 stations within or near to the Severn Basin have more than 20 years of record in the standard period 1941–70. Also, the spatial distribution of these stations is far from uniform, most being located near the edge of the basin, leaving a large central tract lacking in long-term temperature data. Due to these data shortcomings, the limited areal differences in mean annual temperature, and because mean annual temperature has a questionable relevance to hydrological and geomorphological processes (notwithstanding the examples in Table 4.1), a map of spatial pattern is not presented here. Furthermore, national-scale maps of various temperature indices have been produced elsewhere (Meteorological Office, 1975). To help place the basin in a global context, however, some information is appropriate: the mean annual temperature for the whole basin (an average of 21 long-term stations) is 9.3°C, with a standard deviation of 0.55°C. Temperature tends to decline with increasing altitude: Harrison (1977) found for Mid-Wales an average lapse of 6.66 K km$^{-1}$ for maximum temperatures, and 3.65 K km$^{-1}$ for minimum temperatures, while the Meteorological Office, after comprehensive nationwide surveys, uses values of 7.0 K km$^{-1}$, 6.0 K km$^{-1}$ and 5.0 K km$^{-1}$ to 'correct' to sea-level its maps of maximum, mean and minimum temperature respectively (Meteorological Office, 1975, p. 2). Local site variability and seasonal changes in lapse rate, of course, are known to occur (Harding, 1978; Lawler, 1985).

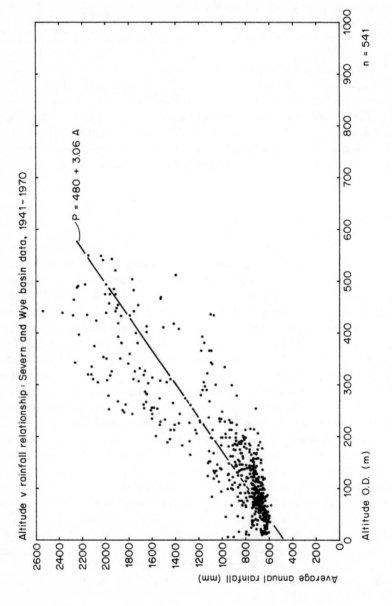

FIGURE 4.6  Relationship between average annual precipitation and altitude in the Severn Basin ($n = 541$).
Data from the *Rainmaster* file of the Meteorological Office, November 1979

**Frost frequency**

In view of the apparent importance of cryergic processes in promoting river-bank and gully erosion within and outside the Severn Basin (e.g. Blacknell, 1981; Bridges and Harding, 1971; Gardiner and Dackombe, 1980; Lawler, 1986, 1987; Thorne, 1978), it is useful to examine spatial patterns of frost activity. However, there are at least three distinct controls on the areal pattern of frost incidence: first, an altitudinal lapse effect, leading to a tendency for higher frequencies of frost on higher ground, second, the well-known occurrence of temperature inversions in hilly terrain (during clear-sky, usually anticyclonic, nights) in which cold air draining katabatically off donor slopes ponds up in valley bottoms, encouraging the maintenance of very low temperatures there and, third, a continental influence, with a reduction in maritime (moderating) influence as one moves inland. All three controls are exercised in the Severn Basin which makes for a great deal of uncertainty in a regional-scale map of air frost frequency based on so few stations (Figure 4.3). Moreover, such maps may not be representative of conditions at the catchment surface, and it should be stressed that one day of air frost is recorded if screen air temperature drops below 0°C in a specified 24-hour period, regardless of the number of 'freeze-thaw' cycles which take place. Nevertheless, Figure 4.10 is the relevant part of the national map constructed by the Meteorological Office (1975), which shows evidence of increases with both altitude and distance from the coast. Given that the efficacy of soil frost action increases in the presence of moisture (Outcalt, 1971; Williams, 1964) then, all other factors being equal, cryergic processes should be most important in the wetter uplands of Wales (compare Figures 4.5 and 4.10). Figure 4.10 does not show the topographically-related 'frost-hollows'—e.g. at Shawbury (Gallagher, 1959) and in the Avon valley (Lawrence, 1952): only a dense network of measurement sites, satellite-imagery (Kalma *et al.*, 1983) or close-range remote-sensing from vehicle-mounted infra-red thermometers (Thornes, 1985) could provide the necessary coverage to quantify these local-scale spatial variations. Statistically based prediction models to handle such effects have also been advanced (e.g. Bootsma, 1976; Myburgh, 1974).

**Potential evapotranspiration**

Potential evapotranspiration (defined as the amount of water evaporated or transpired back into the atmosphere assuming unlimited moisture supplies), depending upon the method of computation used, is related positively to air temperature, solar radiation receipt (or sunshine hours) and wind speed, and negatively to atmospheric humidity. As such it is an estimated, not measured, quantity (see Ward, 1975, for detailed discussion of the concept). Being a

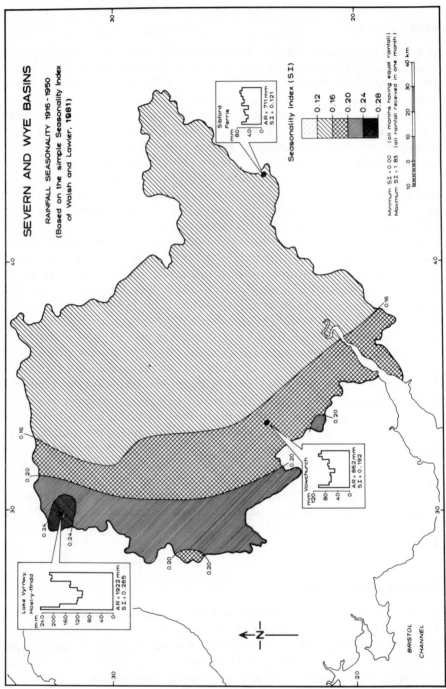

FIGURE 4.7 Rainfall seasonality in the Severn Basin, using the simple index of Walsh and Lawler (1981)

composite variable, the data requirements necessary for calculation are large, and appropriate climatological stations few (Figure 4.3). Nevertheless, the Severn-Trent Water Authority has prepared a map of the Severn Basin which is reproduced in rationalized form in Figure 4.11. Little spatial variation exists, although values of 500 mm or more are recorded in the lower and middle Severn (representing up to around 70 per cent of annual rainfall— compare Figures 4.5 and 4.11), declining to approximately 400 mm in the western upland areas with reduced temperatures and radiation receipts.

## CONCLUSION

The more important variables from a palaeohydrological viewpoint have been distinguished from the large number of other meteorological indices possible and, despite limitations in some datasets (particularly temperature), maps of the basin have been presented. These attempt to show how a few climatic elements may vary widely in magnitude across the catchment and the brief discussion possible has pointed to a number of controls of these spatial patterns. The overall picture at a very general level is one of increasing potential in hydrological and geomorphological activity as one moves west-wards across the Severn Basin from some point on the easterly periphery: such a transect would reveal a steadily increasing annual rainfall receipt, with individual rainfall events achieving greater intensity, a more pronounced concentration of rainfall into the winter half of the year, a possible greater tendency for heavy rainstorms to fall in winter when soil moisture deficits are smallest and contributing source areas more strongly developed, a lower proportion of precipitation evaporated back into the atmosphere, and a greater incidence of frost. If this increasing meteorological dynamism is coupled with the steeper slopes of the western uplands (cf. Newson, 1981) and the readily-available sediment supplies left as a legacy of periglacial activity (Lewin, Cryer and Harrison, 1974) then the potential, at least, for increased hydrological and geomorphological work in these areas becomes evident.

## ACKNOWLEDGEMENTS

I should like to thank the large number of people from the Meteorological Office, Severn-Trent Water Authority and the Welsh Water Authority for supplying much of the raw information on which the figures are based, Mr T. Grogan who drew the maps, Dr T. P. Burt for supplying some data for Table 4.3, and the reviewers for useful comments.

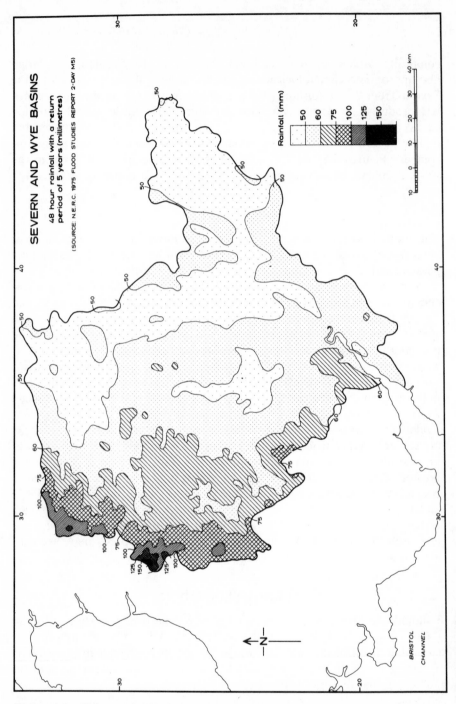

FIGURE 4.8 48-hour rainfall with a return period of 5 years (2-Day M5) in the Severn Basin (after NERC, 1975). Reproduced by permission of Natural Environment Research Council

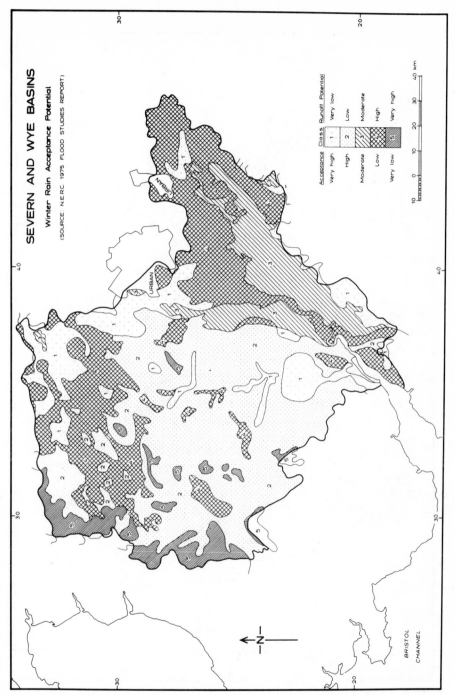

FIGURE 4.9    'Winter Rain Acceptance Potential' in the Severn Basin (after NERC, 1975). Reproduced by permission of Natural Environment Research Council

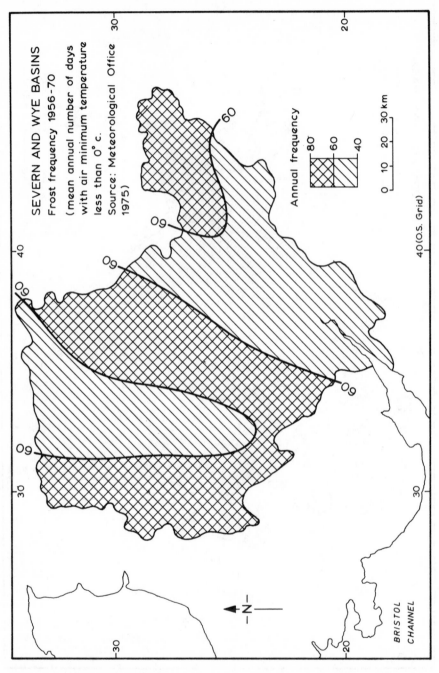

SEVERN AND WYE BASINS

Frost frequency 1956-70

(mean annual number of days with air minimum temperature less than 0°c.

Source: Meteorological Office 1975)

Annual frequency

80
60
40

0  10  20  30 km

40(O.S. Grid)

BRISTOL CHANNEL

FIGURE 4.10    Average annual frost frequency (1956–70) in the Severn Basin (from data supplied by the Meteorological Office, 1975)

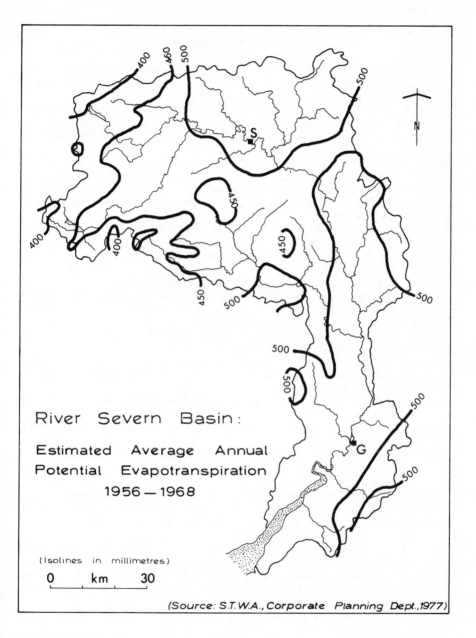

Within the map:

River Severn Basin:

Estimated Average Annual
Potential Evapotranspiration
1956 — 1968

(Isolines in millimetres)

0    km    30

(Source: S.T.W.A., Corporate Planning Dept.,1977)

FIGURE 4.11   Estimated average annual potential evapotranspiration 1956–68 for the Severn Basin (excluding the Avon and Wye catchments). G = Gloucester; S = Shrewsbury. Reproduced by permission of the Severn-Trent Water Authority

## REFERENCES

Anderson, M. G., Bosworth, D. and Kneale, P. E. (1984). Controls on overland flow generation, in Burt, T. P. and Walling, D. E. (eds), *Catchment Experiments in Fluvial Geomorphology*, Geo Books, pp. 21–34.

Arnett, R. R. (1974). Environmental factors affecting the speed and volume of topsoil interflow, in Gregory, K. J. and Walling, D. E. (eds), *Fluvial Process in Instrumented Watersheds, Institute of British Geographers, Special Publication*, **6**, 7–22.

Atkinson, B. W. (1983). Rainshadow—the dynamical factor, *Weather*, **38**, 264–9.

Atkinson, B. W. and Smithson, P. A. (1976). Precipitation, in Chandler, T. J. and Gregory, S. (eds), *The Climate of the British Isles*, Longman, London 129–82.

Ballantyne, C. (1983). Precipitation gradients in Wester Ross, North-West Scotland, *Weather*, **30**, 379–415.

Beaver, S. H. and Shaw, E. M. (1970). The climate of Keele, *Keele, Univ. Library, Occ. Publ. No. 7*, 165 pp.

Beckinsale, R. P. (1934). Climate of the Cotteswold Hills, *Proc. Cotteswold Naturalists' Field Club*, **25**, 155–81.

Blacknell, C. (1981). River erosion in an upland catchment, *Area*, **13**, 39–44.

Bleasdale, A. and Chan, Y. K. (1972). Orographic influences on the distribution of precipitation, in *Distribution of Precipitation in Mountainous areas*, vol. 2, World Meteorological Office No. 326, Geneva, 322–33.

Bootsma, A. (1976). Estimating minimum temperature and climatological freeze risk in hilly terrain, *Agricultural Meteorology*, **16**, 425–43.

Bridges, E. M., and Harding, D. M. (1971). Micro-erosion processes and factors affecting slope development in the Lower Swansea Valley, *Institute of British Geographers, Special Publication*, **3**, 65–80.

Burt, S. D. (1975). The climate of Rugby. Part 1. Temperature, *Journal of Meteorology*, **1**, 9–14.

Chandler, T. J., and Gregory, S. (eds) (1976). *The Climate of the British Isles*, Longman, London, 390 pp.

Derbyshire, E. (ed.) (1976). *Geomorphology and Climate*, Wiley, Chichester.

Derbyshire, E., Gregory, K. J. and Hails, J. R. (1979). *Geomorphological Processes*, Dawson, Folkestone, 312 pp.

Douglas, I. (1976). Erosion rates and climate: geomorphological implications, in Derbyshire, E. (ed.), *Geomorphology and Climate*, Wiley, Chichester, pp. 269–87.

Dreibelbis, F. R. (1949). Some influences of frost penetration on the hydrology of small watersheds, *Transactions, American Geophysical Union*, **30**(2), 279–82.

Dunne, T. (1978). Field studies of hillslope flow processes, in Kirkby, M. J. (ed.), *Hillslope Hydrology*, Wiley, Chichester, pp. 227–93.

Farquharson, F. A. K., Mackney, D., Newson, M. D. and Thomasson, A. J. (1978). Estimation of run-off potential of river catchments from soil surveys, *Soil Survey of England and Wales, Special Survey*, No. 11, 29 pp.

Faulkner, R. and Perry, A. H. (1974). A synoptic precipitation climatology of South Wales, *Cambria*, **1**, 127–38.

Ferguson, R. (1977). Linear regression in geography, *Catmog*, **15**, Norwich, 44 pp.

Finlayson, B. L. (1978). Suspended solids transport in a small experimental catchment, *Zeitschrift für Geomorphologie*, **22**, 192–210.

Fournier, F. (1960). *Climat et érosion*, Presses Universitaires de France, Paris, 210 pp.

Fullen, M. A. and Reed, A. H. (1986). Rainfall, runoff and erosion on bare arable soils in east Shropshire, England, *Earth Surface Processes and Landforms*, **11**, 413–25.

Gallagher, J. G. (1959). Record low temperature at Shawbury, *Meteorological Magazine*, **88**, 15–17.

Gardiner, T. (1983). Some factors promoting channel bank erosion, River Lagan, County Down, *Journal of Earth Sciences, Royal Dublin Society*, **5**, 231–9.

Gardiner, V. and Dackombe, R. V. (1980). Gullying on Cannock Chase, in Doornkamp, J. C. and Gregory, K. J. (eds) *Atlas of Drought in Britain 1975–76*, Institute of British Geographers, p. 46.

Garrett, J. H. (1913). The climate and topography of Cheltenham and its near neighbourhood, *Proc. Cotteswold Naturalists' Field Club*, **18**, 137.

Gloyne, R. W. (1968). Some climatic influences affecting hill-land productivity, *British Grassland Society, Symposium No. 4, Hill Land Productivity*, 9–15.

Gradwell, M. W. (1955). Soil frost studies at a high country station. Pt. 2. *New Zealand Journal of Science and Technology*, Sect B, **37**, 267–75.

Gregory, K. J. (1983). Introduction, in Gregory, K. J. (ed.) *Background to Palaeohydrology*, Wiley, Chichester, pp. 3–23.

Gregory, K. J., and Gardiner, V. (1975). Drainage density and climate, *Zeitschrift für Geomorphologie*, **19**, 287–98.

Gregory, K. J. and Ovenden, J. C. (1979). Drainage network volumes and precipitation in Britain, *Transactions of the Institute of British Geographers*, **4**, 1–11.

Gregory, K. J., and Thornes, J. B. (1980). The Severn basin: sample basin review, *Bulletin de l'Association Française pour l'Etude du Quaternaire*, **1–2**, pp. 61–4.

Gregory, K. J. and Walling, D. E. (1973). *Drainage Basin Form and Process: A Geomorphological Approach*, Arnold, London, 456 pp.

Gregory, S. (1976). Regional climates, in Chandler, T. J. and Gregory, S. (eds) *The Climate of the British Isles*, Longman, London, pp. 330–42.

Hamlin, M. J. (1971). A study of synthetic flow generation techniques using Elan Valley data, *Journal of the Institution of Water Engineers*, **25**, 355–70.

Harding, R. J. (1978). The variation of the altitudinal gradient of temperature within the British Isles, *Geografiska Annaler*, **60A**, 43–9.

Harrison, S. J. (1977). The bioclimatic resources of upland Wales, *South Hampshire Geogr.*, **9**, 1–11.

Harvey, A. M. (1974). Gully erosion and sediment yield in the Howgill Fells, Westmorland, in Gregory, K. J. and Walling, D. E. (eds), *Fluvial Processes in Instrumented Watersheds, Institute of British Geographers Special Publication*, **6**, 45–58.

Henkel, D. J. and Skempton, A. W. (1955). A landslide at Jackfield, Shropshire, in a heavily over-consolidated clay, *Géotechnique*, **5**, 131–42.

Hewlett, J. D. and Hibbert, A. R. (1967). Factors affecting the response of small watersheds to precipitation in humid areas, in Sopper, W. E. and Lull, H. W. (eds), *Forest Hydrology, Proc. Adv. Sci. Seminar of Aug–Sep 1965*, Pergamon, Oxford, pp. 275–90.

Hill, A. R. (1973). Erosion of river banks composed of glacial till near Belfast, Northern Ireland, *Zeitschrift für Geomorphologie*, **17**, 428–42.

Hill, F. F., Browning, K. A. and Bader, M. J. (1981). Radar and raingauge observations of orographic rain over South Wales, *Quarterly Journal of the Royal Meteorological Society*, **107**, 643–70.

Hooke, J. M. (1979). An analysis of the processes of river bank erosion, *Journal of Hydrology*, **42**, 39–62.

Howe, G. M., Slaymaker, H. O. and Harding, D. M. (1967). Some aspects of the flood hydrology of the upper catchments of the Severn and Wye, *Transactions of the Institute of British Geographers*, **41**, 33–58.

Jones, J. A. A. (1978). Soil pipe networks: distribution and discharge, *Cambria*, **5**, 1–21.

Kalma, J. D., Byrne, G. F., Johnson, M. E. and Laughlin, G. P. (1983). Frost mapping in southern Victoria: an assessment of HCMM thermal imagery, *Journal of Climatology*, **3**, 1–19.

Kesel, R. H. and Baumann, R. H. (1981). Bluff erosion of a Mississippi river meander at Port Hudson, Louisiana, *Physical Geography*, **2**, 62–82.

Langbein, W. B., and Schumm, S. A. (1958). Yield of sediment in relation to mean annual precipitation, *Transactions, American Geophysical Union*, **39**, 1076–84.

Lawler, D. M. (1979). Climatology and meteorology of the Severn and Wye basins: a review of literature and data sources, *unpub. Report to the UK Working Party of IGCP Project 158A 'Palaeohydrology of the Temperate Zone'*, 23 pp. (copies available from author).

Lawler, D. M. (1980). Climatic data for the Severn and Wye basins, *unpub. Report to the UK Working Party of IGCP Project 158A 'Palaeohydrology of the Temperate Zone'*, 5 pp. (copies available from author).

Lawler, D. M. (1985). A 1000-metre altitudinal temperature traverse on the Puy-de-Dôme, Massif Central, France, *Journal of Meteorology*, **10**, 285–9.

Lawler, D. M. (1986). River bank erosion and the influence of frost: a statistical examination, *Transactions of the Institute of British Geographers*, **11**, 227–42.

Lawler, D. M. (1987). Bank erosion and frost action: an example from South Wales, in Gardiner, V. (ed.) *International Geomorphology 1986: Proceedings First International Conference on Geomorphology (Manchester, Sept. 1985), Part 1*, Wiley, Chichester, pp. 575–90.

Lawrence, E. N. (1952). Frost investigation, *Meteorological Magazine*, **81**, 65–74.

Lewin, J., Cryer, R. and Harrison, D. I. (1974). Sources for sediments and solutes in mid-Wales, *Institute of British Geographers Special Publication*, **6**, 73–85.

Marsh, T. J. and Littlewood, I. G. (1978). An estimate of annual runoff from England and Wales 1782–1976, *Hydrological Sciences Bulletin*, **23**, 131–42.

Mason, B. J. (1976). Towards the understanding and prediction of climatic variations, *Quarterly Journal of the Royal Meteorological Society*, **102**, 473–98.

McGreevy, J. P. (1982). Frost and salt weathering: further experimental results, *Earth Surface Processes and Landforms*, **7**, 475–88.

Meteorological Office (1975). Maps of mean and extreme temperature over the United Kingdom, 1941–1970, *Climatological Memorandum*, No. 73.

Meteorological Office (1977). *Average Annual Rainfall Map of Southern Britain, 1941–70*, Meteorological Office, Bracknell.

Meteorological Office (1982). The Climate of Great Britain: The Midlands, *Climatological Memorandum*, 132, 16 pp.

Meteorological Office (1983). The Climate of Great Britain: Wales, *Climatological Memorandum*, 140, 20 pp.

Morgan, R. P. C. (1970). Climatic geomorphology: its scope and future, *Geographica*, **6**, 26–35.

Myburgh, J. (1974). An index to relate local topography to mean minimum temperatures, *Agrochemophysica*, **6**, 73–8.

NERC (1975). *Flood Studies Report*, Natural Environment Research Council, 5 vols.

Newson, A. J. (1976). Some aspects of the rainfall of Plynlimon, Mid-Wales, *Institute of Hydrology Report*, **34**, 36pp.

Newson, M. D. (1979). The results of ten years' experimental study on Plynlimon, mid-Wales, and their importance for the water industry, *Journal of the Institution of Water Engineers and Scientists*, **33**, 321–33.

Newson, M. D. (1981). Mountain streams, in Lewin, J. (ed.) *British Rivers*, Allen & Unwin, London, pp. 59–89.

Oliver, J. (1971). Climatology, in Balchin, W. G. V. (ed.) *Swansea and its Region*, University College Swansea, pp. 41–58.

Outcalt, S. I. (1971). An algorithm for needle ice growth, *Water Resources Research*, **7**, 394–400.

Peltier, L. C. (1950). The geographical cycle in periglacial regions as it is related to climatic geomorphology, *Annals of the Association of American Geographers*, **40**, 214–36.

Penman, H. L. (1963). Vegetation and hydrology, *Technical Communication* 53, Commonwealth Bureau of Soils, Harpenden, Commonwealth Agricultural Bureau.

Perry, A. H. (1976). Synoptic climatology, in Chandler, T. J. and Gregory, S. (eds), *The Climate of the British Isles*, Longman, London pp. 8–38.

Perry, A. H. (1979). Climatic variation in West Glamorgan, in Bridges, E. M. (ed.), *Problems of Common Land: the Example of West Glamorgan*, Department of Geography, University College Swansea, 2:2/1–2:2/9.

Pidgeon, J. D. and Ragg, J. M. (1979). Soil, climatic and management options for direct drilling cereals in Scotland, *Outlook on Agriculture*, **10**, 49–55.

Prior, D. B., Stephens, N. and Douglas, G. R. (1971). Some examples of mudflow and rockfall activity in north-east Ireland, in Brunsden, D. (ed.), *Slopes: Form and Process, Institute British Geographers, Special Publication*, **3**, 129–40.

Reed, A. H. (1979). Accelerated erosion of arable soils in the United Kingdom by rainfall and run-off, *Outlook on Agriculture*, **10**, 41–8.

Richards, K. S. (1982). *Rivers: Form and Process in Alluvial Channels*, Methuen, London, 358 pp.

Rodda, J. C. (1962). An objective method for the assessment of areal rainfall amounts, *Weather*, **17**, 54–9.

Rowsell, H. (1963). Rainfall over the catchment area of the Severn River Board, 1916–50, *Meteorological Office Hydrological Memoir*, No. 15, 24 pp.

Saward, B. (1950). Climate, in Wise, M. J. (ed.), *Birmingham and its Regional Setting*, British Association, London, pp. 47–54.

Schumm, S. A. (1965). Quaternary palaeohydrology, in Wright, H. E. and Frey, D. G. (eds), *The Quaternary of the United States*, Princeton University Press, pp. 783–94.

Selby, M. J. (1976). Slope erosion due to extreme rainfall: a case study from New Zealand, *Geografiska Annaler*, **58A**, 131–8.

Severn-Trent Water Authority, Corporate Planning Department. (1977). *River Severn Basin: Report of Survey*, S.T.W.A., Birmingham, 137 pp.

Severn-Trent Water Authority, Directorate of Operations (1977). *Hydrometric Yearbook 1974 & 1975. Part 1—Station Catalogue*, S.T.W.A., Birmingham, 200 pp.

Smith, K. (1972). *Water in Britain*, Macmillan, London, 241 pp.

Soons, J. M. and Greenland, D. E. (1970). Observations on the growth of needle ice, *Water Resources Research*, **6**, 579–93.

Starkel, L. and Thornes, J. B. (eds) (1981). Palaeohydrology of river basins, *British Geomorphological Research Group Technical Bulletin*, **28**, 107 pp.

Stoddart, D. R. (1969). Climatic geomorphology: review and reassessment, *Progress in Geography*, **1**, 159–222.

Sumner, G. N. (1978a). Climate and vegetation, in Thomas, D. (ed.), *Wales: A New Study*, David and Charles, Newton Abbot, pp. 36–69.

Sumner, G. N. (1978b). A drainage density map of Wales, *Cambria*, **5**, 156–66.

Sutcliffe, J. V. (1978). Methods of flood estimation: a guide to the Flood Studies Report, *Institute of Hydrology Report*, **49**, 50 pp.

Taylor, J. A. (1976). Upland climates, in Chandler, T. J. and Gregory, S. (eds), *The Climate of the British Isles*, Longman, London, pp. 264–87.

Thorn, C. E. (1980). Alpine bedrock temperatures: an empirical study, *Arctic and Alpine Research*, **12**, 73–86.

Thorne, C. R. (1978). *Processes of bank erosion in river channels*, unpublished Ph.D. thesis, University of East Anglia, 447 pp.

Thornes, J. B. (1983). Discharge: empirical observations and statistical models of change, in Gregory, K. J. (ed.), *Background to Palaeohydrology*, Wiley, Chichester, pp. 51–67.

Thornes, J. E. (1985). Thermal mapping and road surface temperature, *Proceedings of the 2nd International Road Weather Conference*, Danish Ministry of Transport, Copenhagen.

Unwin, D. J. (1969). The areal extension of rainfall records: an alternative model, *Journal of Hydrology*, **7**, 404–14.

Visher, S. S. (1945). Climatic maps of geological interest, *Geological Society of America Bulletin*, **56**, 713–36.

Walling, D. E. and Webb, B. W. (1983). Patterns of sediment yield, in Gregory, K. J. (ed.), *Background to Palaeohydrology*, Wiley, Chichester, pp. 69–100.

Walsh, R. P. D. (1980). Runoff processes and models in the humid tropics, *Zeitschrift für Geomorphologie*, **36**, 176–202.

Walsh, R. P. D. and Lawler, D. M. (1981). Rainfall seasonality: description, spatial patterns and change through time, *Weather*, **36**, 201–8.

Ward, R. C. (1975). *Principles of Hydrology*, McGraw-Hill, New York, 2nd edn, 367 pp.

Warwick, G. T. (1971). The physical background, in Cadbury, D. A. *et al.* (eds), *A Computer-mapped Flora; A Study of the County of Warwickshire*, Academic Press, London, pp. 4–15.

Williams, P. J. (1964). Unfrozen water content of frozen soil and soil moisture suction, *Géotechnique*, **14**, 231–46.

Wilson, L. (1973). Relationships between geomorphic processes and modern climates as a method in paleoclimatology, in Derbyshire, E. (ed.), *Climatic Geomorphology*, Macmillan, New York, pp. 269–84.

Wischmeier, W. H. (1959). A rainfall erosion index for a universal soil-loss equation, *Soil Science Society of America Proceedings*, **23**, 246–9.

Wolman, M. G. (1959). Factors influencing erosion of a cohesive river bank, *American Journal of Science*, **257**, 204–16.

Palaeohydrology in Practice
Edited by K. J. Gregory, J. Lewin and J. B. Thornes
© 1987 John Wiley & Sons Ltd.

# 5

# The Present-day Hydrology of the River Severn

## T. R. WOOD

*Severn-Trent Water Authority*

### AVAILABLE HYDROLOGICAL DATA

In the nineteenth century a series of locks was constructed on the Severn between Gloucester and Stourport to allow the river to be used as a commercial waterway. The first of these locks was opened in 1836 and a daily record of river level has been kept ever since. Records of this length are unusual in the UK. Despite this long level record there was no continuous flow record on the Severn itself before 1921, although as early as 1886 flows had been measured in the headwaters of the Vyrnwy as an aid to reservoir construction. In 1921 a chart recorder was installed at Bewdley and this is the longest flow record available on any large river in the basin.

Extensive professional hydrometry in the British Isles did not begin until after the 1933–34 drought when the Institution of Civil Engineers and the Royal Society jointly persuaded the then Prime Minister, Ramsey MacDonald, to establish the National Surface Water Survey. The United Kingdom was thus rather slow compared to other countries in establishing a national network of flow measurement stations. Furthermore the early years of the Surface Water Survey were difficult financially and hydrometry had to wait until after the Second World War before substantial expansion of the network occurred.

In the 1950s and 1960s the Severn River Board and subsequently the Severn River Authority constructed a number of flow and level recording stations. A continuous level record from the Severn at Shrewsbury is available from 1950 and Worcester from December 1953. At Montford (SJ 412144, upstream of Shrewsbury) a flow record is available from October 1953 and at Plynlimon from October 1952. Notable additions to the network occurred in the 1960s: Tern at Walcot 1960 (SJ 592123) and the Severn at Abermule 1962 (SO 164958). In the early 1970s, with central government grant aid available to the river authorities, a steady expansion occurred in the network. In particular improvements were made in low-flow measurement with the introduction of purpose-built weirs and low-flow controls.

From 1974 to the present day the hydrometry of the River Severn has been managed by the Severn-Trent Water Authority. During the past 10 years the importance of real time data has increased with the growing need and the increased ability of hydrologists to forecast short-term changes in river conditions both for water-supply purposes and for flood warning. The network has also been reviewed and some stations closed, while others have been modified and improved. Figure 5.1 shows the present river flow and level stations in the Severn and Trent basins.

Technological advances achieved in the past several years have improved the standard of hydrometric measurements and have enabled data to be transmitted reliably and quickly. The development of multipath ultrasonic flow measurement and electro-magnetic flow measurement techniques has offered opportunities of measuring river flow in circumstances where current metering had encountered difficulties. Despite these advances current metering continues to give reliable and cost-effective results from a large number of conventional sites and together with a range of weirs and flumes still forms the backbone of the present network. The high cost of constructing flumes and weirs and the need to allow passage of boats are the main reasons that such structures are confined to tributaries of the Severn. The only flumes or weirs on the main arm of the Severn which have been constructed specifically for hydrometric purposes are in the headwaters at Plynlimon.

In the headwaters a great deal of specialist hydrometry has been introduced by the Institute of Hydrology in their comparative study (Newson, 1979) of the Severn and Wye. With the exception of the Institute's activities in the uppermost 9 km$^2$ of the headwaters, no other agency is active in hydrometry apart from the Water Authority and the integrated management of the whole basin by one authority has allowed the development of a consistent hydrometric policy and eased the problems of network management.

Figure 4.4 shows the associated rainfall network which the Authority also manages and Figure 4.3 shows the climatological network. Much of these networks is nowadays telemetered to enable real time hydrological decisions to be taken. The data are captured in the same manner as the flow and level information. The telemetry system depends upon micro- and mini-computers which poll solid state outstation equipment over British Telecom landlines. Polling is normally undertaken every 24 hours when the last 96 15-minute readings are collected. However, in times of flood or drought polling can take place more frequently. The received data are then processed automatically with the data being printed locally or interrogated remotely. The same

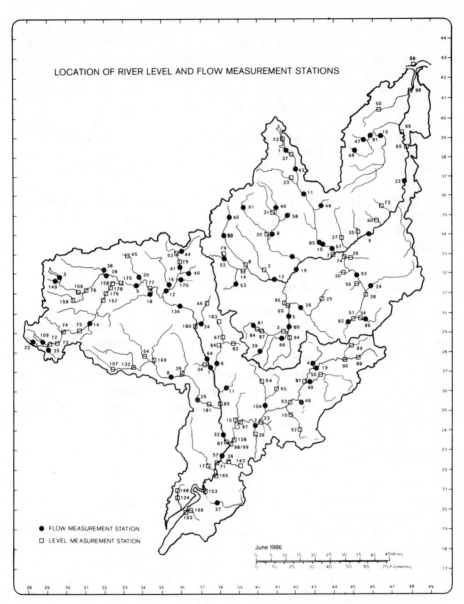

FIGURE 5.1   Location of river level and flow measurement stations

information is also available to a free-standing mini-computer which holds the main forecasting models.

Below Bewdley the 12-metre range of the Bristol Channel tides and the low-gradient thalweg lead to backwater problems at several of the existing gauging stations. Flood gaugings have been taken on many occasions at Gloucester but at lower flows there is frequently no unique relationship between stage and discharge due to tidal interference. In an attempt to overcome these problems a multipath ultrasonic gauge was constructed in 1975 at Ashleworth (SO 819250). However, the lower paths of the Ashleworth gauge encountered high suspended sediment loads due to the strong tidal currents. The system failed to work effectively and flows were no longer processed after 1979. Because of the navigation interests on the Lower Severn a weir could not be constructed and current metering is the only means presently employed of flow measurement at any site below Bewdley. A new ultrasonic flow measurement station is planned, however, at Saxons Lode (SO 863390). Haw Bridge (SO 844279) is the lowermost full-range gauging station on the Severn, monitoring some 9895 km$^2$ of the 11,422 km$^2$ of the Severn Basin, but it is heavily affected by tides for a fortnight in every month.

## GENERAL CHARACTERISTICS OF THE SEVERN BASIN

On a European scale the Severn Basin is small, since the mean discharge of the Severn is less than 1/50th of the flow of the Danube. However, seen in the more limited scale of the British Isles, the Severn is the largest single river basin in England and Wales, and the Severn is the longest river (354 km). The basin embraces several significantly differing topographic and hydrological regimes and the Severn's two major tributaries, the Teme and the Avon, have distinctive individual characteristics. Most visitors characterize the Severn as a natural river and as being substantially unaltered by man. This description belies the fact that the low-flow hydrology of the Severn is modified by river regulation and by abstraction. Nevertheless the basin has a low overall population density and the Severn, above the confluence with the Stour, carries little effluent.

Over the past 100 years the Severn and its tributaries have been extensively developed for water-supply purposes. Liverpool, Birmingham and Coventry, together with the towns of the Black Country, provided for their growing populations either by direct abstraction from the river or by reservoir construction, exporting much of this water permanently from the Severn Basin. However, a significant amount (over 170 Ml/d) is returned as effluent to the Avon (Ledger, 1972). The middle reaches of the Severn are nevertheless deprived both of the natural flow in the river and of the returned effluent. Because industrial abstraction and discharge are small throughout much of the basin, the river appears to be in a natural state. In practice the river is managed very carefully (Wood, 1981) so that the 'natural' appearance is preserved.

The annual flow regime of the Severn is typical of large rivers in temperate climates. The mean discharge is some 10,200 Ml/d at Gloucester and the variance is small. Notably the lowest recorded annual discharge at Bewdley is still 60 per cent of the mean. Nevertheless individual monthly mean flows can fall to suprisingly low values. It is possible to calculate the flow which would have occurred without the effect of abstraction of discharge—the so called 'naturalized' flow. Naturalized flows show that without low-flow augmentation by reservoir releases, individual months could be as little as 12 per cent of the mean. Again that statistic is based on Bewdley, which for such a large catchment (4274 km²), has a high monthly variance.

Figures 5.2–5.5 show four of the principle gauging stations in the Severn basin and indicate the variation in monthly flows that occur. The flows are not naturalized. Naturalization would significantly affect the Evesham and Bewdley hydrographs. In the lower Avon effluent masks the natural variability, while in the Severn the low flows are supported by reservoir releases, preventing their falling below 850 Ml/d at Bewdley. In contrast the lowest naturalized monthly mean flow at Bewdley was some 315 Ml/d and occurred in August 1976.

## LOW-FLOW CHARACTERISTICS

Figures 5.6 and 5.7 show two residual flow diagrams for the River Severn. Figure 5.6 represents a fairly frequently occurring dry-weather condition which approximates to that exceeded 95 per cent of the time. Figure 5.7 shows the lowest recorded flows. At most sites these were experienced in 1976. In those conditions Llyn Clywedog was releasing close to its maximum permitted rate of discharge (500 Ml/d) in support of abstractions in the lower river. Without the support the river would have been unable to withstand the substantial public water supply and spray irrigation abstractions. In severe droughts the river would effectively cease to flow for extensive lengths of its course.

Figure 5.6 demonstrates a number of important features. The upper third of the Severn, upstream of the Vyrnwy confluence, has a particularly low dry-weather flow due to the impermeable strata and consequent absence of groundwater support. Although the catchment of the Vyrnwy is itself impermeable the low flows are augmented by the compensation releases from the dam. Below Montford significant base flow is added to the Severn by the River Perry and the River Tern which drain the Triassic sandstones of North Shropshire. Further downstream the Teme and the Avon, whose topography and character are so different from each other, both contribute substantially to the low flows. Some 440 Ml/d comes from the Avon and 310 Ml/d from the Teme.

Figure 5.7 demonstrates the impact of river regulation during a drought. The Water Authority is obliged to maintain a minimum five-day average flow

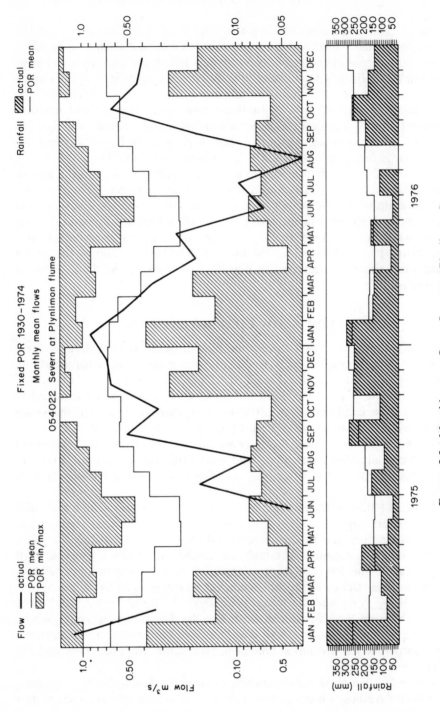

FIGURE 5.2    Monthly mean flows: Severn at Plynlimon flume

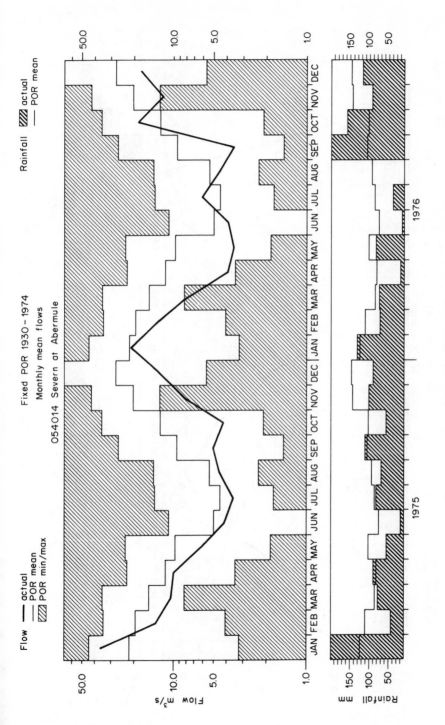

FIGURE 5.3   Monthly mean flows: Severn at Abermule

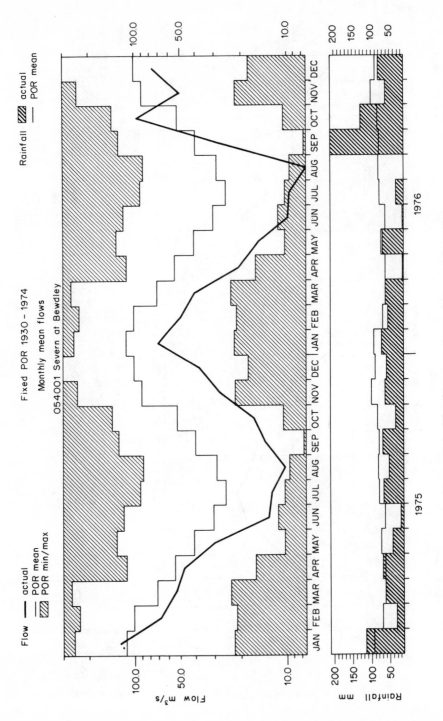

FIGURE 5.4   Monthly mean flows: Severn at Bewdley

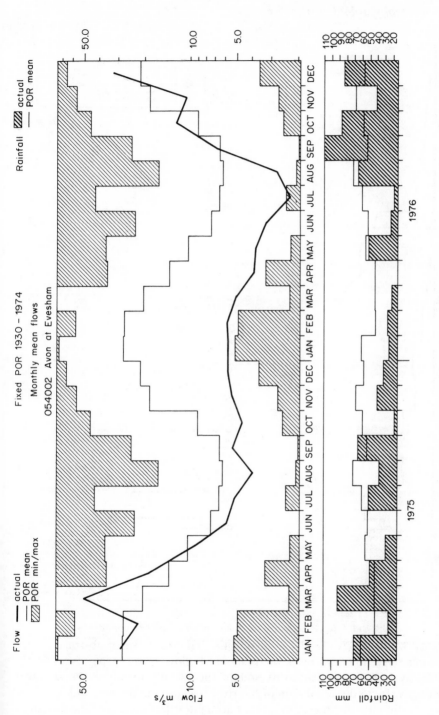

FIGURE 5.5   Monthly mean flows: Severn at Evesham

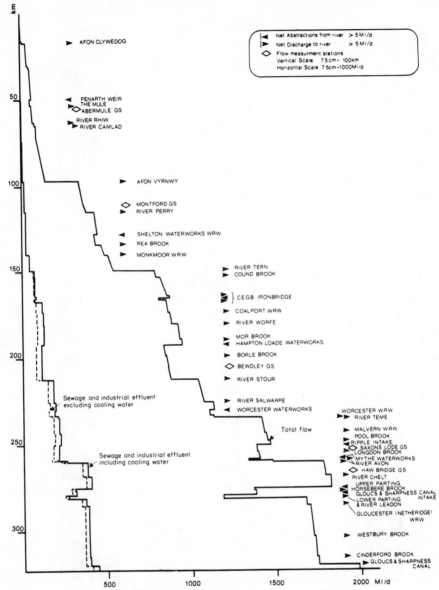

FIGURE 5.6　Residual flow diagram: dry weather conditions

of 850 Ml/d at Bewdley by adjusting the balance between abstraction, the
natural river flow and reservoir releases. Following the experiences of the
1975–76 drought the Authority sought successfully to adjust the statutory
rules governing the regulation of the river because the previous rules had led
to operational difficulties. Temporary drought orders preserved the dwindling

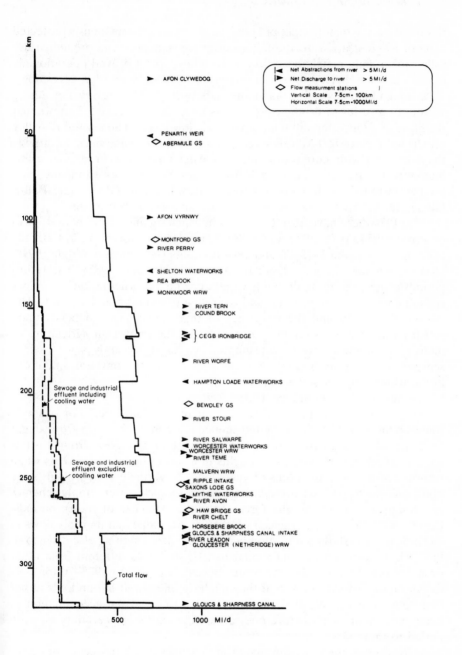

FIGURE 5.7    Residual flow diagram: Minimum flow conditions

resources during that drought but now the river can be permanently protected without as great a risk to the failure of public supplies. The previous rules called for a daily matching of flow at Bewdley, which proved operationally difficult and led in 1975 to wasteful releases of water from Llyn Clywedog.

Although Llyn Clywedog maintained a flow of 730 Ml/d at Bewdley during 1976 until a series of drought orders was introduced, Figure 5.7 shows that the flows at Gloucester still fell to extremely low levels. The revised Bewdley prescribed flow of 850 Ml/d is a recognition of the growing need to provide the lower river with more water. Substantial abstractions occur in Gloucester, indirectly for public water supply to Bristol Waterworks Company, and smaller abstractions also occur between Bewdley and Gloucester. Below Gloucester the river is saline and there are almost no abstractions.

Since 1974 there have been four years in which significant dry periods have occurred (1974, 1975, 1976 and 1984). Considerable discussion has already been devoted to the 1975–76 and 1984 droughts (Hamlin and Wright, 1978; Doornkamp and Gregory, 1980; Institute of Hydrology, 1985). The return periods assigned to these droughts varies according to whether one examines rainfall data or river flow data, and according to duration and location. The impact of drought and the frequency of that impact again depends upon which parameter one examines. For example, the impact on agriculture or industry or domestic water consumption varies significantly since each is sensitive to differing drought characteristics and to the physical limitations of the water supply systems that serve them.

One lesson that should be learnt is that judging the return period of extreme events from short records is fraught with difficulty. For example some of the return periods quoted immediately after the 1975–76 drought have already had to be modified in the light of the 1984 drought. It is dangerous practice to describe droughts in terms of hundreds of years when records are no more than tens of years long. Nevertheless, hydrologists are often obliged to provide drought data against which water supply schemes are designed and are thereby forced to make estimates of return periods. Certainly the Severn experienced some very low individual monthly flows in 1976; flows of 1–90 days' duration being well below anything else on record, but with a six-month minima similar to 1975 and little different from other droughts (see Figure 5.8). The return periods ascribed to 1975–76 for 1–90 day duration flows do look as if they might be measured in hundreds rather than tens of years. We shall, however, have to await the accumulation of many more years' data before one can ascribe with some certainty a return period to these flows.

As to whether there are any significant changes in the frequency of occurrence of low flows, again the short hydrological record leaves one guessing. There is some evidence that UK weather is becoming more variable, with enhanced annual run-off and a tendency to drier summers and wetter

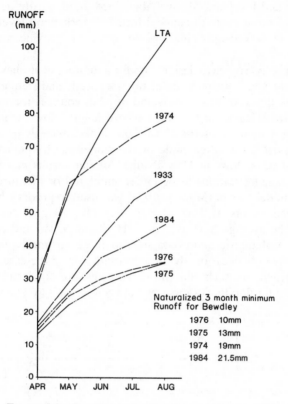

BEWDLEY NATURALIZED CUMULATIVE RUNOFF
APRIL - AUGUST

FIGURE 5.8    Bewdley naturalized cumulative runoff
plot

autumns. However, this is a pattern observed only over the past 10 years and is probably a small perturbation in the long-term cycle of climatic change. Engineering hydrologists might regard palaeohydrology as too academic to be of direct use, but it is increasingly clear that historic records should be used to understand the return periods of extreme events—whether they be the return periods of major floods or of droughts.

## FLOOD FLOWS

Floods have a longer term effect on the landscape than droughts because of their impact on channel form. Hence, one might expect palaeohydrology to offer some information about the relative magnitude of present-day floods. Was, for example, the biggest flood recently recorded (1947), a really rare event? Is there evidence that in the not too distant past some events outpointed it? Sadly, engineering hydrologists have to rely heavily upon the limited flow and level records available since there is little design evidence forthcoming from a study of channel form. Historic flood levels recorded in newspapers or carved into buildings do, however, provide some important information.

In providing hydrological design data for a number of studies in the Severn Basin, estimates have been made of the magnitude and frequency of instantaneous peak flows, of peak levels and of peak volumes for several reaches of the Severn. Because of the short record length, flow records have been generated at a number of sites based on rainfall records or other gauging stations. Figure 5.9 is an example of the magnitude frequency relationship of simulated peak flows at Haw Bridge. Similar results can be seen for a number of gauging stations using either simulated or measured flows. The simulation model Hysim (Manley, 1978) was used to produce both peak and *n*-day volume events at Haw Bridge. Figure 5.10 provides a frequency analysis of the average flow over a 10-day flood period and compares it to peak flows. Volumetric flood data such as these are rarely analysed in the UK since they only occasionally are important, but in the case of the Severn the physiography is such that flood volumes are significant parameters. Prolonged winter flooding of the area between Saxon's Lode and Ashleworth

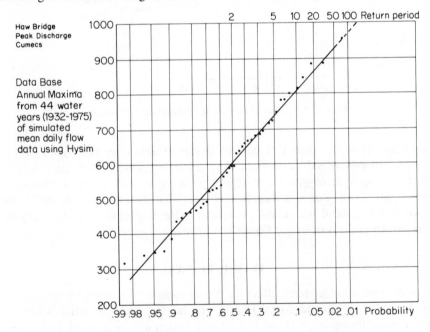

FIGURE 5.9   Normal distribution fitted to flood peaks at Haw Bridge

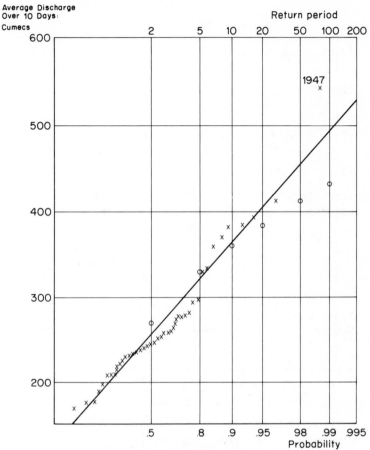

o   10 day volumes calculated from the design hydrograph at
Worcester when the Avon contributes <20% of the design flood at Haw Bridge

x   10 day annual maxima in 44 water years (1932–1975) of simulated data
with the Gumbel Distribution

FIGURE 5.10    Comparison of 10-day volumes at Worcester with volume
calculated from design hydrographs

is commonplace with high volumes rather than high peak flows contributing the majority of the flood water. Upstream of Shrewsbury, similar circumstances prevail, with the area around the confluence of the Vyrnwy subject to long periods of inundation once the earthwork flood defence system (known locally as Argae) is overtopped.

By amalgamating all of the available historic flood-level information from newspaper records, from early historical accounts and from the nineteenth-century lock records, it is possible to produce a diagram such as Figure 5.11. This is unusual in that it has been drawn with two straight lines fitted to the

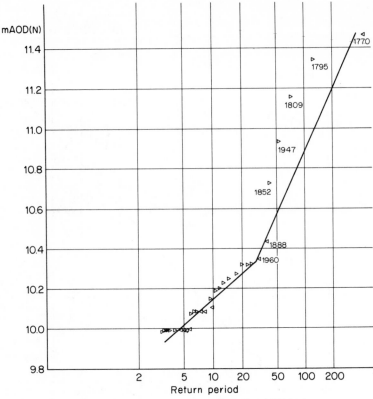

Data base    (i) Levels above 10.35 m  Lock record and adjusted F.S.R. levels
             (ii) Levels below 10.35 m  Record compiled by F. Rowbotham

FIGURE 5.11    Gloucester lock 1770–1976 Adjusted peaks

data, since it was not possible to fit these mixed data to a single distribution. Figure 5.11 shows that although 1947 was the largest flood at Gloucester in recent years, there have been three other floods in the past 230 years which have produced higher levels. It is debatable whether or not flood flows of equal magnitude would today produce the same levels since the conveyance capacity of the river system has been improved. Nevertheless, the historic data do suggest that the 1947 event is not as rare as some hydrologists believed shortly after its occurrence. On the basis of this evidence, the 1947 event is today considered to have a return period of perhaps 100 years or so.

Flood levels at Gloucester do include the effects of tides but there is a limited influence of tidal surges. In the river downstream of Gloucester the effect of tidal surges on peak levels is much more pronounced. Some of the highest recorded surges in that area result from the combination of a spring tide and an active depression moving into the Bristol Channel. A strong onshore south-westerly wind can produce surges in excess of 1 m above predicted tidal levels. Additionally, there is some evidence of the association of tidal surges and fluvial events. Since 1925 there have been some 30 significant fluvial flood peaks in the lower Severn and of these events 12 had surges of over 0.6 m which occurred within five days of the peak. Of these 12, three surges were greater than 1 m. Commonsense tells us that this coincidence is causal since many fluvial events are themselves the product of strong and active depressions. Not only can a surge in combination with a fluvial event cause exceptional flood levels, it often has a significant effect in delaying the evacuation of flood water from the estuary, which again emphasizes the unusual importance of volume-related floods in the lower Severn.

It is again notable that the estimated return period of extreme surges had to be drastically revised after the occurrence on 13 December 1981 of a surge of 1.5 m, which was larger than anything else previously recorded. This underlines some of the points made earlier about short records and rare events.

## DESIGN AND FORECASTING MODELS

Over the past 10 years, several different hydrological and hydraulic models have been used to provide an understanding of the nature of flooding in the Severn Basin. Pre-eminent amongst the hydraulic models are two developed by Hydraulics Research Ltd, namely FLOUT and EMBER. In a series of studies undertaken by Hydraulics Research Limited on behalf of Severn-Trent Water Authority, river reaches such as the lower Avon, the lower Severn and the Vyrnwy confluence have been modelled mathematically. These studies have been aimed at understanding the pattern of present-day flooding and in testing various options in the design of flood protection schemes. (Hydraulics Research Station, 1981).

In addition, EMBER has been modified (Hydraulics Research Station, 1982) to operate in a real time mode, so offering Severn-Trent the possibility of running a forecasting system for the tidal reaches. This model is complex since it explicitly deals with tidal and fluvial forces. In addition it can cater for the computationaly difficult problems encountered in embanked reaches.

The only non-real hydrological simulation model used extensively in the Severn has been Hysim (Manley, 1978). This was used to generate 45 years of daily flows to be input to the hydraulic routing models developed by HR Ltd. Hysim contains its own hydraulic routing models capable of dealing with the fluvial reaches, but does not include facilities for tidal rivers.

In recent years attention has turned to the development of real time hydrological and hydraulic models. In addition to EMBER the Water Authority has itself developed forecasting tools. This development period followed a careful examination of the benefits of developing a flow-based forecasting system (Chatterton, Pirt and Wood, 1979). The models themselves (Douglas and Dobson, 1986) are heavily dependent upon a hydrometric data collection system described by Jones (1980). In essence, there are two models, a simplified conceptual rainfall runoff model and a hydraulic budgeting model which deals with fluvial reaches. These models have been successful in a variety of circumstances and are used daily to provide forecasts of impending flood and to aid river management during low flows. Together with EMBER they will provide the whole river system with an effective forecasting system during the 1980s.

## THE CHANGING LANDSCAPE

Hydrological records are too short to allow one to understand whether the frequency of rare events is changing. Indeed much of this chapter shows that they are not long enough to enable accurate estimates to be made of the frequency of very rare events, even assuming that there are no longer term changes. Nevertheless, there is some evidence available from two directions which shows the need to be conscious of possible changes in the hydrological record. Both of these effects are of secondary importance but both demonstrate that changes in land use could produce potentially serious effects.

Firstly, there is evidence in the Upper Severn that in recent years runoff has been flashier than previously with the time for hydrographs to rise being shortened by as much as 40 per cent and with hydrographs recording a higher percentage runoff. This conclusion is based on the analysis of only eight unit hydrographs and therefore the observations may be due to chance alone. Also, there is no proof that the observed changes are caused by changes in land use. Nevertheless, the reseeding of considerable areas of the Welsh uplands with shorter, finer grasses has been coincidental with the observed change. Such grasses would lead to lower interception and reduced retention times, and as such are the most likely cause. However, a great deal more data are required before any firm conclusion can be offered.

More positive evidence of the effects of land use change comes from the seminal study of Plynlimon by the Institute of Hydrology (Newson, 1979). This has conclusively demonstrated that interception is significantly greater in tree-covered catchments and that there is a major impact on both the quantity and the quality of river water which is attributable to afforestation. This study has been influential in determining the policy of the water industry towards afforestation in reservoired catchments. With the likelihood that up to 2 million hectares could be taken out of agricultural production in the next 10 years in the United Kingdom due to European over-production, and with the continued substantial import of softwood, there is pressure on the government to encourage further afforestation in the uplands. The impact on water management and on the hydrological record could be significant.

## CONCLUSIONS

Although river management and hydrology are today orientated towards real time information to deal with day-to-day problems, the importance of the long-term data at key sites should not be neglected. Many of those records are short and the role of historic evidence is often critical in assessing the return periods of extreme events.

## ACKNOWLEDGEMENT

I am grateful to a number of colleagues, particularly Will Bradford and Richard Douglas, for permission to draw upon their unpublished work. Diagrams 5.2 to 5.5 were plotted by the Institute of Hydrology and I am grateful for their permission to reproduce them.

## REFERENCES

Chatterton, J. B., Pirt, J. and Wood, T. R. (1979). The benefits of flood forecasting, *J. Inst. Wat. Eng Sci*, **33**, No. 3, 237–51.

Douglas, J. R. and Dobson, C. (1986). Real time forecasting in diverse drainage basins, presented Sept. 1985 to Weather Radar and Flood Warning Symposium, Lancaster.

Doornkamp, J. C. and Gregory, K. J. (eds) (1980). *Atlas of Drought in Britain 1975–6*, Institute of British Geographers. 87 pp.

Hamlin, M. J. and Wright, C. E. (1978). The effects of drought on river systems, *Proceedings Royal Society, London*. Series A, **363**, 69–96.

Hydraulics Research Station (1981). *A computational Hydraulic model of the River Severn*, vols I and II, Report No. Ex 945, Wallingford.

Hydraulics Research Station (1982). *Flow forecasting in tidally influenced river reaches*, *Report* No. Ex 1001, Wallingford.

Institute of Hydrology (1985). *The 1984 Drought*, Wallingford, 74 pp.

Jones, H. H. (1980). An overview of hydrological forecasting in a multi-functional Water Authority, *International Association for Scientific Hydrology Publication No. 129*, 195–202.

Ledger, D. C. (1972). The Warwickshire Avon: a case study of water demands and water availability in an intensively used river system, *Transactions Institute British Geographers*, **55**, 83–110.

Manley, R. E. (1978). The use of a hydrological model in Water Resources Planning, *Proceedings Institute of Civil Engineering*, **65**, 223–35.

Newson, M. D. (1979). The results of ten years' experimental study on Plynlimon, Mid-Wales, and their importance for the water industry, *J. Inst. Wat. Eng. Sci.*, **33**, 321–33.

Wood, T. R. (1981). River Management, in J. Lewin (ed.), *British Rivers*, Allen & Unwin, London, pp. 170–95.

Palaeohydrology in Practice
Edited by K. J. Gregory, J. Lewin and J. B. Thornes
© 1987 John Wiley & Sons Ltd.

# 6

# *Climatic Change over the Last Millenium in Central Britain*

D. M. LAWLER

*Department of Geography, University of Birmingham*

## INTRODUCTION

Although temporal change and variability in climate has long been a subject of enquiry (e.g. Brooks, 1926), it is only over the last 30 years that an explosion of interest has occurred, resulting in a massive literature, including contributions scientific (e.g. Lamb, 1977; Gribbin, 1978; as well as the journals *Climatic Change* and *Palaeogeography, Palaeoclimatology and Palaeoecology*), popular (e.g. Weiss, 1978) and fantastic (e.g. Ballard, 1978). In relation to palaeohydrology, the study of climatic change may feature in at least two ways. First, if independent assessments of climatic change are available they may be used to assist inference of hydrological changes. Second, climatic change may itself be inferred from changes in hydrology, sediments or channel form, uncovered through, for example, stratigraphical analysis or palaeohydraulic reconstruction. Thornes (1983) has stressed the need for *independent* assessments of climatic change. The dangers of circularity in argument, also, should be recognized: for example, it is clearly important that a climatic change deduced from biological evidence, such as *coleoptera* data (Coope, 1977), should be used cautiously in the explanation of other biologically-related changes (e.g. vegetation history).

This chapter reviews the main shifts of climate over the last millenium identified in Central Britain and particularly the Severn Basin. Emphasis is very much placed on the definition of patterns of change and variability, rather than on their causes or implications. The review draws on the acknowledged 'classic' work on climatic change of the last 25 years, but also attempts to bring to the attention of those researchers from a large range of disciplines with interests in palaeohydrology some important recent work now appearing in climatological journals and which may not be generally familiar. In addition, some original material on precipitation variability is presented. The chapter attempts to build on Chapter 4 and to provide background material to the debates conducted elsewhere in this volume on the relative impact of man and climatic change on hydrological, sedimentological and geomorphological changes. The climatological variables explored here are based on the suite of precipitation and temperature indices discussed in Chapter 4 and which are of most potential relevance to palaeohydrologists. More extensive reviews of climatic change in general, for a large range of timescales and areas, can be found elsewhere (e.g. Bradley, 1985; Flohn and Fantechi, 1984; Goudie, 1983; Gribbin, 1978; Lamb, 1977; Mörner and Karlén, 1984; Wright, 1976).

## CLIMATIC RECONSTRUCTION

The study of palaeoclimatology has generated a huge methodological litera-ture, and substantial technical advances, at all timescales of interest, have been made in recent decades. The whole field is reviewed in excellent detail by Bradley (1985) and Lamb (1977, 1982). Within the area of recent climatic reconstruction using historical records, H. H. Lamb has been an energetic pioneering influence. A useful review of available sources of information is Lamb (1982, pp. 67–81) in which the most commonly used sources (e.g. diaries, chronicles, annals, grain-price records, ships' logs) are briefly exam-ined. Ingram, Underhill and Farmer (1981) also produce a neat methodolog-ical summary. Lamb (1982) claims that 'from the mid-sixteenth century onwards there are daily reports of the weather from somewhere or other in western or central Europe with few years omitted' (p. 79) and that 'every season of any kind of dramatic character in Europe since about AD 1100 is probably either known already or could finally be determined' (p. 74). Atten-tion is drawn to the limitations of such datasets, particularly in terms of transcription errors, confusion over different calendar systems, language vari-ations and deliberate falsification of records. Indeed, the uncertainty of grain-yield and crop-failure records, especially, has been the subject of recent vigorous debate by Dury (1984, 1985) and Osmaston (1985).

Early instrumental records are often extremely useful (e.g. see Oliver, 1965) especially if details of site exposure and equipment design, accuracy and precision are known. Although there are occasional references to extremely ancient instrumental records (e.g. the fourteenth-century Korean raingauge noted by Biswas, 1970), the age of 'serious' meteorological obser-vation, at least in the British Isles, is considered to have begun in the seventeenth century. The longest rainfall and temperature datasets available all date from this period (see Craddock, 1976; Craddock and Craddock, 1977; and Lewis, 1977 for historical reviews of early attempts at meteorological monitoring). For instance, 'rainfall measurements have been made at three or more sites in the British Isles since 1725' (Craddock, 1979, p. 332) although organization and routine publication of data began in the 1860s (with *English Rainfall* and, later, *British Rainfall*). Howells (1983) provides a useful summary of long-running rainfall stations in the British Isles. Such records are used in the section below in an exploration of climatic change and variability in central Britain over the recent period (approximately the last 300 years): discussion then briefly focuses on historical climatology—the period prior to the era of direct measurement when changes have to be deduced from indirect evidence or descriptive reports.

## THE INSTRUMENTAL PERIOD

### Precipitation

It is appropriate within a palaeohydrology enquiry to examine first the change and variability over time of the basic input to the system. Sites mentioned in the text are located in Figure 6.1, and are generally within the Severn Basin or roughly peripheral to it. An examination of annual rainfall changes in the British Isles has become possible largely through the efforts of workers at the Climatic Research Unit (CRU) of the University of East Anglia and the Meteorological Office. Over the last 15 years J. M. Craddock, P. D. Jones and T. M. L. Wigley especially, very much in the scholarly traditions of H. H. Lamb and G. Manley, have painstakingly pieced together fragments of rainfall data from a number of gauges (each often with a different height above ground, orifice diameter, exposure, etc.) within a given area, and homogenized these disparate records into single, lengthy, internally consistent, monthly and annual precipitation series. They have published a collection of papers discussing methodology and results and including useful data-tabulations (see *Climate Monitor*, **14**(2), May 1984, for a bibliography of recent CRU work). Typical homogenization procedures are fully described in Craddock (1977, 1978, 1979), while datasets for Oxford (1767–1814), for example, can be found in Craddock and Craddock (1977) and for Cirencester (1844–1977) in Jones (1980). A new homogeneous England and Wales precipitation series (1766–1980), superseding that of Nicholas and Glasspoole (1932), has recently been produced by Wigley, Lough and Jones (1984). A list of available homogenized series is given by Jones (1983) and Wright and Jones (1982).

Annual rainfall changes over the last 100–250 years for thirteen sites (many based on CRU-derived data) have been brought together for comparison in Figure 6.2. Weak, positive trends, nothing like the pronounced changes observable at some tropical sites (e.g. see Stoddart and Walsh (1979) and Walsh and Lawler (1981)), can be discerned at some stations, notably the

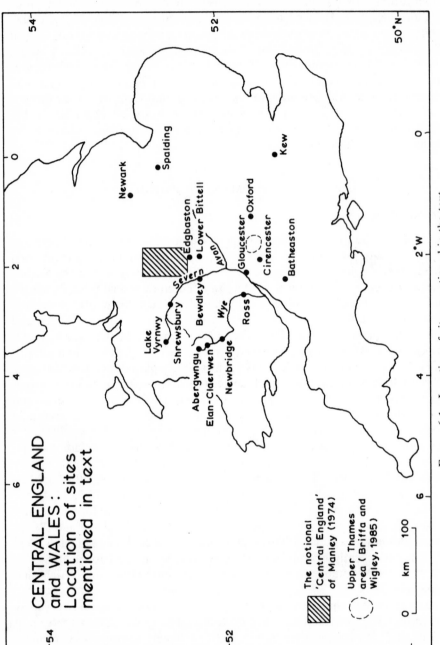

CENTRAL ENGLAND and WALES: Location of sites mentioned in text

The notional 'Central England' of Manley (1974)

Upper Thames area (Briffa and Wigley, 1985)

Newark
Spalding
Kew
Edgbaston
Lower Bittell
Oxford
Gloucester
Cirencester
Batheaston
Severn
Avon
Lake Vyrnwy
Shrewsbury
Bewdley
Wye
Ross
Abergwngu
Elan-Claerwen
Newbridge

km

FIGURE 6.1  Location of sites mentioned in the text

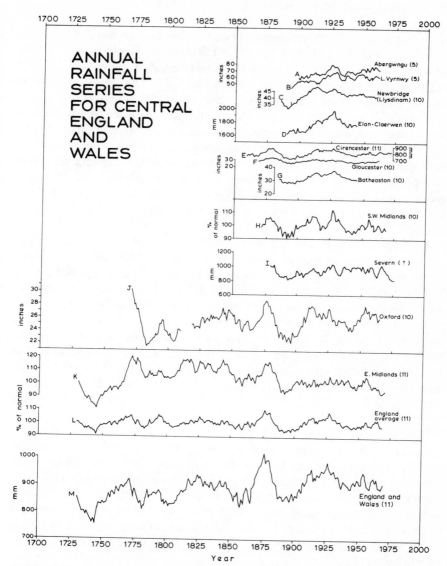

FIGURE 6.2    Annual rainfall series, over the instrumental period, for selected sites in Central England and Wales (the number in parentheses represents the moving-average term for the given series). Various sources have been used: A and B (Howe, Slaymaker and Harding, 1967); C (annual volumes of *British Rainfall*); D (plotted from data in Risbridger and Godfrey, 1954); E (plotted from data in Jones, 1980); F (annual volumes of *British Rainfall*); G (Gregory, 1956); H (plotted from data in Craddock, 1976); I (Thornes, 1983); J (first part from Smith, 1974; second part from Craddock and Craddock, 1977); K (plotted from data in Craddock and Wales-Smith, 1977); L (plotted from data in Craddock, 1976); M (Rodda, Sheckley and Tan, 1978)

upland sites of the basin headwaters (Figure 6.2A–D), the composite Severn series (Figure 6.2I), and the England and Wales run (Figure 6.2M). Almost no trend emerges at the remaining (lowland) easterly locations. The recent slight increase in annual rainfall may be related to the onset of warming following the so-called Little Ice Age: Burroughs (1980) and Mason (1976) have demonstrated a (weak) positive relationship between yearly precipitation totals and mean annual temperature for Britain—(but cf. Smith (1967) on Oxford)—principally through, they argue, a greater energy availability for oceanic evaporation. Peaks and troughs seem to be more accentuated in the uplands, too. The picture broadly reflects the east–west regional variations in rainfall trends proposed by Gregory (1956, his Figure 3) for this part of the country. At most sites, precipitation peaks can be identified around 1770, 1790, 1825, 1875 and 1925 although the moving average technique employed in Figure 6.2 makes difficult the precise timing of rapid fluctuations. With almost all datasets, the peaks of 1875 and 1925 are the strongest. 'A faint suggestion of a six year cycle' has been suggested for the Severn dataset (1875–1975) by Thornes (1983, p. 63) while the author's own autocorrelation analysis of the Cirencester record showed no marked periodicity other than an 18-year cycle significant at $p = 0.05$. Year-to-year variability is generally high: for example at Cirencester the mean annual rainfall (1844–1977) is 810 mm with a (sample) standard deviation of 143 mm and a coefficient of variation of 17.6 per cent (Figure 6.3). When separate measures of dispersion were calculated for concatenated 22-year periods, there was no evidence of any great changes in annual rainfall variability over time.

However, although information on annual totals (and their reliability) is adequate for many water resource planning purposes, there is a need to examine records at monthly, seasonal and daily timescales within many hydrological or palaeohydrological studies. By way of example, individual monthly trends are shown for Cirencester in Figure 6.3. It will be noted that, as Smith (1974) discovered for the Oxford record, little consistency emerges in overall pattern between months, even adjacent ones. One of the main features is the decline in October rainfall (Figure 6.3), defined by time-trend regression as

$$OP = 489 - 0.214 \, YR \qquad (6.1)$$

where *OP is October precipitation (mm) and YR* is year, in which coefficients of correlation ($r = -0.194$) and regression are statistically significant at $p = 0.05$. This month's precipitation has therefore been decreasing at a rate of approximately 0.2 mm per annum or roughly 0.25 per cent per annum over the 1844–1977 period. This decline, however, is almost counterbalanced by an increase in December precipitation (*DP*) similarly obtained as (Figure 6.3)

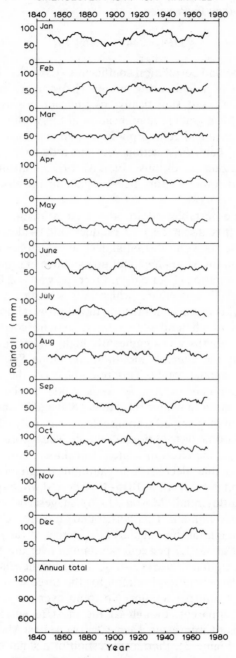

FIGURE 6.3   Eleven-year moving averages
of monthly and annual rainfall at Cirencester: 1844–1977 (plotted from data given
in Jones, 1980)

$$DP = 244 + 0.169\ YR \qquad\qquad (6.2)$$

where, again, slope and correlation coefficients ($r = +0.16$) are significant at $p = 0.05$. Scrutiny of neighbouring months (November, January, February) reveals only a weak and statistically insignificant positive trend, but the general tendency for a gradual amplification of the winter rainfall maximum may have hydrological and geomorphological implications. For example, the efficacy of soil and bank frost action very much depends on moisture availability during freezing conditions (Jumkis, 1977; Outcalt, 1971) and this might be expected, other factors being equal, to increase slightly through time.

The longer series established for the East Midlands (1726–1975), the reliability of which is discussed by Craddock and Wales-Smith (1977) and Tabony (1980), also shows a substantial decline in October rainfall since 1850 (and a similar trend for September) but no evidence of sympathetic increases in December precipitation (Figure 6.4). Over the whole 250-year period only June rainfall showed a decrease which was statistically significant ($r = -0.164, p = 0.01$). It was also interesting that it proved impossible to identify groups of months which tended to covary, and no correlation coefficient between *adjacent* months was significantly high. Interestingly, too, analysis demonstrated that the *summer* month totals (June–September) are most closely related to annual totals. This re-emphasizes the point that annual rainfall information may be a very weak guide to the important flood-producing precipitation in winter (cf. Howe, Slaymaker and Harding, 1967).

This introduction of 'grouped response' leads naturally into a consideration of the changing seasonal distribution of rainfall. The record at Oxford (Smith, 1974) shows sizeable increases in winter (December, January and February) rainfall and noticeable decreases in autumn (September, October and November) precipitation over the 1815–1970 period while spring and summer seasons exhibit little trend. Mason (1976) suggests a similar pattern for England and Wales rainfall. This was found, too, for Cirencester (Figure 6.3) where winter rainfall has significantly increased (at $p = 0.05$) by 0.35 mm per annum or about 1.7 per cent per annum over the 1844–1975 period. A clear winter precipitation rise is also obtained for the Elan–Claerwen data: over the period 1887–1949 total rainfall for the three winter months rose by 2.2 mm or 0.4 per cent per annum. No trend over the longer term for the East Midlands is apparent, though. If the most recent 50–60-year period only is considered at Cirencester, Elan–Claerwen, East Midlands and Oxford (Smith, 1974, his Figure 3), then a suggestion of a slight decrease in winter rainfall can be noted and this concurs with the statement of Rodda, Sheckley and Tan (1978, their Figure 2) for England and Wales as a whole (although they used a six-month definition of 'winter'). Nevertheless, it should be stated that in none of the examples detailed above was a simple, consistent decrease

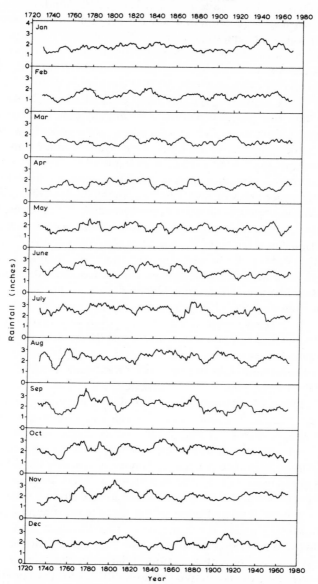

FIGURE 6.4    Eleven-year moving averages of monthly rainfall in the East Midlands: 1726–1975 (plotted from data given in Craddock and Wales-Smith (1977): the corrections given in Tabony (1980) have not been incorporated). The base site is Pode Hole, near Spalding (see Figure 6.1)

since 1920 noted, as Rodda, Sheckley and Tan (1978) found: instead, Cirenc-ester and Oxford showed pronounced decreases from around 1915 to 1950, then rises in winter precipitation in the 1950–75 period.

Arguably the most important indicator of changing input to the system, however, is the nature of heavy daily precipitation. Howe, Slaymaker and Harding (1966, 1967) were perhaps the first workers to demonstrate that, in Mid-Wales, heavy rainfalls (>63.5 mm, or 2.5 in.) had apparently increased since 1940, and this partly explained a recent increase in flooding of the River Severn at Shrewsbury. This problem, understandably, has since found expression in the engineering literature (e.g. Butters and Vairavamoorthy, 1977). It was not clear at the time how widespread this type of change was, and a number of workers have since been investigating this aspect of climatic variability for other areas of Great Britain: these studies are summarized in Table 6.1. While some examinations revealed negligible changes, it appears that a recent increase in storm rainfall frequency has been detected at Kew, Oxford, Birmingham (Edgbaston) and in South Wales. Nevertheless, Table 6.1 shows that only southern and western Britain has been looked at in any detail, apart from McEwen's (1986) analysis of Scottish data, and a variety of methods to identify and describe the nature of the change has emerged. In addition, Lawler (1985) has demonstrated that, in a number of respects, the picture is far from simple. First, it was shown that two sites, only 12.5 km apart and with an altitude difference of just 24 m, yielded quite different results: Figure 6.5 shows for Edgbaston (Figure 6.1) an increase in the frequency of substantial falls of rain (>45.7 mm) since the late 1940s, yet at nearby Lower Bittell Reservoir in Worcestershire (Figure 6.1), little pattern emerges (Figure 6.6). Second, the precise daily rainfall magnitude selected for the conventional frequency plot through time could also significantly affect conclusions drawn: in Figure 6.6, for example, little trend emerges with the 25 mm and 38 mm thresholds, a suspicion of a recent increase since the 1960s can be identified for the 51 mm events, and evidence of a recent *decrease* is noted when falls of >58 mm are considered. Because of the small number of events involved, great caution has to be exercised when identifying such trends. Third, a related point is that there is no general agreement on which thresholds to use: the same daily rainfall event may have vastly different hydrological consequences in differing catchments and would, of course, represent a different magnitude relative to the rainfall regime of a particular site. Fourth, no simple synchroneity across the country exists in any changes identified and the methods used to identify the timing and significance of such increases are themselves open to controversy (see Perry and Howells, 1982; Jolliffe, 1983). Fifth, changes in heavy rainfall frequency appear to have been accompanied by changes in the seasonal incidence of events. Thus, at Lake Vyrnwy in upland Mid-Wales, and, to a lesser extent at Edgbaston (Figure 6.1), the timing of heavy falls appears to have shifted

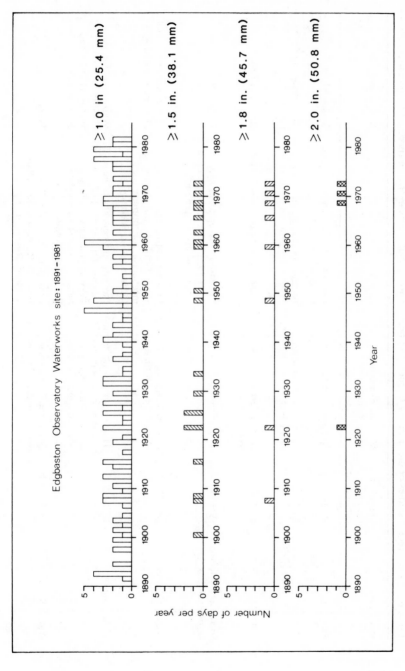

FIGURE 6.5 Frequencies of daily rainfall totals of different magnitudes at Edgbaston Observatory (Birmingham): 1891–1981 (Lawler, 1985)

TABLE 6.1 Some recent studies of the changing frequency of heavy daily rainfall events in Britain (updated from Lawler, 1985)

| Study area | Period of record considered | Methods of analysis/ presentation | Results | Reference |
|---|---|---|---|---|
| Kew | 1881–1975 | Annual exceedances analysis | Intensity of heavy daily rainfalls increased since 1951 | Rodda (1978) |
| Oxford | 1861–1967 | Decadal totals of daily rainfalls ⩾1 in. and ⩾2 in. | Increase in falls ⩾2 in. since 1950; not significant increase of falls ⩾1 in. | McFarlane and Smith (1968) |
| Oxford | 1881–1965 | Return period analysis using partial duration series | Increases in heavy falls, 1941–65 | Rodda (1970) |
| Surrey and selected British stations | 1956–1971 | 'Exceptional falls' defined as ⩾30 mm in 2 consecutive days: 5-year totals compared | Increase in 'exceptional falls' since 1956 in SE England; most other stations showed increase since 1967 | Finch (1972) |
| Wales. Central | 1892–1964 | Daily rainfalls ⩾2.5 in. plotted through time | Increase in frequency 'since about 1940' and was 'triggering mechanism' or 'primary cause' for increase in flooding | Howe, Slaymaker and Harding (1966, 1967) |
| Wales. South | 1875–1979 | 'Heavy rainfalls' (⩾76 mm in a day or 102 mm in 2 days) plotted through time; moving averages | Increase in frequency of heavy rainfalls for periods 1929–33 and 1958–79; increases 'a major factor' in increased flooding in the Swansea Valley | Walsh and Hudson (1980); Walsh, Hudson and Howells (1982) |
| Wales. South | 1878–1980 | 5-year total frequencies of daily rainfalls ⩾25 mm | Peak frequency 1915–20 at Beacons Reservoir; little change at Gnoll Reservoir | Perry (1982) |

| Location | Period | Method | Findings | Reference |
|---|---|---|---|---|
| Wales, South | 1901–1980 | Frequencies of heavy daily rainfalls plotted through time; return period analysis based on partial duration series | 1926–80 period characterized by more frequent heavy rains than pre-1926 | Perry and Howells (1982) |
| West Midlands (and Mid-Wales) | 1881–1981 | Frequencies of heavy rainfalls plotted through time; return period analysis with partial duration series; 3-D magnitude–frequency–time surface constructed | Lake Vyrnwy: continuation of increase in the frequency of rainfalls $\geq 2$ in. and $\geq 2.25$ in.. but not for falls $\geq 2.5$ in.; increase in heavy daily falls since 1959 at Edgbaston; lack of any real trend at nearby Lower Bittell | Lawler (1985) |
| South Wales, East Anglia, Ireland and Scotland (also mid-Wales, Southwest, Northern, Central and Southern England) | c. 1870–1980 | 65 long records examined; frequencies of heavy rainfalls (various thresholds) plotted through time; trend and breakpoint tests; return period analysis | Marked increases observed in heavy daily rainfall frequencies, since the mid 1920s, in South and Central Wales and Southwest England (less clear pattern elsewhere); such increases 'accompanied by major changes in flood magnitude-frequency'; *decreases* in heavy rainfall frequency noted for some easterly districts | Howells (1985) |
| Sussex (especially Balcombe) | 1909–1973 | Heavy rainfall frequencies plotted through time; running means and polynomial curve fitting used to investigate quasi-regular fluctuations | No long-term trend found, but cyclic fluctuations noted: wet cycle (1936–52); dry cycle (1919–35); wet cycle (1953–69); cycles may be connected to the lunar years | Ghayoor (1986) |
| Wales, Central (Upper Severn) | 1899–1982 | Heavy rainfall frequencies plotted through time; breakpoint analysis; return period analysis | Periods of higher frequencies are: late 1920s, late 1930s and mid 1960s; main breakpoint for Lake Vyrnwy is 1925; upland and lowland stations show different trends and seasonal distributions in storm events | Higgs (1987) |

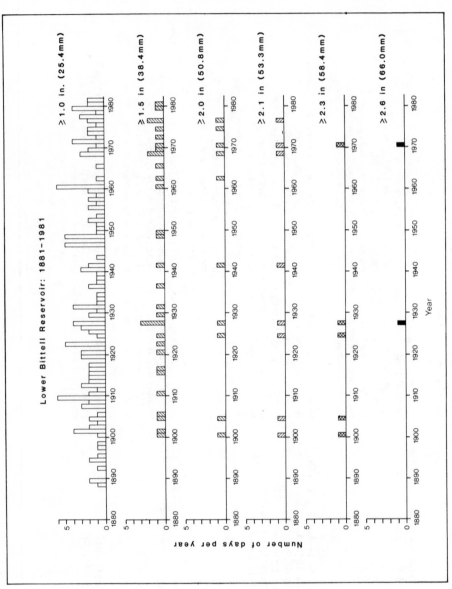

FIGURE 6.6  Frequencies of daily rainfall totals of different magnitudes at Lower Bittell Reservoir (Worcester-shire): 1881–1981 (Lawler, 1985)

from the summer months early in the century to become an autumn or winter phenomenon in the 1930s and 1940s before a gradual drift back to the late summer months took place in the post-war period (Figure 6.7). These shifts in event timing, regardless of magnitude/frequency changes, are likely to influence strongly any hydrological impact (see Chapter 7). Lastly, recent work has suggested a reversal, since around 1968, of the trend of increasing storm rainfalls, at least in Mid-Wales (Severn Area Unit, 1980; Lawler, 1985; Higgs, 1987), and flood frequency has appeared to decrease in sympathy (Severn Area Unit, 1980).

In the light of these and other areas of uncertainty, and the lack of any substantial information on the spatial coherence of recent changes in storm precipitation, Howells (1985) conducted a detailed analysis of 65 long-term daily rainfall records drawn from South Wales, East Anglia, Ireland and Scotland, including a small number of gauges in the Mid-Wales/Borderlands area, Southwest England, Northern England and Central and Southern England. She showed that pronounced increases in storm rainfall from around the mid 1920s were largely confined to South and Mid-Wales, and Southwest England, while, with one or two exceptions, no coherent pattern was identifiable elsewhere. There is a weak tendency for the perceived increase to become less notable in an easterly direction across the country. The scale of change was apparently more marked in upland areas. The timing of change also showed considerable variation (e.g. Kew and Oxford displayed marked post-1950 heavy-rainfall increases—see Table 6.1). For the most part, too, daily rainfall trends tended to be masked by unchanging annual totals, and were not related in any simple way to changing frequency of weather types, e.g. as given in Lamb (1972). In summary, then, despite the need for more detailed regional analyses recognized by Howells (1985), we now know that, in answer to the question posed by Rodda (1970, p. 58) and Goudie (1983, p. 170), the recent increase in heavy rainfall events does not appear to be a nationwide phenomenon. However, in the affected areas, such changes may be more important in the explanation of recent flood-frequency increases than changes in land use (Higgs, 1987; Howe, Slaymaker and Harding, 1966; Thornes, 1983; Walsh, Hudson and Howells, 1982), although this issue of relative impact is exceedingly complex (see Chapter 7).

## Temperature

In an examination of recent climatic change in Central Britain it is fortunate to be able to draw upon the classic, and oft-cited, monthly mean and annual temperature series for Central England (1659–1973) assembled by Manley (1953, 1974) and subsequently updated in issues of *Climate Monitor*. These data, meticulously drawn together from disparate instrumental records, and

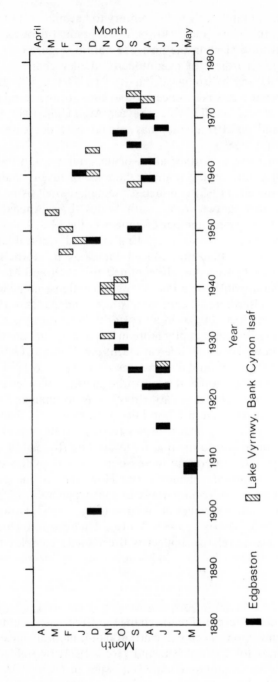

FIGURE 6.7  Monthly distribution of heavy daily rainfall events through time. For Lake Vyrnwy, Bank Cynon Isaf (1905–80), events ⩾2.45 in. (62 mm) are plotted; for Edgbaston Observatory, Birmingham (1881–1981), events ⩾1.5 in. (38 mm) are plotted (Lawler, 1985)

then checked and corrected, have since been the subject of a number of analyses over the last 30 years. By 'Central England', Manley (1974, p. 390) suggests that the temperatures given are representative of 'open rural surroundings in the lowlands of Staffordshire, Shropshire, North Warwickshire (52° 30' to 53 °N, 1° 45' to 2° 15'W) at 100 to 200 feet above sea level' (Figure 6.1). Records from Lancashire and Oxford form the basis of the dataset. The values are probably accurate to within 0.2 K back to 1720, and to within 1.0 K before then (Lamb, 1982, p. 68). The decadal moving averages of the annual temperature series (Figure 6.8), show that a gradual warming over the 315-year period has taken place, which Dyer (1976) found to be statistically significant (at $p = 0.01$). The standard errors of these 10-year means have been estimated to be 'no more than ±0.2 °C in the 1660s and 1670s, and little over ±0.1 °C in the 1690s' (Lamb, 1985, p. 106). Furthermore, Figure 6.8 suggests that almost all of this annual increase is accounted for by temperature increases in winter and spring, with no overall trend apparent in the summer and autumn series. In fact, the correlation analysis of individual months through time by Dyer (1976) demonstrates statistically significant positive trends for the winter half of the year, from October through to April, with non-significant results for the May–September period.

Other significant features of the series have been debated on numerous occasions and include, for the annual data, a pronounced low around 1700 and a marked peak around 1740 (both of which are evident in all four seasons), and the well-documented gradual warming in the first half of this century, followed by a cooling from 1950–70, to be replaced by a notable temperature increase over the last fifteen years (Figure 6.8). This last feature has been ascribed to increases in atmospheric $CO_2$ and has been the subject of vigorous debate in recent scientific literature: e.g. see Wigley, Jones and Kelly (1980), Bach (1984) and Perry (1984). It is notable, though, that the annual picture is not always representative of what is happening to individual seasons: witness the substantial drop in winter temperature from 1930 to 1970, compared to the annual trend over the same period. Dyer's (1976) time series analysis of Manley's data failed to identify clear cycles, other than a peak at 94 years in the spectrum for the annual data, although Lamb (1982, p. 73) mentions a 23-year oscillation which may be associated with the 'double sunspot cycle' (Mason, 1976, p. 478).

It is clearly beyond the scope and purpose of this short chapter to review the massive literature on the possible *causes* of pattern and trend in the Manley temperature series (but see Mason, 1976, and Gribbin, 1978), although no simple explanations appear adequate in the face of a multiplicity of possible influences. In a multivariate statistical study, which proved far from conclusive, for instance, Miles and Gildersleeves (1978, p. 202) (a) suggested that 'the effect of volcanic dust appears therefore to be small, say

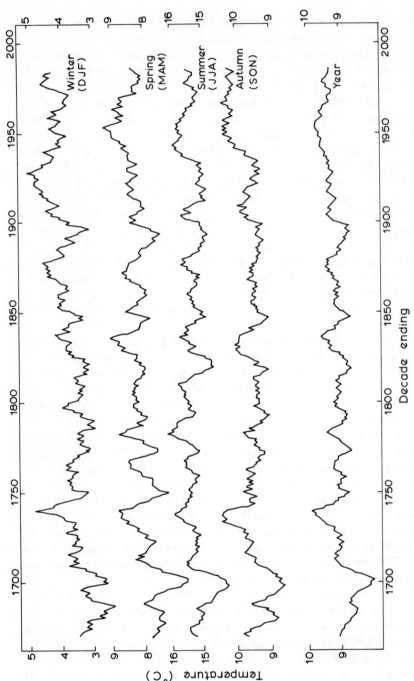

FIGURE 6.8  Ten-year running means of temperatures in Central England, for each season and the whole year, 1659–1982. Originally devised by Manley (1953, 1974), this updated version has been redrawn from Lamb (1985). Reproduced by kind permission of H. H. Lamb and the editors and publishers of *The Climatic Scene* (Allen and Unwin, 1985)

less than 0.1 to 0.2 K for major series of eruptions', (b) found an equally limited relationship between solar activity (as measured by the Wolf sunspot number) and temperature (noting 'a depression of no more than 0.1–0.2 K during the time of zero sunspots'), and (c) stated that 'the carbon dioxide effect is the one which emerges most consistently, and provides the best explanation of the enhanced linear trend after 1870'.

**Derived or related climatic variables**

Mean monthly or annual temperature changes, in themselves, may not be very meaningful in palaeohydrological studies but, in conjunction with other variables, they might be expected to influence, for example, evapotranspiration rates (Penman, 1963), and vegetation and crop growth (Dennett, Elston and Diego, 1980; Parry and Carter, 1985) (and hence soil moisture regimes, interception losses, catchment surface roughness), and to reflect, however weakly, more geomorphologically relevant indices such as frost frequency/intensity patterns. It is these related variables which are now briefly examined. Wales-Smith (1977) has produced a long series (1698–76) of monthly potential evaporation (PE) totals representative of grassland at Kew, using Penman's (1948) procedure for the post-1870 period, and temperature data alone for years prior to 1870, when data on, for instance, sunshine hours and wind speed were generally unavailable. The annual trend is a slight positive one (Figure 6.9), with a marked increase this century which Wales-Smith (1977) largely ascribes to a longer duration of sunshine and higher temperatures since around 1911. When the pattern is analysed on a seasonal basis, enhancement of *summer* evaporation only is found, with limited trend in the other seasons. All other factors being equal, this may mean an increasing delay in the autumnal growth of runoff source areas and hence a slight reduction in the magnitude of early winter flow peaks. At other times of the year, the hydrological impact of such PE changes may be less important. Soil moisture reconstructions from tree rings have recently been made for the Upper Thames catchment by Briffa and Wigley (1985), and this represents, despite considerable simplification of the problem, an interesting development in dendroclimatological method. Davis (1972) has found for Oxford a substantial secular variation in the onset of spring since 1869. The first day of spring was defined as the end of the first five-day period of the year within which average daily maximum screen temperature reaches 10°C (representing reinitiation of plant growth), and was found to occur anytime between the 10th and 23rd pentad (*c.* 19 February–25 April). Around 1920, however, an abrupt and statistically significant change (at $p = 0.005$) to an earlier onset of spring took place. In the post-war period, though, spring has tended to drift back to a later date (Figure 6.10). Lyall's (1970) study of Newark data supports this. This follows the pattern of mean temperatures

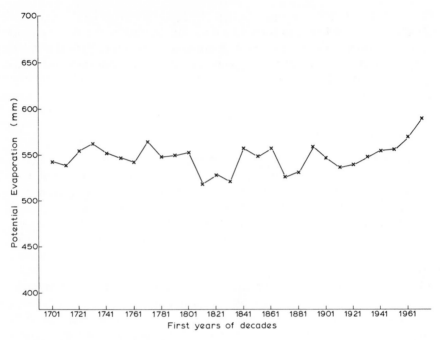

FIGURE 6.9    Trend of potential evaporation at Kew: 1698–1976 decadal averages (from Wales-Smith, 1977). From data supplied by the Meteorological Office

given by Manley (Figure 6.8), and Davis (1972) argues that it reflects changing circulation types, the amount of ice in the East Greenland Sea and North Baltic areas (which influences the temperature of, respectively, northerly and easterly airstreams over the UK), and the sea temperature pattern in the North Atlantic.

Finally, some climatologists have commented on the apparent increase in recent years of variability or extremes in the British weather. If significant, this might make the recent period rather unrepresentative for the calibration of rainfall-runoff models or for stochastic flow simulation exercises in which data sequences are generated which preserve the statistical structure of observed time series (e.g. see Kottegoda, 1980). Smith (1967), for example, has drawn attention to a post-war increase in the severity of winter at Oxford (as measured by various indices of temperature and snow cover). Lamb (1977) has listed a sequence of extremes that have affected the British Isles since 1960, and a tabulation of significant climatic hazards experienced in the UK from 1968 to 1978 has been presented by Perry (1981, p. 16). Morris and Marsh (1985, p. 329) analysed drought sequences and magnitudes, calculated from a bulked England and Wales precipitation dataset, and drew attention to the 'erratic nature of post-1974 rainfall' and suggested 'how volatile the

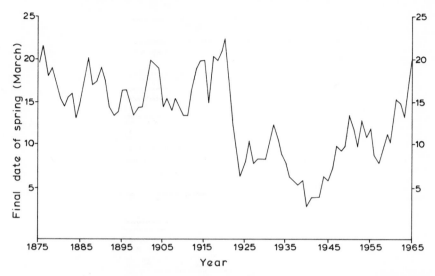

FIGURE 6.10   Final date of the onset of spring at Oxford: ten-year running means (from Davis, 1972). Reproduced by permission of the Royal Meteorological Society

climate has been over the last ten years'. Their plotting of 'drought index' values (measured as a departure from long-term means) from 1766–1984 (Figure 6.11) revealed not only an increased drought severity over the last 100 years, but also the dominance over the whole record of the 1975–76 event (see Doornkamp, Gregory and Burn (1980) for a detailed assessment of drought magnitude and impact here). Wigley and Jones (in press) have also demonstrated a statistically significant increase in the frequency of extreme dry summers and wet springs over the last 10 years. Equally, however, there have been workers who have challenged this view of a recent increase in weather extremes, e.g. Ratcliffe *et al.*, (1978), cited in Perry (1982). Furthermore, a recent review by Perry (1982) entitled 'Is the climate becoming more variable?' could only conclude that, because of the equivocal nature of the evidence, there still is 'disagreement as to whether that variability is greater today than in the past' (Perry, 1982, p. 111).

Such controversy is by no means unusual in investigations of climatic change over the instrumental period. It mainly arises due to a common lack of widespread spatial coherence in change (and hence different regions exhibiting variable responses), the presence of undetected heterogeneity in some climatic records, different methods being used to detect the timing and magnitude of change, the scrutiny of annual totals alone, which may mask change or variability in more sensitive indicators at shorter timescales, and the difficulties of separating trend from noise in the climatic signal. These

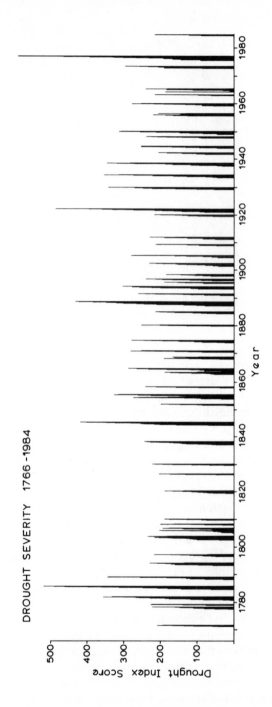

FIGURE 6.11  Drought index values for England and Wales: 1766–1984 (from Morris and Marsh, 1985). Reproduced by permission of Artetech Publishing Co.

problems are often exacerbated by further difficulties when we shift attention to the historical period before instrumental records began.

## THE HISTORICAL PERIOD

Prior to the period of 'serious' observations using meteorological equipment, climate has to be inferred from other sources, including, for instance, surviving diaries, chronicles, journals, phenological records and ships' logs, in addition to the quite different kind of evidence provided by tree-ring and ice-core studies. Clearly, then, rather than a progressive and gradual decline in the reliability of climatic information as one goes back in time, there exists, in Britain at least, a sharp drop in both accuracy and precision of data around the late seventeenth century, at the transition from direct measurement to indirect estimation. A natural starting point for this period of indirect written evidence is the end of the Dark Ages in Britain (*c*. AD 800). If G. Manley was the pioneering figure in work on early instrumental records, H. H. Lamb emerges as the individual who has probably done more than any other to reconstruct climates in the pre-instrumental era. Lamb has largely made use of written sources—see Lamb (1977, 1982) for details of techniques and their reliability—and the reconstructions have been tested against actual meteorological measurements for the post-1700 period (Lamb, 1982, p. 77) and against other European reconstructions for medieval times (Lamb, 1982, p. 78). Although decadal values of temperature and rainfall have been produced, Lamb (1982, p. 77) suggests an integration over longer periods is more appropriate: 'taken in blocks of fifty years, the results are thought to be reliable as a first approximation to the temperature and rainfall history of England sincd AD 1100'. Lamb, Lewis and Woodruffe, (1966, p. 175) consider that, back to 1400, standard errors of derived 100-year means are less than 0.15 K. His classic work is summarized in a large number of papers, monographs and books (e.g. Lamb, 1965, 1977, 1982, 1984a).

A recent statement of temperature trend in central England since AD 800 appears in Lamb (1984a) and is presented here as Figure 6.12. The main features, which are broadly similar to those outlined earlier by Lamb, Lewis and Woodruffe (1966), are as follows:

1. A warming phase, 900–1150. Increased anticylonic activity here gave warmer summers and cooler winters.
2. The medieval warm epoch, climaxing between 1150 and 1300 (Lamb, 1984b, p. 35). Average annual temperatures considered to be 0.5–0.8 K above the 1900–50 mean and 1.2–1.4 K above Little Ice Age conditions (Lamb, 1984b, p. 38).
3. A cooling phase, 1300–1450, heralded by particularly stormy weather,

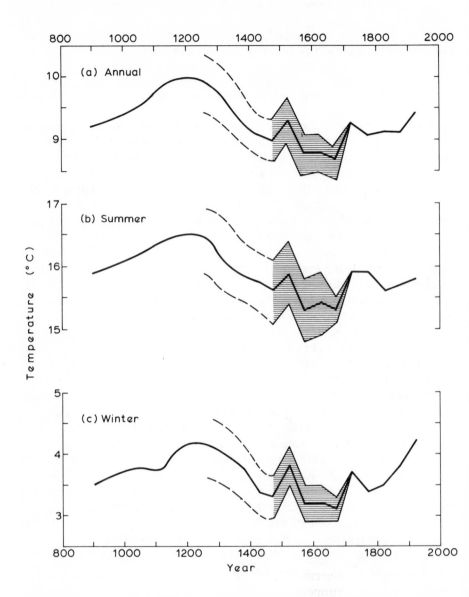

FIGURE 6.12  Estimated temperatures in Central England since AD 800. Probable 50-year averages: (a) for the whole year; (b) for July and August; (c) for December, January and February. A range of uncertainty (approximately three times the standard error of the estimates) is indicated (from Lamb, 1984a). Reproduced by permission of D. Reidel Publishing Company

which is held to be responsible for the sand-dune invasions of the South Wales coast (e.g. Lees, 1982).

4. A cold epoch, commonly referred to as the 'Little Ice Age', 1450–1700, but interrupted by a short-lived amelioration around 1500. The climax came around 1550–1700, but the Little Ice Age has been considered to extend to 1850, when maximum expansion of Alpine glaciers was recorded (Lamb, 1984b, p. 45). Numerous reports of sea ice around the coast of Britain and the freezing of the London Thames emanate from this period. Mean annual temperatures were around 0.7–0.8 K below the 1900–50 average, although differences of around 1.2 K are recorded in winter (Figure 6.12).

5. A warming phase, 1700–1950, defined by instrumental records and described above (Figure 6.8).

Further details of the characteristic weather patterns (and their impact on European society) of all these phases and epochs (especially the Little Ice Age) can be found in Lamb (1977, 1982, 1985).

Rainfall reconstructions for the pre-instrumental era are a little more problematic. Lamb has presented a tentative derivation of 50-year averages of annual, and summer, precipitation from AD 1100 (and 100-year averages back to AD 800) for England and Wales (Lamb, Lewis and Woodroffe, 1966; Lamb, 1984b), expressed as a percentage of the 1916–50 mean (Figure 6.13). These are again based on the Lamb summer and winter indices, obtained as a function of the ratio of reports of mild/cold winters and wet/dry summers in the written records available, which are then calibrated against post-1700 observations (Lamb, 1982, p. 77). These descriptive reports, however, largely reflect the wetness of the ground surface, which depends on the *balance* between inputs and losses (particularly evapo-transpiration—controlled by temperature, wind speed, humidity, radiation and vegetation cover), rather than on precipitation alone. As explained by Lamb (1965) and Lamb, Lewis and Woodroffe (1966) use was also made of regression equations which predict rainfall from temperature estimates, taking advantage of the moderately strong positive correlations typically obtained for British locations (e.g. see Burroughs, 1980). Strong similarities therefore exist between the annual trends in Figures 6.12 and 6.13. Thus, the medieval warm epoch was characterized by around 3 per cent increases in annual rainfall totals, while, for the Little Ice Age, 7–10 per cent decreases are noted, deficits which rise to 15–20 per cent in the very dry 1740–1760 period (Lamb, 1984b). However, this conceals an apparently strong seasonal variation in response: for example, Figure 6.13 shows that summer precipitation dropped dramatically during the medieval warm epoch, with a compensatory increase in the other months of the year. Summer rainfall is also shown to be particularly *variable*, and Lamb, Lewis and Woodroffe

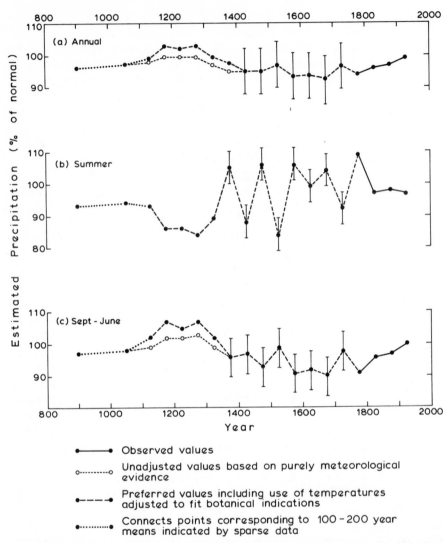

FIGURE 6.13   Estimated precipitation over England and Wales since AD 800. Probable 50-year averages, expressed as percentages of twentieth-century values: (a) for the whole year; (b) for July and August; (c) for the rest of the year (September–June inclusive). A range of uncertainty (±3 times the standard error) is indicated (from Lamb, 1984b). Reproduced by permission of D. Reidel Publishing Company

(1966, p. 190) point to the 'greater wetness of the British summers in the second halves of each century since 1300' (Figure 6.13).

The limitations of the climatic data presented in this section, as well as the contrasts in accuracy and resolution at different timescales and with different variables, pose great difficulties for palaeohydrological studies. One promising complementary line of enquiry, though, is through dendrological work: recently, for example, researchers at the Climatic Research Unit have produced, from tree-ring studies, an *annual* soil moisture series back to the early 1800s for the Upper Thames catchment (Briffa and Wigley, 1985), a spatially averaged precipitation and temperature series back to 1830 for England (Briffa *et al.*, 1983) and even riverflow series for basins in Southern Britain (Jones, Briffa and Pilcher, 1984). This latter study is believed to be the first such dendrohydrological reconstruction in Europe. Longer-term retrodictions may well be possible at this finer resolution. More recently, Jones, Ogilvie and Wigley (1985) have produced an extremely interesting volume containing further flow reconstructions using the Central Water Planning Unit model of Wright (1978)—cited in Jones, Ogilvie and Wigley (1985)—for ten catchments in England and Wales. It also includes an excellent tabulation of historical flood and drought information for England and Wales back to 1556 and supersedes, in comprehensiveness and reliability, the useful report by Potter (1978). These, and other studies now emerging, provide a substantial platform for further progress in assessing the forcing role of climatic change within historical hydrology.

## ACKNOWLEDGEMENTS

I am very grateful to the staff of the Meteorological Office, Severn-Trent Water Authority, Welsh Water Authority for the provision of data and other assistance in the preparation of this paper. I also thank Mr J. M. Craddock, Dr K. A. Howells, Dr P. D. Jones (CRU), Dr M. L. Parry (University of Birmingham) and Dr R. P. D. Walsh (Swansea University) who drew my attention to some useful references, the editors for their helpful comments on an earlier draft, Mr T. Grogan (University of Birmingham) for drawing the figures, and Dr H. A. Lawler for assistance with MS production.

## REFERENCES

Bach, W. (1984). Carbon dioxide and climatic change: an update, *Prog. Phys. Geog.*, **8**, 83–93.
Ballard, J. G. (1978). *The Drought*, Triad/Panther Books, St Albans, 188 pp.
Biswas, A. K. (1970). *History of Hydrology*, North-Holland, Amsterdam.
Bradley, R. S. (1985). *Quaternary Palaeoclimatology*, Allen & Unwin, Boston, 472 pp.
Briffa, K. R., Jones, P. D., Wigley, T. M. L., Pilcher, J. R. and Baillie, M. G. L. (1983). Climate reconstruction from tree rings: Part 1, Basic methodology and preliminary results for England, *J. Climatology*, **3**, 233–42.
Briffa, K. R. and Wigley, T. M. L. (1985). Soil moisture reconstruction using tree rings, *Climate Monitor*, **14**, 106–13.
Brooks, C. E. P. (1926). *Climate through the Ages*, Benn, London.

Burroughs, W. J. (1980). Average temperature and rainfall figures in British winters, *Weather*, **35**, 75–9.

Butters, K. and Vairavamoorthy, A. (1977). Hydrological studies on some river catchments in Greater London, *Proc. Instn. Civ. Engrs*, Part 2, **63**, 331–61.

Coope, G. R. (1977). Fossil coleopteran assemblages as sensitive indicators of climatic changes during the Devensian (last) cold stage, *Phil. Trans. R. Soc. Lond. B.*, **280**, 313–40.

Craddock, J. M. (1976). Annual rainfall in England since 1725, *Quart. J. R. Met. Soc.*, **102**, 823–40.

Craddock, J. M. (1977). A homogeneous record of monthly rainfall totals for Norwich for the years 1836 to 1976, *Met. Mag.*, **106**, 267–78.

Craddock, J. M. (1978). Methods of comparing rainfall records, *Climate Monitor*, **7**, 64–7.

Craddock, J. M. (1979). Methods of comparing annual rainfall records for climatic purposes, *Weather*, **34**, 332–46.

Craddock, J. M. and Craddock, E. (1977). Rainfall at Oxford from 1767 to 1814, estimated from the records of Dr Thomas Hornsby and others. *Met. Mag.*, **106**, 361–72.

Craddock, J. M. and Wales-Smith, B. G. (1977). Monthly rainfall totals representing the East Midlands for the years 1726 to 1975, *Met. Mag.*, **106**, 97–111.

Davis, N. E. (1972). The variability of the onset of spring in Britain, *Quart. J. R. Met. Soc.*, **98**, 763–77.

Dennett, M. D., Elston, J. and Diego, R. (1980). Weather and yields of tobacco, sugar beet and wheat in Europe. *Agric. Met.*, **21**, 249–63.

Doornkamp, J. C., Gregory, K. J. and Burn, A. S. (1980). *Atlas of Drought in Britain 1975–76*, Inst. Br. Geogr., London 82 pp.

Dury, G. H. (1984). Crop failures on the Winchester manors, 1232–1349, *Trans. Inst. Br. Geogr.*, **9**, 401–18.

Dury, G. H. (1985). Crop failures on the Winchester manors, 1232–1349: pseudo-controversy as unnecessary, *Trans. Inst. Br. Geogr.*, **10**, 501–3.

Dyer, T. G. J. (1976). An analysis of Manley's central England temperature data: I, *Quart. J. R. Met. Soc.*, **102**, 871–88.

Finch, C. R. (1972). Some heavy rainfalls in Great Britain 1956–1971, *Weather*, **27**, 364–77.

Flohn, H. and Fantechi, R. (1984). *The Climate of Europe: Past, Present and Future*, D. Reidel, Dordrecht, 356 pp.

Ghayoor, H. A. (1986). Temporal fluctuations of daily rainburst and the cycle of the lunar year, *J. Climatology*, **6**, 83–95.

Goudie, A. S. (1983). *Environmental Change*, Oxford University Press, 2nd edn, 258 pp.

Gregory, S. (1956). Regional variations in the trend of annual rainfall over the British Isles, *Geog. J.*, **122**, 346–53.

Gribbin, J. (ed.) (1978). *Climatic Change*, Cambridge University Press, 280 pp.

Higgs, G. (1987). Environmental change and hydrological response: flooding in the Upper Severn catchment, in Gregory, K. J., Lewin, J. and Thornes, J. B. (eds), *Palaeohydrology in Practice*, Wiley, Chapter 7 (this volume).

Howe, G. M., Slaymaker, H. O. and Harding, D. M. (1966). Flood hazard in mid-Wales, *Nature*, **212**, 584–5.

Howe, G. M., Slaymaker, H. O. and Harding, D. M. (1967). Some aspects of the flood hydrology of the upper catchments of the Severn and Wye, *Trans. Inst. Br. Geogr.*, **41**, 33–58.

Howells, K. A. (1983). Long-running rainfall stations in the British Isles, *Swansea Geographer*, **20**, 33–49.

Howells, K. A. (1985). *Changes in the magnitude-frequency of heavy daily rainfalls in the British Isles*, unpublished Ph.D. thesis, University of Wales, 501 pp.

Ingram, M. J., Underhill, D. J. and Farmer, G. (1981). The use of documentary sources for the study of past climates, in Wigley, T. M. L., Ingram, M. J. and Farmer, G. (eds), *Climate and History*, Cambridge University, Press, pp. 180–213.

Jolliffe, I. T. (1983). Large falls of rain in Wales—a simple statistical case study, *Weather*, **38**, 103–6.

Jones, P. D. (1980). A homogeneous rainfall record for the Cirencester area, 1844–1977, *Met. Mag.*, **109**, 249–58.

Jones, P. D. (1983). Further composite rainfall records for the United Kingdom, *Met. Mag.*, **112**, 19–27.

Jones, P. D., Briffa, K. R. and Pilcher, J. R. (1984). Riverflow reconstruction from tree rings in Southern Britain, *J. Climatology*, **4**, 461–72.

Jones, P. D., Ogilvie, A. E. J. and Wigley, T. M. L. (1985). Riverflow data for the United Kingdom: reconstructed data back to 1844 and historical data back to 1556, *Climatic Research Unit Research Paper*, **8**, 166 pp.

Jumkis, A. R. (1977). *Thermal Geotechnics*, Rutgers University Press, 375 pp.

Kottegoda, N. T. (1980). *Stochastic Water Resources Technology*, Macmillan, London, 384 pp.

Lamb, H. H. (1965). The early medieval warm epoch and its sequel, *Palaeogeography, Palaeoclimatology, Palaeoecology*, **1**, 13–37.

Lamb, H. H. (1972). British Isles weather types and a register of the daily sequence of circulation patterns 1861–1971, *Met. Off. Geophys. Memoir*, **16**, No. 116, 85 pp.

Lamb, H. H. (1977). *Climate: Present, Past and Future, vol. 2: Climatic history and the future*, Methuen, London, 835 pp.

Lamb, H. H (1982) *Climate, History and the Modern World*, Methuen, London, 387 pp.

Lamb, H. H. (1984a). Climate and history in northern Europe and elsewhere, in Mörner, N.-A. and Karlén, W. (eds), *Climatic Changes on a Yearly to Millenial Basis*, D. Reidel, Dordrecht, pp. 225–40.

Lamb, H. H. (1984b). Climate in the last thousand years: natural climatic fluctuations and change, in Flohn, H. and Fantechi, R. (eds), *The Climate of Europe: Past, Present and Future*, D. Reidel, Dordrecht, pp. 25–64.

Lamb, H. H. (1985). The Little Ice Age period and the great storms within it, in Tooley, M. J. and Sheail, G. M. (eds), *The Climatic Scene*, Allen & Unwin, London, pp. 104–31.

Lamb, H. H., Lewis, R. P. W. and Woodroffe, A. (1966). Atmospheric circulation and the main climatic variables between 8000 and O BC: meteorological evidence, *Proc. International Symp. on World Climate 8000 to 0 BC, London 18–19 Apr 1966*, Royal Meteorological Society, pp. 174–217.

Lawler, D. M. (1985). Changes in the frequency of heavy daily rainfalls: some preliminary observations from Wales and the West Midlands, *University of Birmingham, Dept. of Geography, Working Paper*, 27, 15 pp.

Lees, D. J. (1982). The sand dunes of Gower as potential indicators of climatic change in historical time, *Cambria*, **9**, 25–35.

Lewis, R. P. W. (1977). Rainfall recording in the British Isles, 1677–1977, *Met. Mag.*, **106**, 378–80.

Lyall, I. T. (1970). Recent trends in spring weather, *Weather*, **25**, 163–5.

Manley, G. (1953). The mean temperature of central England, 1698–1952, *Quart. J. R. Met. Soc.*, **79**, 242–61.

Manley, G. (1974). Central England temperatures: monthly means 1659 to 1973, *Quart. J. R. Met. Soc.*, **100**, 389–405.

Mason, B. J. (1976). Towards the understanding and prediction of climatic variations, *Quart. J. R. Met. Soc.*, **102**, 473–98.

McEwen, L. J. (1986). *River channel planform changes in upland Scotland, with specific reference to climatic and land use changes over the last 250 years*, unpublished Ph.D. thesis, University of St Andrews.

McFarlane, D. and Smith, C. G. (1968). Remarkable rainfall in Oxford, *Met. Mag.*, **97**, 235–45.

Miles, M. K. and Gildersleeves, P. B. (1978). A statistical study of the likely influence of some causative factors on the temperature changes since 1665, *Met. Mag.*, **107**, 193–204.

Mörner, N.-A. and Karlén, W. (eds) (1984). *Climatic Changes on a Yearly to Millenial Basis*, D. Reidel, Dordrecht, 667 pp.

Morris, S. E. and Marsh, T. J. (1985). United Kingdom rainfall 1975–1984: Evidence of climatic instability?, *J. Meteorology*, **10**, 324–32 (see also *J. Met.*, **11**, (1986), p. 25 for an important corrigendum of this paper).

Nicholas, F. J. and Glasspoole, J. (1932). General monthly rainfall over England and Wales, 1727–1931, *British Rainfall (1931)*, 299–306.

Oliver, J. (1965). A weather register for Haverfordwest, Pembrokeshire, 1724–7, *Weather*, **20**, 364–9.

Osmaston, H. (1985). Crop failures on the Winchester Manors 1232–1349 AD; some comments, *Trans. Inst. Br. Geogr.*, **10**, 495–500.

Outcalt, S. I. (1971). An algorithm for needle ice growth, *Water Resources Res.*, **7**, 394–400.

Parry, M. L. and Carter, T. R. (1985). The effect of climatic variations on agricultural risk, *Climatic Change*, **7**, 95–110.

Penman, H. L. (1948). Natural evaporation from open water, bare soil and grass, *Proc. Roy. Soc., A*, **193**, 120–45.

Penman, H. L. (1963). Vegetation and Hydrology, *Technical Communication*, 53, Commonwealth Bureau of Soils, Harpenden, Commonwealth Agricultural Bureau.

Perry, A. H. (1981). *Environmental Hazards in the British Isles*, Allen & Unwin, London, 191 pp.

Perry, A. H. (1982). Is the climate becoming more variable?, *Prog. Phys. Geog.*, **6**, 108–114.

Perry, A. H. (1984). Recent climatic change—is there a signal amongst the noise?, *Prog. Phys. Geog.*, **8**, 111–17.

Perry, A. H. and Howells, K. A. (1982). Are large falls of rain in Wales becoming more frequent?, *Weather*, **37**, 240–3.

Potter, H. R. (1978). The use of historic records for the augmentation of hydrological data, *Inst. Hydrol. Report*, 46, 59 pp.

Risbridger, C. A. and Godfrey, W. H. (1954). Rainfall, run-off and storage: Elan and Claerwen gathering grounds, *Proc. Inst. Civ. Engrs*, **3**, 345–88 (Discussion, 389–408).

Rodda, J. C. (1970). Rainfall excesses in the United Kingdom, *Trans. Inst. Br. Geogr.*, **49**, 49–60.

Rodda, J. C. (1978). Discussion of: Hydrological studies on some river catchments in Greater London, *Proc. Inst. Civ. Engrs*, Part 2, **65**, 469–70.

Rodda, J. C., Sheckley, A. V. and Tan, P. (1978). Water resources and climatic change, *J. Inst. Wat. Engrs and Scientists*, **32**, 76–83.

Smith, C. G. (1967). Winters at Oxford since 1815, *Oxford Magazine*, 265–8.

Smith, C. G. (1974). Monthly, seasonal and annual fluctuations of rainfall at Oxford since 1815, *Weather*, **20**, 2–16.

Severn Area Unit (1980). Weather conditions in the Upper Severn catchment before and after 1968, *Severn-Trent Water Authority Report*, 5 pp.

Stoddart, D. R. and Walsh, R. P. D. (1979). Long-term climatic change in the western Indian Ocean, *Phil. Trans. R. Soc. Lond., B.*, **286**, 11–23.

Tabony, R. C. (1980). A revised rainfall series for Spalding, Lincolnshire, *Met. Mag.*, **109**, 152–7.

Thornes, J. B. (1983). Discharge: empirical observations and statistical models of change, in Gregory, K. J. (ed.) *Background to Palaeohydrology*, Wiley, Chichester, pp. 51–67.

Wales-Smith, B. G. (1977). An analysis of monthly potential evaporation totals representative of Kew from 1698 to 1976, *Met. Mag.*, **106**, 297–313.

Walsh, R. P. D. and Hudson, R. N. (1980). *The floods of the River Tawe of late December 1979*, unpublished report commissioned by Inco Europe Ltd, Clydach, South Wales, 47 pp.

Walsh, R. P. D., Hudson, R. N. and Howells, K. A. (1982). Changes in the magnitude-frequency of flooding and heavy rainfalls in the Swansea valley since 1875, *Cambria*, **9**, 36–60.

Walsh, R. P. D. and Lawler, D. M. (1981). Rainfall seasonality: description, spatial patterns and change through time, *Weather*, **36**, 201–8.

Weiss, M. E. (1978). *What's Happening to our Climate?*, Messner, New York, 93 pp.

Wigley, T. M. L. and Jones, P. D. (in press). England and Wales precipitation: a discussion of recent changes in variability and an update to 1985, *J. Climatology*

Wigley, T. M. L., Jones, P. D. and Kelly, P. M. (1980). Scenario for a warm, high-$CO_2$ world, *Nature*, **283**, 17–21.

Wigley, T. M. L., Lough, J. M. and Jones, P. D. (1984). Spatial patterns of precipitation in England and Wales and a revised, homogeneous England and Wales precipitation series, *J. Climatology*, **4**, 1–25.

Wright, C. E. and Jones, P. D. (1982). Long period weather records, droughts and water resources, in *Optimal Allocation of Water Resources, Proc. Exeter Symposium (July 1982) of the Int. Assn. Hydrol. Sci., IAHS Publ.*, **135**, pp. 89–100.

Wright, P. B. (1976). Recent climatic change, in Chandler, T. J. and Gregory, S. (eds), *The Climate of the British Isles*, Longman, London, pp. 224–47.

Palaeohydrology in Practice
Edited by K. J. Gregory, J. Lewin and J. B. Thornes
© 1987 John Wiley & Sons Ltd.

# 7

# Environmental Change and Hydrological Response: Flooding in the Upper Severn Catchment

G. HIGGS

*Department of Geography, University College of Wales, Aberystwyth*

## INTRODUCTION

The importance of contemporary process studies for palaeohydrological estimation has often been stressed (Gregory, 1983). Changes in catchment
characteristics, both of natural and man-made origin, have important hydrological and sedimentological consequences for rivers. Factors which are
important on a historical timescale incorporate climatic effects such as
changes in the amount and distribution of precipitation, as well as land
use changes such as afforestation, pasture improvement, urbanization and
reservoir construction.

One of the major tools for identifying the hydrological consequences of
land use change, especially since the International Hydrological Decade,
has been that of the experimental catchment. Despite criticisms of such
experiments, focusing in particular on costs, leakages and unrepresentative-
ness (Ackermann, 1966), much of our current knowledge is based on them
and on so-called 'representative basins' which have been set up to provide
data 'typical' of a particular hydrological region.

One of the earliest approaches was to observe a research catchment before
a treatment such as afforestation or deforestation was applied, and then to
continue observations to identify the degree of change caused. An example
is the Hubbard Brook Experimental Forest in New England. A 16 ha catchment was cleared of hardwood forest cover, and regrowth was prevented
with herbicides. Annual water yields were seen to have increased for the
first two water years after treatment. Instantaneous peaks also increased
(Hornbeck, Pierce and Federer, 1970). A similar approach has been

conducted for the Coal Burn catchment in Northumberland, whereby the hydrological and sedimentological characteristics of the catchment have been compared prior to, and following, forestry drainage (Robinson, 1981). Analysis of changes in catchment runoff, from the scale of the annual water balance to individual storm characteristics, reveals a great increase in the magnitude of hydrograph peaks (by up to 40 per cent) suggesting an increase in flood flows within the catchment.

A second approach has involved the comparison of the hydrological responses of two or more adjacent catchments within a similar hydroclimatic region with differing land use covers (for example, forested and non-forested basins). The use of such 'paired' or 'multiple' catchments to study the hydrologic relationships of a single type of cover is preferable to the single watershed approach since the latter does not take into account any possible climatic variation during the period of measurement which may have a bearing on the conclusions drawn. An early example of the 'paired' catchment approach was the work of Bates and Henry (1928) from 1911 to 1926, at Wagon Wheel Gap. Two small forested catchments of 80 ha were instrumented. After a period of 8 years one catchment was deforested, and measurements of rainfall and runoff continued for another 7 years. The streamflow pattern of the two catchments before and after deforestation was compared and the conclusion drawn that deforestation had led to an average increase in runoff of 15 per cent over the unforested catchment. A British example has been the Institute of Hydrology's study of the forest/grassland comparison for the Plynlimon Experimental Catchments. By comparing the forested Severn catchment and the grassland Wye catchment in terms of components of the hydrological cycle, important conclusions have been reached as to the role of afforestation in increasing evapotranspiration and water yield reduction and the reduction in flood peak discharges (Newson, 1979).

There are inherent problems in both experimental approaches. For example, in most cases the experimental catchments are kept small because of difficulties in obtaining adequate uniformity over larger areas, while there are complications introduced at larger scales by variations in rainfall and evaporation within the catchment area itself. There are difficulties in extrapolating the results of small catchment experiments, often with short-term data sets, to the interpretation of larger catchment responses to climatic and man-made changes (Pilgrim, 1983). There are also dangers in extrapolating results from one geographical or climatic region to another. The consequences of land management practices are influenced by particular site conditions of geology, climate, soil and vegetation. Thus land use hydrology is, to a large extent, restricted to being a regional science.

An important consideration in studies of the hydrological consequences of catchment changes is, therefore, the research scale. This has to be clearly defined at the commencement of a study since the results of a land use change on peak discharges are likely to vary according to the scale involved. For example, at the smaller plot scale the effect of agricultural drainage may be effectively to lower the water table and create more storage capacity so that peak flows are decreased. Thus peak flows were lower for two agriculturally drained plots in Mid-Wales, one on a clay soil and the other on a peaty hill soil, seemingly suggesting a moderation of the flood risk downstream of farmland drainage schemes (Newson and Robinson, 1983). However, the problem is to extend these conclusions to the larger catchment scale, where other variables appear to be important. Here the effect may be the opposite, i.e. an increase in peak flows, dependent on the location of the drained plots within a catchment. Howe, Slaymaker and Harding (1967), for example, suggested that the drainage of afforested land in the upper reaches of the Severn, together with an increase in the frequency of heavy rainfall events, led to an increase in the frequency of flooding in the post-1940 period, as compared to the period 1911–40. Subsequent analysis has shown that the effects of drainage are complex and can be reversed according to soil and rainfall characteristics of the area in question, the type and efficiency of drainage system and the location within the catchment (Robinson, 1981).

The problem, therefore, is to find the appropriate scale at which to study and to draw conclusions. The experimental plot scale may be too small for any meaningful conclusions to be made with regard to peak flows further downstream. At the other extreme, at the basin-scale, it may be difficult to detect a change in runoff patterns where land use changes are slow, or where several changes have opposite effects so that one is masked by another. Current methods such as unit hydrograph analysis and return period analysis may not be sensitive enough to detect responses to changes in catchment land use. Importantly, it may be changes affecting hydrologically responsive areas within a catchment that will be the most important with regard to quickflow variations (Pilgrim, 1983). Thus it is the influence of catchment factors within these areas that must be determined, and good correlations overall between hydrological variables and variations in the characteristics of the complete drainage basin seem unlikely. Furthermore, areas undergoing changes in land use but with low drainage densities may not show up in runoff variations downstream, and so 'the drainage network is a key to the way in which land use effects are expressed in downstream discharges' (Gregory and Madew, 1982), both in terms of the extent of the network and the size of the channels. The relation of drainage networks to the main areas of land use change within the catchment must be carefully examined when looking at the hydrological consequences of land use change.

It is evident from what has been discussed so far that studies have to overcome problems of scale, lack of hydrological sensitivity amidst complete land use changes, and regional and scale variations. It is often difficult to compare similar studies in differing parts of the world. This is compounded by a lack of long-term land use and especially flow data. Three examples of successful integrated basin studies are those of the Red River of the North on the North Dakota–Minnesota border (Hammen, 1980), the Mississippi at St Paul, Minnesota (Knox *et al.*, 1975) and 11 Illinois catchments, ranging from 1000 to 70,000 km² in basin area (Changnon, 1983). One of the few studies outside the USA has been that of the Upper Hunter Valley, New South Wales (Bell and Erskine, 1981). Flood frequencies and annual runoff were found to have increased since 1946 in the catchment. In an attempt to evaluate the role of land use versus that of climate on the flood hydrology, the annual runoff values were classified according to catchment rainfall indices for various time subdivisions at the most downstream station. The fact that differences in the magnitude of annual runoff changes were largely non-significant between periods of similar rainfall conditions suggested that changes in land use were not important in this catchment and that increases in flood frequencies and runoff were attributable to changes in rainfall. Similar results were obtained for the River Tawe in South Wales (Walsh, Hudson and Howells, 1982) in that changes in flood magnitude/frequency since the late 1920s were attributable to a similar increase in the frequency of heavy daily rainfall events in the catchment. Land use change was deemed to have had no effect on the flooding history of the Tawe. This is in direct contrast to the findings of Howe, Slaymaker and Harding (1967) for the Upper Severn.

The Upper Severn is adopted in this present study as a study catchment. A first aim is to try to detect changes in the flood magnitude, frequency and seasonality of the Severn. Changes in land use and climatic variations within the catchment are then examined, in so far as records allow. Finally, an evaluation is then made as to the role of man versus that of natural change on the flood hydrology of the Severn.

## FLOOD CHANGES IN THE UPPER SEVERN CATCHMENT

The physiographic, climatic and hydrologic characteristics of the Severn catchment have been outlined in previous chapters. With regard to the flood hydrology of the Severn, Figure 7.1 illustrates the locations of the flow and raingauge records analysed. Four of the five flow gauges are sited on the Upper Severn with the fifth—Meifod—sited on the Vyrnwy, a tributary of the Severn. In addition, the record at Bewdley has been analysed since it represents the longest continuous flow record for the upper or middle reaches of the river system.

The synoptic causes of flooding on the Severn have been previously analysed by Howe, Slaymaker and Harding (1967) and subsequently by Harding (1972). Widespread floods were found to be most commonly related to the occurrence of intense depressional storms and sequences of storms.

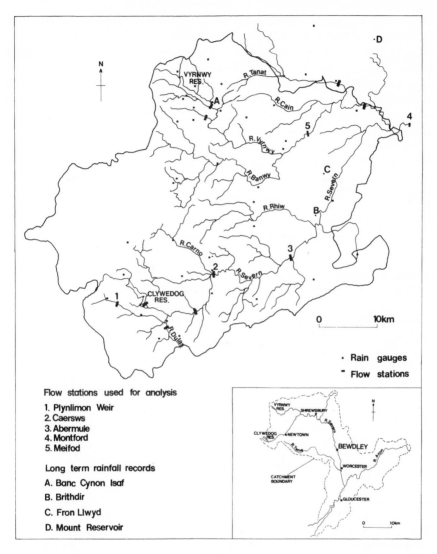

Flow stations used for analysis
1. Plynlimon Weir
2. Caersws
3. Abermule
4. Montford
5. Meifod

Long term rainfall records
A. Banc Cynon Isaf
B. Brithdir
C. Fron Llwyd
D. Mount Reservoir

· Rain gauges
‾ Flow stations

FIGURE 7.1   Map of catchment

The importance of heavy orographic rainfall associated with warm sectors or depressions in the Severn and Wye catchments was emphasized. However, as we have seen, different parts of the catchment may respond differently to the same synoptic situation, so that what may be a flood-causing situation in the uplands may not necessarily appear as major factor on hydrograph shape in, for example, the Severn–Vyrnwy confluence area. Thus localized, summer convective storms such as that of 15 August 1977 (Newson, 1982) may have important consequences for flooding at Plynlimon Weir, whereas at Montford general frontal situations may be more important for flooding.

Table 7.1 summarizes those factors important for flooding on the Upper Severn, together with the effects of changes in such factors. There are basically three sets of factors which are deemed important: climate, surface and channel (the latter including the possible effects of flood protection schemes with regard to the acceleration in the arrival of flood peaks further downstream). The relative roles of these factors are discussed in later sections.

TABLE 7.1   Factors influencing annual flooding on the Upper Severn

| Factor | Change | Effect | Evidence where available |
|---|---|---|---|
| Climate | Increased daily rainfall | Positive | |
| | Increased long-period rainfall | Positive | Severn-Trent Water Authority (1980) |
| | Increased short-period intensity | Positive | |
| | Increased snowmelt | Positive | |
| Surface | Increased mature forest cover | Negative | Newson (1975) |
| | Pasture improvement | Positive | |
| | Increased urban cover | (Positive) | |
| | Increased open drainage (forest established) | (Positive) | |
| | Increased underdrainage | Controversial | |
| Channel | Arterial drainage | Controversial | |
| | Flood protection upstream | Positive | |
| | Reservoir control-post 1968 (Clywedog) | Negative | |
| | Increased sedimentation of reach/reduction in channel capacity | Positive | |

( ) = Those factors unlikely to have had an influence on recent trends (i.e. post-1968).

Four aspects of the flood hydrology of the Severn have been analysed: frequency, magnitude and seasonality variations as well as the changes in the nature of the flooding. The longest flood chronology for the Upper Severn is the flood level record for Welsh Bridge, Shrewsbury extending back to 1840. This was analysed by Howe, Slaymaker and Harding (1967). They identified three main periods: 1840–80 and 1940–64 were recognized as ones of higher flood levels and frequency, with a period of reduced flood levels with fewer floods from 1880 to 1940. In addition it was shown that the 25–year flood level had increased from 5.10 m (for the period 1911–40) to 5.94 m (1940–64).

The flood chronology for Bewdley (1921–83) has been compiled from the partial duration series of floods exceeding 4.20 m above ordnance datum (approximately 268 cumecs). It is evident that the increase in flood levels noted for the mid 1960s by Howe, Slaymaker and Harding has not been sustained into the period 1968–82. Since 1968 there have been no floods exceeding 400 cumecs at Bewdley. With regard to flood frequency, the mid 1970s represented a general lull, although the late 1970s–early 1980s witnessed an increase in the number of floods (Figure 7.2).

The general decline in flood levels has since 1968 coincided with the construction and operation of Clywedog Dam, which was built to maintain a minimum dry weather flow at Bewdley. Comparisons of the flood frequency and the mean annual flood discharge prior to and following 1968 for the Severn at Plynlimon Weir, Caersws, Abermule, Montford and Bewdley and the Vyrnwy at Meifod, reveal the effects of Clywedog on the Severn (Table 7.2). At Plynlimon Weir, where the Severn is unregulated, there has been an increase in flood frequency in the latter period as well as a 14.5 per cent increase in the magnitude of the mean annual flood. At Caersws, however, there has been a decline in the average number of events per year, which is significant at the 5 per cent level, as well as a 25 per cent decrease in the mean annual flood discharge. The table also reveals that downstream the effects of the reservoir decline, so that at Bewdley, for example, there has been an increase in the flood frequency in the period 1968–83, at least for the threshold considered. The effects of Clywedog, therefore, appear beneficial at least at Caersws and Abermule, but decline downstream so that at Bewdley the reservoir has negligible effects on the flood hydrology.

Changes in the magnitude and frequency of flood events on the Upper River Severn can also be analysed through return period analysis; changes in the recurrence intervals of flood events can be gauged by splitting the peaks over threshold record for each of the stations into two or more shorter periods. Data were split into pre- and post-1968 periods for four stations on the Severn and one on the Vyrnwy. The results are presented in Table 7.3 and suggest an increase in the latter period in the discharges associated with each of the return periods for Plynlimon Weir but a decline in those for Caersws, Montford and Bewdley, all downstream of Clywedog. This decline is greater at Caersws than at the other two stations. The Bewdley record was also split into pre- and post-1940 time periods (Figure 7.3). The latter period has witnessed an increase in the discharges associated with all return periods confirming the interpretations of Howe, Slaymaker and Harding (1967), and suggesting that even including the post-1968 decline in flood levels, the period since 1940 as a whole has witnessed a general increase in magnitudes and frequencies on the Severn.

TABLE 7.2 Flood events on the Rivers Severn and Vyrnwy 1946–83

| River | Catchment area km² | Res. % | | Pre-Clwedog | Post-Clwedog | Total |
|---|---|---|---|---|---|---|
| Severn | | | | 1951–Mar 1968 | Apr 1968–83 | |
| Plynlimon Weir | 8.05 | — | No. events >68 cumecs | 90 | 102 | 192 |
| | | | Av. events/year | 5.00 | 6.80 | 5.81 |
| | | | Summer floods % | 24.4 | 24.5 | 24.5 |
| | | | Mean annual flood | 13.10 cumecs | 15.00 cumecs | |
| Severn | | | | 1946–Mar 1968 | Apr 1968–83 | |
| Caersws | 375 | 13.0 | No. events >115 cumecs | 60 | 27 | 87 |
| | | | Av. events/year | 2.61 | 1.80 | 2.28 |
| | | | Summer floods % | 8.30 | 7.40 | 8.00 |
| | | | Mean annual flood | 192 cumecs | 144 cumecs | |
| Severn | | | | 1960–Mar. 1968 | Apr 1968–83 | |

**Abermule** — 580, 8.4

| | 1952–Mar 1968 | Apr 1968–83 | |
|---|---|---|---|
| No. events >157 cumecs | 16 | 12 | 28 |
| Av. events/year | 2.00 | 0.75 | 1.16 |
| Summer floods % | 6.20 | 16.66 | 10.7 |
| Mean annual flood | 272 cumecs | 200 cumecs | |

**Severn Montford** — 2025, 2.41

| | 1923–Mar 1968 | Apr 1962–83 | |
|---|---|---|---|
| No. events >200 cumecs | 81 | 71 | 152 |
| Av. events/year | 4.76 | 4.73 | 4.74 |
| Summer floods % | 14.81 | 11.26 | 13.6 |
| Mean annual flood | 272 cumecs | 249 cumecs | |

**Severn Bewdley recorder** — 4325, 1.08

| | 1948–Mar 1968 | Apr 1968–83 | |
|---|---|---|---|
| No. events >210 cumecs | 223 | 91 | 314 |
| Av. events/year | 4.84 | 6.06 | 5.15 |
| Summer floods % | 12.10 | 7.69 | 10.3 |
| Mean annual flood | 345 cumecs | 302 cumecs | |

**Vyrnmy Meifod** — 405, 18.3

| | 1948–Mar 1968 | Apr 1968–83 | |
|---|---|---|---|
| No. events >160 cumecs | 25 | 19 | 44 |
| Av. events/year | 1.19 | 1.26 | 1.22 |
| Summer floods % | 8.00 | 16.00 | 11.0 |
| Mean annual flood | 230 cumecs | 210 cumecs | |

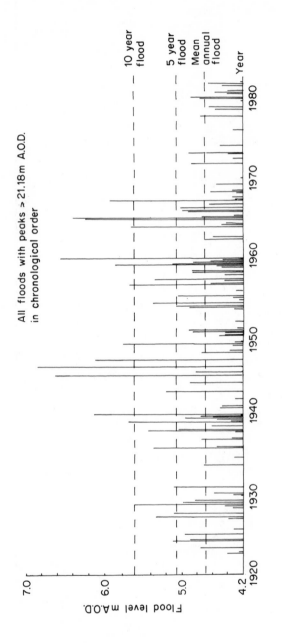

FIGURE 7.2 Flood chronology at Bewdley, 1923–83

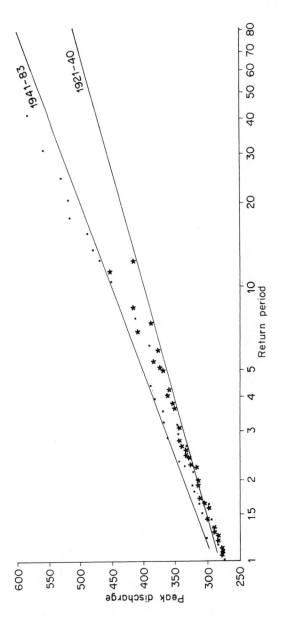

FIGURE 7.3    Flood frequency variations at Bewdley, 1921–40 and 1941–83

TABLE 7.3    Changes in the magnitude/frequency of flood flows in the Upper Severn

| Station | River | Years of record | Return period (years) | | | | | |
|---------|-------|-----------------|------|------|------|------|------|------|
| | | | 1 | 2 | 5 | 10 | 20 | 40 |
| Plynlimon Weir | Severn | 1951–83 | | | | | | |
| | | Pre-1968 | 10.6 | 12.7 | 15.7 | 17.9 | 20.1 | 22.3 |
| | | Post-1968 | 11.2 | 14.4 | 18.7 | 21.9 | 25.2 | 28.4 |
| | | Change % | +6 | +13 | +19 | +22 | +24 | +27 |
| Caersws | Severn | 1946–83 | | | | | | |
| | | Pre-1968 | 122 | 163 | 220 | 262 | 304 | 348 |
| | | Post-1968 | 122 | 142 | 169 | 189 | 210 | 230 |
| | | Change % | — | −13 | −23 | −28 | −31 | −34 |
| Montford | Severn | 1953–83 | | | | | | |
| | | Pre-1968 | 227 | 261 | 306 | 340 | 375 | 410 |
| | | Post-1968 | 236 | 246 | 260 | 272 | 283 | 294 |
| | | Change % | +4 | −6 | −15 | −20 | −25 | −28 |
| Bewdley | Severn | 1923–83 | | | | | | |
| | | Pre-1968 | 284 | 330 | 394 | 444 | 492 | 544 |
| | | Post-1968 | 285 | 300 | 320 | 335 | 351 | 365 |
| | | Change % | +0.4 | −9 | −19 | −25 | −29 | −33 |
| Meifod | Vyrnwy | 1948–83 | | | | | | |
| | | Pre-1968 | 186 | 217 | 260 | 292 | 325 | 357 |
| | | Post-1968 | 176 | 204 | 240 | 267 | 295 | 324 |
| | | Change % | −5 | −6 | −8 | −9 | −9 | −9 |

All flows in cubic metres per second (cumecs)

In addition to the flood magnitude and frequency variations noted, an attempt has been made to establish whether or not the characteristics of the river's response to rain falling in the catchment has varied in recent years, i.e. whether there is a change in the nature of flooding response to unit rainfall impulses. This involves deriving 1 hr 10 mm unit hydrographs for storm events from the mid 1960s to the early 1980s using an Institute of Hydrology computer programme kindly supplied through M. Robinson. By way of example, the unit hydrographs for Caersws are plotted in Figure 7.4. These illustrate doubling of the magnitude of the peak flow during the time period. The times to peak flow have generally declined over the same period, suggesting therefore a more rapid hydrological response at Caersws, with earlier and larger peak flows in the late 1970s and early 1980s. This more 'flashy' pattern of storm response is similar to that noted by Robinson (1981) for experimental plots at Coalburn following peat drainage prior to afforestation, in which the peak magnitudes were 40 per cent greater and the time to peaks 25 per cent shorter following drainage operations. It appears, therefore, even allowing for the limitations of the unit hydrograph technique, such as its sensitivity to rainfall profiles, there has been a change in the catchment

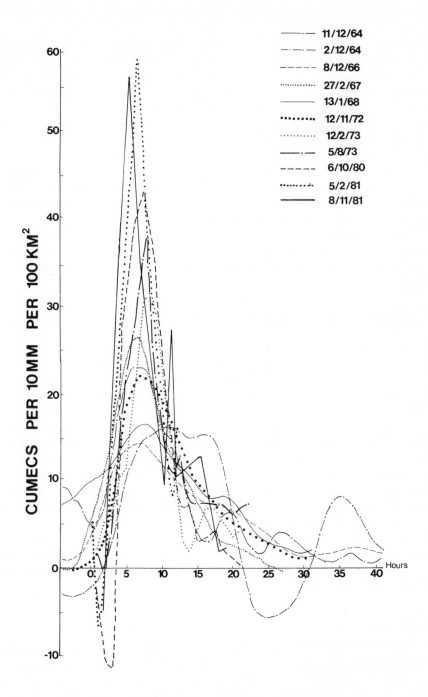

FIGURE 7.4   Unit hydrograph variations at Caersws, 1964–81

behaviour upstream of Caersws, which may be accounted for by such land use changes as pasture improvement, perhaps subsequent to land drainage.

Another flood characteristic which may be subject to variation is that of seasonality. Plots of the seasonal incidence of floods in the catchment for each of the stations revealed no large-scale departure from the November to March predominance of floods associated with widespread frontal rain. There has therefore been no real shift in the seasonality of flooding on the Severn during the study period.

These analyses so far show that there have been statistically significant changes in both the magnitude and frequency of flooding on the Upper River Severn, particularly since 1968, with the late 1970s witnessing the return to a more 'normal' flooding regime characteristic of the mid 1960s after a relative lull in the period 1968–76. In addition the nature of flooding has changed, with earlier and larger peak flows for storm events in the late 1970s compared with those of the 1960s and early 1970s. Possible explanations for these variations can now be suggested.

## ENVIRONMENTAL CHANGE IN THE UPPER SEVERN CATCHMENT

That the characteristics of magnitudes and frequencies of floods are highly sensitive to variations in climate circulation systems and changes in land use, is well established (Knox *et al.*, 1975; Knox, 1984). However, the relative importance of these two factors and their complex interactions is much less clear. Thus the fluvial responses to land use changes in a catchment are often analogous to fluvial responses produced by certain types of climate changes. In addition, the effects of land use changes on flood regimes are likely to be less important for those extreme rainfalls producing the largest floods when the vegetation and soils of the catchment are unable to absorb rainfall of such intensities. Inevitably, therefore, it may not be possible to separate and distinguish the effects of meteorological and catchment variations on flooding characteristics. The important land use changes in the Upper Severn catchment are outlined below, followed by those meteorological variations anticipated to have an affect on flood flows. Finally, tentative conclusions are drawn as to the relative importance of these factors in the case of the Upper Severn flood history.

### Land use change

Those catchment-wide changes likely to have important consequences for flooding in the catchment are shown in Figure 7.5. The most important land use change within the catchment since the First World War has been that of afforestation. Approximately 14,800 ha (or 9.5 per cent of the catchment area to Shrewsbury) have been afforested in this time, with peaks just after

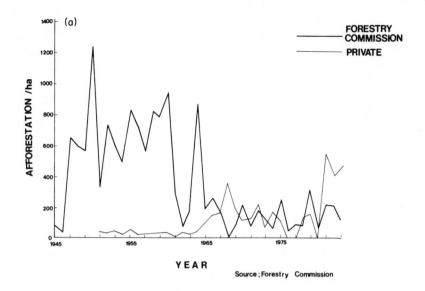

Source; Forestry Commission

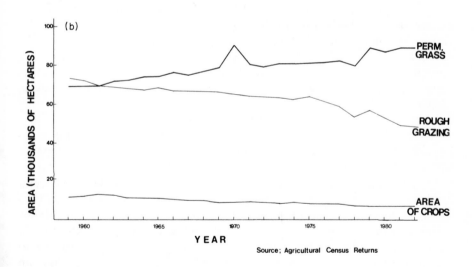

Source; Agricultural Census Returns

FIGURE 7.5   Land use changes in the Upper Severn catchment
(a)  Afforestation 1945–83
(b)  Land use change 1959–82

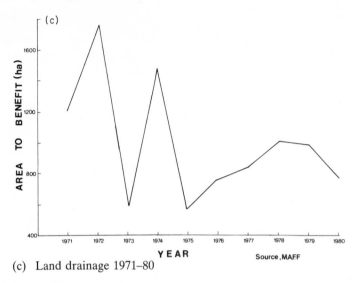

(c)   Land drainage 1971–80

the Second World War of up to 1200 ha/year, falling to 200 ha/year in the late 1970s; see Figure 7.5(A). This afforestation has predominantly taken place on land previously used for rough grazing. Associated with these forest plantations have been forestry drainage practices which were suggested by Howe, Slaymaker and Harding (1967) to have increased flood peaks and reduced time lags in the catchment in the period 1940–64. However, no quantitative data exist as to the extent of this type of drainage in the catchment. The decline in afforestation rates from the mid 1960s, however, suggests that the hydrograph changes noted at Caersws since this period have not been a direct response to forestry practices.

A second important type of land use change has been that of pasture improvement. The conversion of moorland to pasture and arable cropping within the catchment may, through a reduction in retention storage capacities, lead to the hydrological response becoming more rapid. The main trend has been an increase in area under permanent grass from 70,000 ha in 1959 to 90,000 ha in 1982; see Figure 7.5(B). In the same period the area under rough grazing has declined by 4 per cent with vast areas in the north and west of the catchment being converted to forestry. These correspond to the main runoff-producing areas of the catchment, such that there was an approximate 10 per cent decline in rough grazing areas in the catchment to Caersws. These factors may be important for changing hydrograph shape at Caersws.

There has been an enormous post-war spread of field tile drainage and newer plastic drains in Mid-Wales (Green, 1979). The first suggestions that such drainage in the Upper Severn catchment may have altered flooding characteristics followed the major floods of the early and mid 1960s. By this time nearly 4000 ha had been the subject of grant-aided-scheme work (hill gripping, ditching and underdraining) in the years since 1940. This amounted to 9 per cent of the total catchment. The increased level of grant aid for drainage schemes was followed by increases in drainage rates. This may account for the increased peak and severity of flooding on the Severn. Annual rates of drainage activity, as well as the 10-year (1971–80) total, for parishes within the Upper Severn catchment have been calculated through information supplied by MAFF (Ministry of Agriculture, Fisheries and Food). Due to a change in the grant procedure no data exist from 1980 onwards, and data only exist on a divisional basis prior to 1970. The catchment witnessed exceptional increases in underdrainage activity in the 1970s with 1972–73 being the peak year; see Figure 7.5(C). Much of this drainage has been associated with the western upland parishes. Drainage is especially concentrated within areas of heavy soils in the centre of the catchment. There has been a decrease in the areas drained through the 1970s following the early peak years. The total area drained in the parishes in the period 1971–80 amounted to 8500 hectares (or 6 per cent of the catchment). This, together with the clearing of much of the rough pasture and bracken and the planting of improved grasses, has speeded up the rate at which flood runoff reaches the Severn and may thus account for the variations in unit hydrographs noted at Caersws.

### Climatic variations

In order to study those meteorological variations which have an effect on flooding characteristics of the Upper Severn, four long-term records of daily rainfall were analyzed. The Banc Cynon Isaf gauge is situated in the northern uplands of the catchment whereas the remaining three (Brithdir, Fron Llwyd and Mount Reservoir) are situated in the lowland east (Figure 7.1). The records date from 1905, 1925, 1910 and 1899 respectively. Three main types of analysis have been utilized: simple time series analysis on the annual variations in the frequency of intense storm events; breakpoint analysis on these threshold exceedance frequencies; and thirdly, return period analysis to determine changes in the rainfall magnitude/frequency relationships. Variations in the seasonality of extreme rainfalls are analysed to differentiate the trends in winter and summer rainfall. There has been a dearth of very heavy winter one-day events since 1968, especially in the upland sites (Severn Area Unit, 1980). Moderate winter events are also seen to decrease since 1968. However, there is a slight increase in the frequency of very heavy summer events, although this may not have an effect on the seasonality of floods since for a flood to occur in summer requires extremely heavy rainfall to overcome low runoff factors. Trends have differed according to the threshold considered when defining heavy rainfall. Thus falls exceeding 2 in. (50.8 mm) but less than 2.25 in. (57.2 mm) seemed to increase in frequency into the 1970s but falls greater than 2.5 in. (63.5 mm) declined in frequency (Lawler, 1985).

Plots of the variations in storm frequencies for Banc Cynon Isaf (30 mm events) and Mount Reservoir (20 mm events) show the general trends for heavy rainfall variations within the catchment (Figure 7.6). The trends in 5-year moving averages have been plotted as well as the regression plots for the whole time period, and for 1968 to 1982. The mid 1930s, mid 1940s and mid 1970s stand out as being periods of lower storm frequencies whereas the late 1920s, late 1930s and mid 1960s were periods of higher frequencies. Regression plots for the whole time period suggest a steady climatic state with a statistically non-significant positive trend in exceedance frequencies for both gauges. The regression plots for the 1968–82 period for each station reveal, however, that while frequencies have decreased in this period for Banc Cynon Isaf, they have tended to increase at Mount Reservoir. One possible explanation for this is that the decrease in frequencies at Banc Cynon Isaf corresponds to a decline in those winter events which form the bulk of the storms at this upland site. In comparison, Mount Reservoir has a regime dominated by summer events which have increased in frequency since 1968. There has not been a significant increase in winter events at this site (Figure 7.7) whereas summer event frequencies have increased. Analysis of the variations in storm magnitudes at the two sites reveals a return in the late 1960s–early 1970s to those storm magnitudes of the mid 1910s and early 1930s, especially for summer events at Mount Reservoir. This corresponds to an increase in summer flooding in the period 1968–82 on the River Severn. Variations in those 2-day rainfall events considered important for flooding on the Severn reveal similar trends with a decline of winter events in the period since 1968, but an increase in the frequency of summer 2-day events.

Statistical analysis of the variations in frequencies of storm events either side of given break points for the four gauges in the catchment suggests that of the four breakpoints analysed by the Cox and Lewis (1966) test statistic, which compares the frequencies of a number of differing thresholds using the normal approximation to the binomial distribution, there has been a significant change in the frequency of heavy rain events either side of a 1925 breakpoint for Banc Cynon Isaf—for both the 40 mm and 60 mm thresholds (Table 7.4). However, the changes either side of a 1940, 1950 and 1968 breakpoint were not significant. This agrees with the findings of Howells (1985) that there had been a significant increase in storm frequencies above the 64 mm threshold of 380 per cent for Banc Cynon Isaf, for the period 1925–80 compared with that of 1905–25. There is also a pattern of increase from the 1940s to the present, confirming the findings of Howe, Slaymaker and Harding (1967). However, these changes were not significant at the 10 per cent level and a mid 1920s increase was more evident. The remaining three records indicated a variety of breaks for various thresholds. For example, the record for Mount Reservoir suggests an increase in annual frequencies of storms exceeding 20 mm which is significant at the 5 per cent level for both the 1925 and the 1940 breakpoints. There has also been an increase in frequencies of 40 mm events since 1968 (significant at the 2 per cent level). This corresponds to an increase in the frequency of summer events in the lowland east of the catchment since 1968, and therefore validates the conclusions of time series analysis.

TABLE 7.4  Significant variations in heavy storm events in the Upper Severn catchment

| Gauge | Time period | Breakpoint | Threshold (mm) | Significance level |
|---|---|---|---|---|
| Banc Cynon Isaf | 1905–82 | 1925 | 40 | Increase in average annual frequencies significant at 10% level |
| | | | 60 | Increase, significant at 1% level |
| Mount Reservoir | 1899–1982 | 1925 | 20 | Increase, significant at 5% level |
| | | 1940 | 20 | Increase, significant at 5% level |
| | | 1968 | 40 | Increase, significant at 2% level |
| | | | 50 | Increase, significant at 5% level |
| Fron Llwyd | 1910–80 | 1940 | 20 | Increase, significant at 1% level |
| | | 1968 | 20 | Decrease, significant at 10% level |

NB The record for Brithdir (1925–82) revealed no significant variations either side of the 1940, 1950 or 1968 break points.

Return period analysis on daily rainfalls above specified thresholds for each of the four gauges suggests that rainfalls increased from 1926 onwards at Banc Cynon Isaf, for the whole range of recurrence intervals (Figure 7.8). Increases ranged from 17 per cent for the 2-year fall (45.5–53.2 mm) to 31 per cent for the 20-year fall (66.5–87.0 mm). Increases at Mount Reservoir, however, were smaller (1 per cent or less). At Banc Cynon Isaf, rainfalls associated with a particular return period are heavier in the post-1940 period (increase range from 15 to 17 per cent). Similar increases since 1940 were noted for Brithdir and From Llwyd. In the post-1968 period the Banc Cynon Isaf record shows a decrease in those rainfalls for the whole range of recurrence intervals considered (from 1.5 per cent for the 2-year return period to 7 per cent for the 20-year return period). Conversely, Mount Reservoir

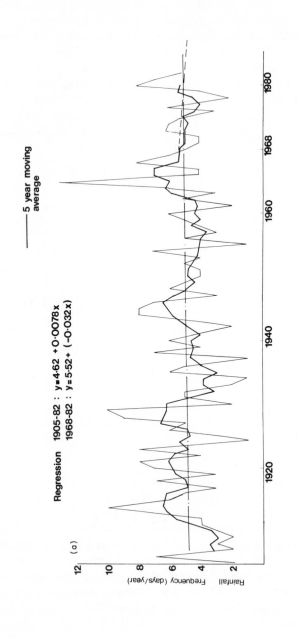

Regression   1905-82 :  y = 4·62 + 0·0078 x
             1968-82 :  y = 5·52 + (-0·032 x)

—— 5 year moving
   average

(a)

Rainfall    Frequency (days/year)

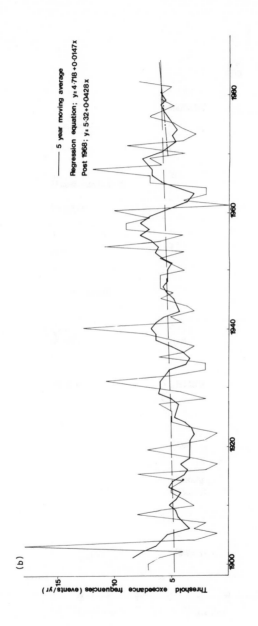

FIGURE 7.6  Variations in the frequency of storm events
(a) Banc Cynon Isaf 30 mm threshold
(b) Mount Reservoir 20 mm threshold

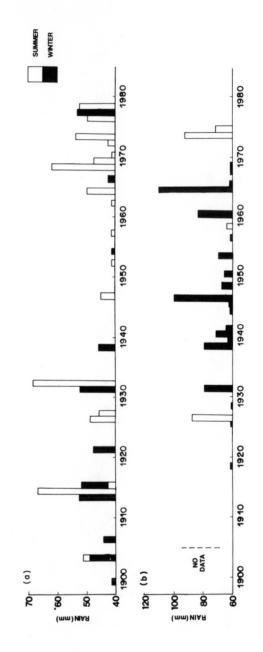

FIGURE 7.7  Incidence of heavy storm events in the Severn catchment
(a)  Mount Reservoir
(b)  Banc Cynon Isaf

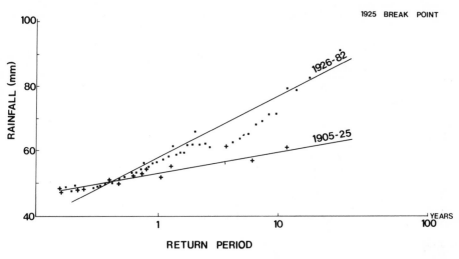

FIGURE 7.8   Variations in the return periods of daily rainfalls at Banc Cynon Isaf 1905–82

has a steeper magnitude/frequency relationship after 1968 than before; for example, the 20-year return period rainfall increased from 50.8 mm to 56.8 mm (12 per cent). This is mirrored in the Fron Llwyd and Brithdir records and reflects the summer dominance of extreme storm events in the east of the catchment.

It is evident, therefore, that trends in rainfall frequency and magnitude are not uniform between upland and lowland stations, between winter and summer events, or between different rainfall thresholds. The upland/lowland division within the catchment can be explained in terms of the differing heavy rainfall-producing mechanisms, the former being dominated by winter frontal rainfall and the latter by convective summer storms. Whereas winter events have declined in frequency in the mid 1970s, there has been an increase in heavy summer rainfall events. In addition, it appears that the increase in heavy rainfall events over the catchment noted by Howe, Slaymaker and Harding (1967) stems from the mid 1920s rather than 1940.

### Synoptic variations and flooding history of the Severn

The correlation between heavy rainfall event frequencies and floods at Bewdley can be shown through the relationship of 5-year running means (Figure 7.9). Thus the high frequencies of floods in the mid 1960s correspond to a period of high storm event frequencies. Possible explanations for such variations can be suggested through Lamb's synoptic classification (1972).

The synoptic classification of each rainfall day on the peaks over threshold

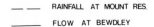

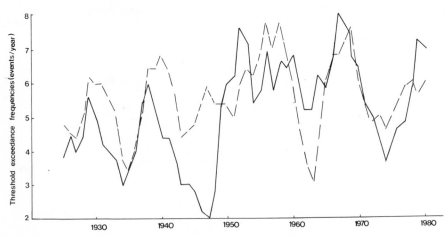

FIGURE 7.9   Correlation of flood frequency at Bewdley and heavy storm event
frequency at Mount Reservoir

series for seven gauges in the catchment revealed the highest percentage of
falls at each site were associated with cyclonic and westerly weather situations
(Table 7.5). At Hafren Lodge, Abercannon Farm and Banc Cynon Isaf
(upland stations) westerly weather was the dominant cause of heavy falls. At
the remaining four gauges (lowland), cyclonic weather was more important.

TABLE 7.5   Distribution of largest falls with differing airflow types

| Station | Threshold (mm) | Percentage of airflow type days with large falls | | | | | | | |
|---------|----------------|------|------|------|------|------|------|------|------|
| | | A | C | W | NW | N | E | S | Un |
| Banc Cynon Isaf 1905–82 | 30 | 0.14 | 2.88 | 3.03 | 0.21 | 0.05 | 0.52 | 0.88 | 1.68 |
| Hafren Lodge 1945–80 | 30 | 0.20 | 3.16 | 4.56 | 1.38 | — | 0.41 | 0.76 | 1.99 |
| Abercannon Farm 1949–82 | 30 | 0.31 | 3.13 | 4.56 | 1.81 | 0.22 | 0.76 | 0.75 | 1.11 |
| Brithdir 1925–82 | 20 | 0.19 | 2.71 | 1.29 | 0.28 | 0.86 | 1.69 | 0.89 | 2.04 |
| Fron Llwyd 1910–80 | 20 | 0.20 | 2.71 | 0.91 | 0.04 | 0.39 | 1.48 | 0.89 | 1.99 |
| Mount Reservoir | 20 | 0.29 | 4.12 | 1.19 | 0.24 | 0.38 | 2.01 | 1.52 | 3.27 |
| Llanyblodwel 1949–82 | 20 | 0.27 | 3.00 | 1.60 | 0.43 | 0.45 | 1.86 | 0.91 | 3.10 |

Un = unclassified weather days

Together the cyclonic and westerly types accounted for over 50 per cent of the heavy rainfalls at each station. By splitting the record, any change in the likelihood of a particular weather type producing a heavy daily rainfall may be analysed by comparing the percentage of airflow type days producing large falls either side of two breakpoints—1925 and 1968—for which changes in heavy rainfall regimes were noted. Both the Banc Cynon Isaf and Mount Reservoir records showed a notable increase in the chance of a cyclonic day producing a heavy rainfall in the period 1926–82 compared to the period prior to this (Table 7.6). Analysis of a 1968 breakpoint suggests a decrease in the chances of both the westerly andcyclonic types producing falls in the period 1969–82 at Mount Reservoir and Banc Cynon Isaf. This is despite no major changes in the annual frequencies of either weather type. It appears, therefore, that variations in heavy rainfalls and ultimately flooding character-istics, can be explained on the basis of changes in the likelihood of weather types producing heavy rainfall.

Finally, a comparison has been made between the flood frequency vari-ations at Bewdley and those of the PSCM indices of Murray and Lewis (1966) which are based on Lamb's synoptic classification. Thus there is a significant correlation between 5-year running means of flood frequency and the Cyclonicity Index (Figure 7.10). As the index increases (i.e. the weather tends to be more cyclonic), the number of floods increases (for example, the mid 1960s). In comparison, the more anticyclonic period of the early–mid 1970s witnessed a decrease in flood frequencies.

It is suggested, therefore, that variations in flooding on the Severn can be correlated with those of heavy storm frequencies, variations of which can be explained in synoptic terms.

TABLE 7.6 Changes in the likelihood of a synoptic weather type producing a heavy rainfall either side of 1925 and 1968

| Airflow type | Banc Cynon Isaf | | Mount Reservoir | |
|---|---|---|---|---|
| | 1925/26 | 1968/69 | 1925/26 | 1968/69 |
| A | + | ˘ | ˘ | − |
| C | + | −− | ++ | −− |
| W | + | −− | + | − |
| NW | − | − | −− | − |
| N | ˘ | ˘ | + | ++ |
| E | + | − | ++ | −− |
| S | − | ˘ | ++ | − |
| Unclassified | − | − | −− | −− |

++Large increase (>0.50%)
+ Increase (0.10% to 0.49%)
˘ No change (+/− to 0.09%)
− Decrease (−0.09% to −0.49%)
−−Large decrease (>−0.50%)

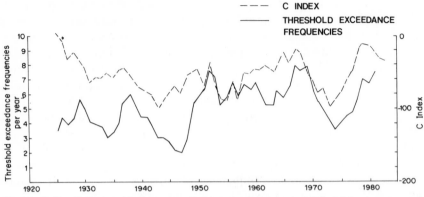

FIGURE 7.10  Relationship between 5-year running means of flood frequency and the Cyclonicity Index

## CONCLUSION

The number of flood events in the Severn catchment has exhibited a general increase during the past 60 years. This is directly related to variations in heavy rainfall in the catchment stemming from the mid 1920s. Since 1968, however, there has been a decline in flood frequencies and magnitude, related to a decrease in winter precipitation, but also to a change in the nature of flooding in the upper reaches, with a tendency to a more 'flashy' regime. Land use changes, and in particular agricultural drainage and pasture improvement in the catchment upstream of Caersws, can be suggested as possible reasons for a more rapid upstream hydrological response. The characteristics of flood magnitudes and frequency are highly sensitive to climatic variations, and in particular to changes in circulation regimes, as well as catchment changes, and the effects of the latter will vary according to the areas involved and the location of change within the catchment. Thus in larger catchments, for example, drainage schemes in the headwater regions may have different consequences on the flooding regime to those of floodplain schemes, through the effects on the sychronization of flood waves. It has also been suggested that up to 20 per cent of a catchment has to be altered for the effects of a particular land use change to be detected in the flow record (Riggs, 1985). In many cases there is a lack of data at the basin scale to cover the range of catchment changes which may have hydrological consequences, such as forestry drainage.

By establishing the relationship between contemporary climatic and land use changes and hydrological response, evaluations of past responses of river systems to environmental changes can be attempted (Knox, 1983). We have seen here that such relationships are often very complex. The effects of land use change vary according to the return periods of the flood considered, with negligible consequences for the less frequent, larger magnitude floods. Climatic change may be important through direct consequences on the recurrence intervals of floods or, on longer timescales, through the resulting displacement of vegetation formations. The sensitivity to changes in catchment factors will vary according to climatic region, such that arid and semi-arid regions—where vegetation is more sensitive to climatic change—often have a more rapid hydrological response to land use change. Knox (1983) has established the responses of rivers in the United States to Holocene climates through examination of environmental associations in six unique regions. He concluded that Holocene climatic changes had a significant effect on the magnitude and frequency of floods through effects on displacements of major vegetative formations. Climate was considered, therefore, to be of overriding importance in long-term variations.

Results from the present study suggest that fluctuations in flood frequency in the historical time period are the norm and that such fluctuations accompany a continuous variation in climatic and catchment factors demonstrating that stability or temporal equilibrium is far from obvious, and that 'fine tuning' of palaeohydrology estimates to such factors is by no means straightforward.

## ACKNOWLEDGEMENTS

I would like to acknowledge the help provided during the study leading up to this chapter of my supervisors, Professors J. Lewin and M. D. Newson and also the staff at the Institute of Hydrology, MAFF, the Forestry Commission, the Public Records Office, the Meteorological Office and the Severn-Trent Water Authority for provision of data. The research was funded by a Natural Environment Research Council Studentship.

## REFERENCES

Ackermann, W. C. (1966). *Guidelines for Research on Hydrology of Small Watersheds*, US Dept. Interior, OWRR, Washington DC, 26 pp.

Bates, C. G. and Henry, A. J. (1928). Forest and streamflow at Wagon Wheel Gap, Colorado. Final Report, *Monthly Weather Review Supplement*, **30**, 1–79.

Bell, F. C. and Erskine, W. D. (1981). Effects of recent increases in rainfall on floods and runoff in the Upper Hunter Valley, *Search*, **12**, 82–3.

Changnon, S. A. (1983). Trends in floods and related climate conditions in Illinois, *Climatic Change*, **5**(4), 341–63.

Cox, D. R. and Lewis, P. A. W. (1966). *The Statistical Analysis of Series of Events*, Methuen, London; Wiley, New York, 285 pp.

Green, F. H. W. (1979). Field drainage in Europe; a quantitative survey, *Institute of Hydrology Report 57*, 78 pp.

Gregory, K. J. (ed.) (1983). *Background to Palaeohydrology*, Wiley, London, 486 pp.

Gregory, K. J. and Madew, J. R. (1982). Land use change, flood frequency and channel adjustments, in Hey, R. D., Bathurst, J. C. and Thorne, C. R. (eds), *Gravel Bed Rivers*, Wiley, Chichester, pp. 757–81.

Hammen, J. (1980). Flooding of the Red River of the North; Multivariate analysis of climatic and land use parameters, *Great Plains–Rocky Mountain Geographical Journal*, **9**, 49–58.

Harding, D. M. (1972). Floods and droughts in Wales, unpublished Ph.D. thesis, University of Wales.

Hornbeck, J. W., Pierce, R. S. and Federer, C. A. (1970). Streamflow changes after forest clearing in New England, *Water Resources Research*, **6**, 1124–32.

Howe, G. M., Slaymaker, H. O. and Harding, D. M. (1967). Some aspects of the flood hydrology of the upper catchments of the Severn and Wye, *Transactions of the Institute of British Geographers*, **41**, 33–58.

Howells, K. A. (1985). Changes in the magnitude-frequency of heavy daily rainfalls in the British Isles, unpublished Ph.D. thesis. University of Wales.

Knox, J. C. (1983). Responses of river systems to Holocene climates, in Porter, C. S. and Wright, H. E., jr (eds), *Late Quaternary Environments of the United States*, University of Minnesota Press, pp. 26–41.

Knox, J. C. (1984). Fluvial responses to small scale climate changes, in Costa, J. E. and Fleisher, P. J. (eds), *Developments and Applications of Geomorphology*, Springer-Verlag, Berlin, Heidelberg, 318–42.

Knox, J. C., Bartlein, P. J., Hirshboeck, K. K. and Muckenhirn, R. J. (1975). The response of floods and sediment yields to climatic variation and land use in the Upper Mississippi Valley, *Centre for Geographic Analyses, University of Wisconsin-Madison, Institute for Environmental Studies. I.E.S. 52.*

Lamb, H. H. (1972). British Isles weather types and a register of the daily sequence of circulation patterns 1861–1971, *Met. Off. Geophys. Memoir*, **116**, 85 pp.

Lawler, D. M. (1985). Changes in the frequency of heavy daily rainfalls; some preliminary observations from Wales and the West Midlands, *Department of Geography, University of Birmingham. Working Paper Series*, 27, 15 pp.

Murray, R. and Lewis, R. P. W. (1966). Some aspects of the synoptic climatology of the British Isles as measured by simple indices, *Meteorological Magazine*, **95**, 193–203.

Newson, M. D. (1975). *The Plynlimon floods of 5–6 August 1973*, Institute of Hydrology Report No. 26.

Newson, M. D. (1979). The results of ten years experimental study on Plynlimon, Mid-Wales, and their importance for the water industry, *Journal Inst. Water Engrs and Scientists*, **33**, 321–33.

Newson, M. D. (1982). Mountain streams, in Lewin, J. (ed.), *British Rivers*, George Allen & Unwin, London, pp. 59–89.

Newson, M. D. and Robinson, M. (1983). Effects of agricultural drainage on upland streamflow; case studies in Mid-Wales, *Journal of Environ. Management*, **17**, 333–48.

Pilgrim, D. H. (1983). Some problems in transferring hydrologic relationships between small and large drainage basins and between regions, *Journal of Hydrology*, **65**, 49–72.

Riggs, H. C. (1985). Streamflow characteristics, *Developments in Water Science*, 22, Elsevier, Amsterdam, 249 pp.

Robinson, M. (1981). The effects of pre-afforestation drainage upon the streamflow and water quality of a small upland catchment, *Institute of Hydrology Report 73*, 69 pp.

Severn Area Unit (1980). *Weather conditions in the Upper Severn catchment before and after 1968*, Severn-Trent Water Authority, Birmingham.
Walsh, R. P. D., Hudson, R. N. and Howells, K. A. (1982). Changes in the magnitude/frequency of flooding and heavy rainfalls in the Swansea Valley since 1875, *Cambria*, **9**(2), 36–60.

Palaeohydrology in Practice
Edited by K. J. Gregory, J. Lewin and J. B. Thornes
© 1987 John Wiley & Sons Ltd.

# 8

# Historical River Channel Changes

## J. Lewin

*University College of Wales, Aberystwyth*

### INTRODUCTION

For studying river channel change, historical maps and related kinds of documented information cover a most useful timespan which falls between the observable present and the very many millenia spanned by the geological record. Present-day observations are of course essential to the understanding of hydrological processes, but they often cover a period which is so brief that it restricts their value in interpreting evolving forms. River channels take a long time to develop, and one during which environmental constraints themselves are liable to historical change. Conversely, the geological record (including evidence from both sediments and palaeochannels) is often difficult to interpret, not least in terms of the timespans required for observed sedimentary bodies and palaeoforms to develop. These usually necessitate knowledge both of fluvial hydraulics and of hydrological regime, that is, both of the magnitude and processes of formative events and of their frequency and duration, and this is difficult to provide except by using the analogue of historical data. Such historical data on channel change can, potentially at least, be linked to the kind of analyses of climatic, hydrological and other environmental changes that have been fully described in previous chapters so that such important considerations as the spatially variable response of channel change to environmental change in the historic present, and therefore in the past, may be better understood. This chapter considers the historical information on channel change for the River Severn, illustrating both the spatial and the temporal variability in the changes documented, and it suggests the value and the limitations of such information both in looking back at an interpretable past and perhaps forward also to a more or less predictable future.

Sources for historical information on the River Severn channel are graphi-

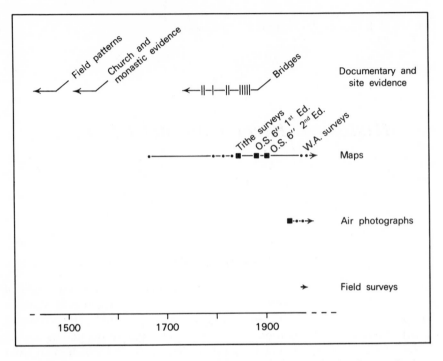

FIGURE 8.1    Historical sources for river channel changes in the Severn Basin

cally illustrated in Figure 8.1. They are necessarily fragmentary. The most recent data come from surveys undertaken by the Severn-Trent Water Authority and by academic researchers. Thus the Authority has commissioned surveys of channel planforms and cross-sections in a number of areas, particularly in association with flood protection schemes and at river gauging sties. A number of areas have similarly been surveyed in the field in conjunction with geomorphological studies (e.g. Dury, 1984; Dury, Sinker and Pannett, 1972; Hey, 1975; Lewis and Lewin, 1983; Newson, 1986; Thorne and Lewin, 1979).

Air photography has been available generally since the 1940s, and in some areas may be used to plot planform changes. For example, at Welshpool study of the recent stages of a long history of channel change is very usefully aided by the availability of large-scale photography flown in 1946, 1973 and 1981. Large-scale maps of good accuracy are also generally available back into the last century, including the editions of Ordnance Survey 1:10560 maps and tithe and estate maps of earlier date. These last can be of greater or lesser reliability. Our example of Welshpool here is unusual in that the Severn channel was well mapped on an estate survey as far back as 1663 so that channel planforms at least may here be examined for upwards of 300 years.

Other documentary information is patchily available and may provide only an indication of channel type or location. Bridges are often dated; they do of course also constrain channel movement once they are built. Medieval ridge and furrow patterns can provide an indication of the length of time for which floodplain areas have *not* been reworked by lateral migration, while other historical information on land and river use (such as fish weirs or navigation works) and archaeological finds can provide clues as to the nature and location of the channel at particular times, or to the limits to recent river migration.

## PATTERNS OF CHANGE

Selected examples of channel pattern change are shown in Figure 8.2. These illustrate the considerable variety in change patterns. In 70 years at Penstrowed, a high-sinuosity channel has been replaced by one that is relatively straight, although the steepened recent channel appears to be one of bedform instability. Cutoff development seems in part to have involved a natural process of chute development, but human agency is also involved in the infilling of the old cutoffs and possibly in the later cutoffs.

By contrast, at Llandinam the major outline of two bends has remained more or less fixed for a century, but detailed examination of the plots for 1902, 1948 and 1975 shows that the locations of gravel accumulation and of the low-flow channel have shifted. Minor phases of erosion and sedimentation have accompanied the downstream migration of mobile gravel bars, as at b in Figure 8.2. The same can be said of the channel at Llanidloes, except that field boundaries and the floodplain topography show signs of a pre-1846 loop which has now been short-circuited.

Historical changes at Maesmawr have been analysed previously (Thorne and Lewin, 1979). Channel migration has involved down-valley shift (bend a), lateral expansion (b, c, d), and chute cutoff of a particularly sharp bend (e). This is one of the relatively few UK sites at which observations of flow hydraulics, of sediment transport (on bend d) and of bank erosion processes have all been conducted (Bathurst, 1979; Bathurst, Thorne and Hey, 1979; Thorne and Lewin, 1979).

Similar complexity can be seen at Welshpool for which the unusual long sequence of records available has already been mentioned. Meander bend development can be appreciated because of this good record of historical material. A sinuous channel in 1775 was replaced by a straighter one involving braiding, probably as a result of an extreme flood known to have occurred in 1795. At that time, the channel below the site of Leighton Bridge was also straightened. This century has seen contrasted growth and

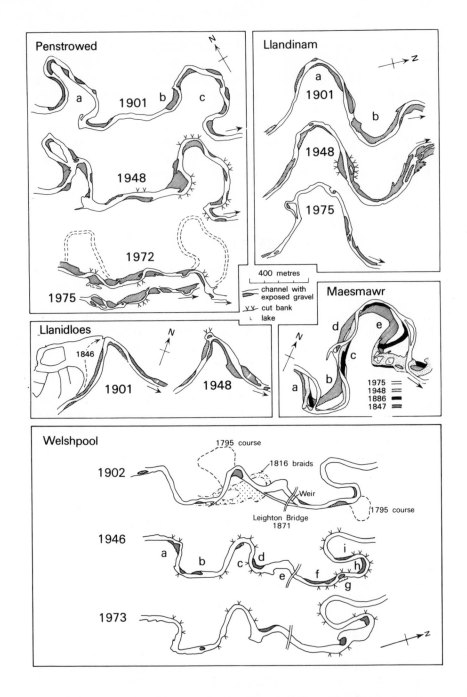

FIGURE 8.2  Channel changes on selected reaches on the Upper Severn

development in the various loops of the sinuous meander pattern, partly constrained by the construction of Leighton Bridge in 1871 (Figure 8.2). Thus a has translated down-valley, b altered rather little, c and d have expanded, e and f have been constrained by a weir and channel, while g, h and i have been involved in the development of a 'classic' sine-generated curve form (Langbein and Leopold, 1966).

These five examples are all of actively developing channels—some showing very considerable loop development, some cutoffs and changes in channel pattern, and other migratory bedforms within more or less stable courses. Other areas, if they show such changes at all, may also more closely follow human activity than natural change. For example, the lower part of the Trannon, a 72 km² left-bank tributary of the Severn, has been converted from a rather stable low-gradient meandering channel in the early nineteenth century (Figure 8.3) to one in which straight and sinuous segments alternate, and where erosion is now a problem (Newson, 1986). Part of the earlier 'natural' system appears to have been anastomosing (in the sense of Smith, 1983) in a multithread pattern. The channelization schemes were initially associated with railway development and in purpose designed to alleviate a flood problem. Though records are not widely available for schemes of early date, it seems likely that many sites on main river courses have been affected by *ad hoc* channelization works of one kind or another, while the extent of recent channel works is considerable (Brookes, Gregory and Dawson, 1983).

In taking the number of examples of historical pattern change on the Severn to six, not all the possible range of planform developments has been covered. Attention has been drawn elsewhere to counterpoint sedimentation in which the channel recedes from a cut bank on a concave bend (e.g. of

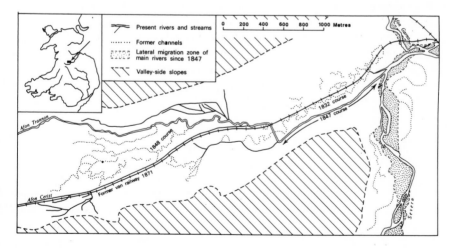

FIGURE 8.3   The Lower Trannon, showing present and former courses

the Teme, see Lewin, 1983, Figure 13.1), and to vertical aggradation and incision in recent times (Lewin, Bradley and Macklin, 1983) or more generally in the Holocene (Macklin and Lewin, 1986). The latter may considerably constrain lateral mobility, with undercut terraces providing a large component of local bedload and also distorting channel planforms on a local scale. More generally, some wonderment has been expressed at the fact that the classic model of meander development and point bar sedimentation on a simple curving channel has lasted so long (Carson, 1986): the field examples from the Severn so far given reinforce the need for a broader approach to the complexities of channel change.

## MODELLING CHANGE IN SPACE AND TIME

High rates of historical channel shift are to be found widely in the upper parts of the Severn and Wye basins. Figure 8.4 shows sites at which historical rates of change have been estimated over about 100 years. Rates are expressed in long-term reach averages, involving several individual bends. Spatially and temporally averaged rates of 1 m$^2$ m$^{-1}$ yr$^{-1}$ are not uncommon. The map also shows alluvial valley floors over 200 m wide together with confined 'gorge' reaches, for it is only in the former that there is the opportunity for sustained lateral migration. The present largely incised course of the Upper Wye, for example, limits the potential for channel change, while elsewhere the glacial history of the region has left a pattern of discontinuous alluvial basins separated by reaches with narrower valley floors and only shallow fills. Borehole evidence suggests that these scoured and infilled basins can be of considerable depths. The role of such features in sediment storage is discussed elsewhere in this volume by A. G. Brown (Chapter 14).

A further word of caution needs to be expressed about Figure 8.4: absence of data points does not necessarily mean low rates of migration, but rather that observations have not yet been compiled. However, below Welshpool on the main channel of the Severn rates have in fact been low. This point will be returned to again when discussing change rates on the Lower Vyrnwy for the area indicated in Figure 8.4.

### Overall change

A variety of possible ways is available for modelling change rates. In the broadest terms, rates appear to be large where stream power is largest (Lewin, 1983), and in particular in the middle reaches of the river systems shown in Figure 8.4. Thus neither very small tributaries nor the Severn itself below Welshpool shift very considerably, though there are additional factors of channel confinement in the former and more cohesive sediments on the latter also to be considered. But for discriminating amongst those channels

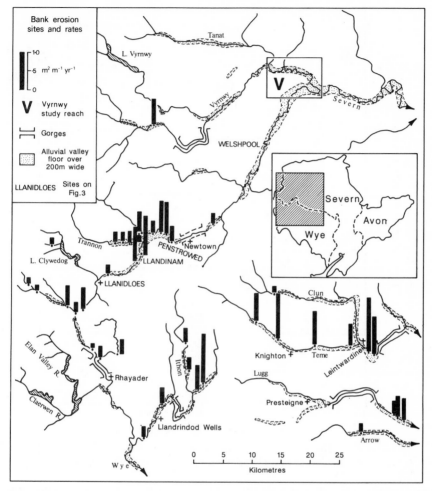

FIGURE 8.4    The Upper Severn and Wye showing average historical change rates

that *are* actively shifting, stream power alone does not appear to explain relative rates of change very well. Figure 8.5 plots bankfull specific stream power (standardized according to channel width) against shift rates, expressed in dimensionless terms as a percentage of channel width. Within this group of middle-reach channels, this suggests that both large and small streams which have low specific powers, shift as much as ones of intermediate size which have higher powers.

An alternative is to consider the development of individual bends. Figure 8.6 plots historical change rates against two measures of loop development. These loops may be both expanding (with path length increasing between

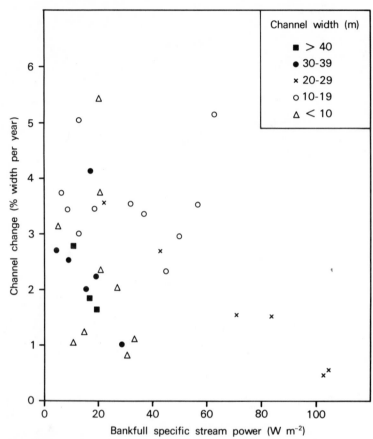

FIGURE 8.5   The relationship between historical channel change rate,
stream power and channel width for 35 Mid-Wales sites

cross-over points) and translating (shifting down-valley), with change rate
being slightly better related to the latter. The scatter is considerable,
however, and it has to be remembered that other processes such as cutoffs
(Lewis and Lewin, 1983) are also important.

Both these examples, involving averaged change rates and the development of individual loops, suggest that a multivariate approach to predicting
change rates is required. D. A. Hughes, C. Blacknell and the writer have
undertaken an extensive multiple regression exercise attempting to predict
historical channel change rates using data for 38 mobile-channel sites in
Wales and the borderland. For this, map data (change rates, catchment area)
were combined with field survey data (channel slope and dimensions at
bankfull), and with data computed following laboratory analysis (sediment
size) or further desk calculation (relative roughness, discharge per unit

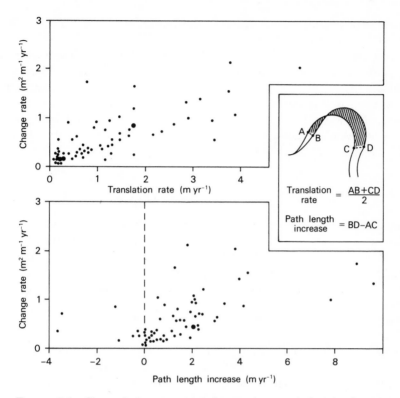

FIGURE 8.6   Channel change rate and meander translation (top) and extension (bottom) for 64 loop changes. Rates for the same loops for different time periods are included.

channel width at bankfull). These variables may usefully be collapsed into alternative ones of channel change which may be related to ones of channel shape, discharge, energy gradient and sediment resistance. For the latter, a dimensionless index was derived (dimensionless or scaled indices are preferred here in general) in which both cohesive fine sediments and coarse materials gave higher values than non-cohesive material of intermediate sizes. The most satisfactory regression results explained change rate expressed as a percentage of channel width ($C$) in terms of the sediment index ($SI$), bankfull discharge per unit width ($Qb/w$), and slope ($S$):

$$\log_{10}C = 0.17 \ SI - 0.45 \ Qb/w - 0.10S + 0.87$$

This gave an $R$ (multiple correlation coefficient) value of 0.73, and an $R^2$ value (adjusted to take account of the loss of degrees of freedom resulting from using more independent variables) of 0.48. Introduction of a 'shape' variable (*w/d* ratio) did not improve the prediction, while taking the logarithm of *C* marginally improved the $R^2$ value.

Although the derived relationship is interesting, it does pose a number of difficulties in interpretation. There is also a problem in that discharge and sediment indices are not statistically independent ($r = -0.67$); independence of variables is very difficult to achieve generally in such studies though it is important statistically. Nonetheless the multiple regression exercise does show, as a starting point, that the incorporation of both power and 'resistance' variables into a channel change model could be useful.

## HISTORICAL VARIATION IN CHANGE RATES

One complication that has to be taken into account in predicting change rates lies in the possibility that change rates vary between historical time periods. As G. Higgs showed in the previous chapter, flood magnitude and frequency have varied historically, and whatever the hydrological cause, the result could be a change in the rate of channel change, particularly with lower rates occurring in periods marked by a decrease in the frequency of floods of channel-forming magnitude.

Table 8.1 is of some interest in this respect. For a set of 21 sites in Mid-Wales it shows average change rates for three time periods. It will be observed that for the first half of this century, rates were lower on average than for the periods before or after. It is unfortunate that surveys are not available on more frequent occasions so that change rate/flood frequency data could be more closely examined, but there is nonetheless a broad 'circumstantial' correspondence between less frequent high-magnitude floods and lower change rates in the early twentieth century.

TABLE 8.1   Historical changes in migration rates

| $n$ | Years | Time period | Site rankings | | | $m^2m^{-1}yr^{-1}$ | |
|---|---|---|---|---|---|---|---|
| | | | 1 | 2 | 3 | $\bar{x}$ | $\sigma$ |
| 21 | 14–17 | (1) 1885–7 to 1901–3 | 8 | 2 | 11 | 0.395 | 0.412 |
| 21 | 44–47 | (2) 1901–3 to 1946–8 | 5 | 11 | 5 | 0.282 | 0.177 |
| 21 | 27–33 | (3) 1946–8 to 1975–8 | 8 | 8 | 5 | 0.346 | 0.221 |

The average picture does, however, disguise some complexities from site to site. Table 8.1 also lists how many sites achieved first, second or third rankings on each of the survey periods. Thus 8 sites achieved their most rapid change in period (1), 5 in period (2) and the remaining 8 in period (3). Further examination shows that over half the sites (11) achieved their *least* rate of change in period (1) when the average change rate for the whole data set was *greatest*. This shows that site-by-site variability over time is considerable and that average behaviour may conceal this.

It is a matter of observation that individual bends may behave differently in response to large floods; the extreme case is of chute cutoffs which may be achieved on suitably sinuous bends, but not on others. It appears, therefore, that site response to flood history may be rather varied and that average behaviour requires a relatively large number of sites for its definition.

## DOWN-CHANNEL SEQUENCES

So far we have considered sites or reaches as if they were spatially independent. On the smallest scale this is known not to be so, and it appears worthwhile to consider this problem more fully than has hitherto been done in the literature. Just as position on a bend and the hydraulics of flows within a meandering reach may determine the loci for bank erosion and sedimentation, so the importance of the transmission of process effects over longer distances deserves to be considered.

For example, on the Maesmawr reach (Figure 8.2), expansion of bends c and d was preceded by that of bend b upstream; the cutoff of bend e led to the development of a whole new bend downstream. At Welshpool, cutoffs similarly had effects, both upstream and downstream, as a result of channel realignment and incision.

Sequences of channel forms and dimensions can be examined spatially and historically on a wider scale. Figure 8.7 shows the lowest 20 km of the River Vyrnwy and its junction with the Severn. Good coverage by historical maps, and a set of ground survey data kindly made available by the Severn-Trent Water Authority allow down-valley trends in channel pattern, sections and rates of change to be examined (Figure 8.8). To an extent, this short reach encapsulates some of the trends in the Severn Basin more widely, though with the advantage that apart from the River Morda, the reach is varied through the interplay of gradient and sediment size and transport, with virtually the same discharges having to be accommodated down the reach for any flood event.

Major characteristics of the reach are as follows (Figure 8.8). Channel change is active in the uppermost third of the reach where the gradient is relatively steep and gravel bed material is moved frequently. By contrast, the lower part of the reach is characterized by extremely low gradients (a former glacial lake is reported for this area; Thompson, 1982), rather stable channels, and banks dominated by fine silty sediments with only minor fine gravel or granule-gravel bedforms. Lateral channel stability is also indicated here by ridge and furrow field patterns extending close to the present channel.

An interesting feature of the reach as a whole is that while bankfull channel capacity remains similar (at around 140 m$^2$), width tends to decline, and

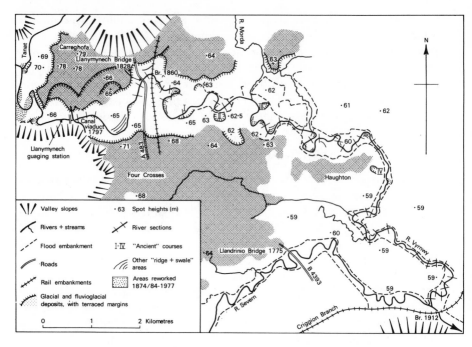

FIGURE 8.7    The Lower Vyrnwy

depth (either mean or maximum) increases. With decreasing gradient downstream, the conveyance capacity of the channel would appear to decrease, unless there is a corresponding decline in channel roughness. Such roughness is difficult to estimate since it must include effects of bedforms, channel planform, and channel vegetation (this appears to be greater in extent in the finer sediments and more stable channel downstream). But the known flood history of the area suggests that within-channel water conveyance downstream is less than it is upstream, where flows can be contained within the channel that lower down pass overbank.

This has important implications for palaeohydrology as well as contemporary management. Channel capacity is not uniquely related to a discharge of given frequency; downstream channels in finer sediments may be adjusted to convey a greater proportion of their high flows overbank, thus of course facilitating the deposition of those same fine sediments comprising a larger proportion of floodplain sediments. Similarly the extent of overbank flood storage must affect downstream flood discharges, and as in engineering flood routing procedures, so spatial studies of downstream changes in channel forms really need to consider out-of-channel water movement as well.

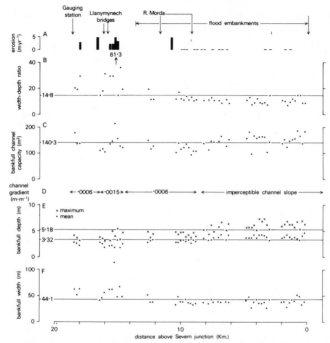

FIGURE 8.8   Change rates and channel morphology on the Lower Vyrnwy. Sites, are shown according to distance above the Severn junction. Average values are shown by the horizontal lines, numerical values being shown towards the left of the diagram

## DISCUSSION

This examination of historical channel change in the Upper Severn Basin has pointed to the considerable variety in planform changes that have taken place, and to the regional disparities in such changes. As suggested previously (Lewin, 1983) what might be called the middle reaches of rivers, or the 'piedmont zone' (Newson, 1981), show the greatest degree of channel migration historically. This is broadly interpretable in terms of available streampower and available alluvial valley floor. However, precise relationships between change rate and streampower, or change rate and evolving pattern, are less than satisfactory. Multivariate models are an improvement, but even these are less satisfactory than is desirable.

Reasons are not hard to find. Hydraulic processes within reaches are demonstrably complex, and it may be that historical change patterns are at too coarse a time and space scale for hydraulic parameters to be appropriately applied. The effects of large floods, occurring between survey dates, are difficult to take into account. It is highly desirable that now survey data are becoming more frequently available, the links between climatic and flood data on the one hand and channel change on the other should continue to be explored over extended timescales.

One further difficulty here lies in the engineering changes to channels themselves, as was discussed for the Trannon. Modification to river channels is widespread and it is often difficult to sort out natural from artificial effects. For example, bank stability is likely to be much affected by the presence or absence of riparian trees; tree growth policies have changed this century, and older historical information is hardly available at all. To what extent are change-rate variations due to the uneven presence of riparian trees?

A final problem concerns down-channel process linkages, because it now seems that historical changes ought not to be considered at sites as if they were spatially independent. This applies both at the reach and the basin scale, and attention has been drawn to the case of the Lower Vyrnwy where, within 20 km, there are systematic and associated changes in pattern forms and changes, channel cross-sectional geometry, gradient, sediments and sedimentary processes, and the extent of within- and overbank flows. There is a clear research need for considering spatial changes in river systems much more systematically and precisely.

The implications for palaeohydrology are important. Historical data show that spatial variety over quite short distances in both rates and patterns of change is expectable and that this is beginning to show the potential for being predictable. Between-site variations are also greater, and certainly less equivocal by way of explanation in terms of environmental change, than are at-a-site changes. Partly this is a function of the inadequacy (in terms of frequency) of survey data, but it is to be hoped that over the decades an improved body of channel pattern and change data will emerge to complement that available for climatic and discharge data. The study of palaeochannel development may in this way achieve a better observational basis.

What do historical studies of channel change show for the palaeohydrology of the last 15,000 years in the Severn Basin? They can have little relevance to interpretations of the late-Devensian, but for the Holocene there is no strong reason to assume other than that the historical picture has broadly prevailed for a very long time. This is one of progressive removal of valley fills from the uplands by migrating rivers. The lowlands, by contrast, have been characterized by much more stable channels and by overbank sedimentation rather than lateral accretion deposits. Rates of change will certainly have varied as a result of changes in landcover and human activity; it also seems likely that the record of the earlier Holocene has been largely removed by river reworking in middle-course areas, but that vertical sequences in finer sediments may much better record fluctuating environmental constraints. It will be helpful if future work is able to put these qualitative statements on a more satisfactory quantitative footing than is at present possible.

## REFERENCES

Bathurst, J. C. (1979). Distribution of boundary shear stress in rivers, in Rhodes, D. D. and Williams, G. P. (eds), *Adjustments of the Fluvial System*, Kendall-Hunt, Dubuque, Iowa, pp. 95–116.

Bathurst, J. C., Thorne, C. R. and Hey, R. (1979). Secondary flows and shear stresses at river bends, *Journal of the Hydraulics Division, American Society of Civil Engineers*, **105**, HY10, 1277–95.

Brookes, A., Gregory, K. J. and Dawson, F. H. (1983). An assessment of river channelization in England and Wales, *The Science of the Total Environment*, **27**, 97–112.

Carson, M. A. (1986). Characteristics of high-energy 'meandering' rivers: The Canterbury Plains, New Zealand, *Geological Society of America Bulletin*, **97**, 886–95.

Dury, G. H. (1984). Abrupt variation in channel width along part of the River Severn, near Shrewsbury, Shropshire, England, *Earth Surface Processes and Landforms*, **9**, 485–92.

Dury, G. H., Sinker, C. A. and Pannett, D. J. (1972). Climatic change and arrested meander development on the River Severn, *Area*, **4**, 81–5.

Hey, R. D. (1975). Design discharges for natural channels, in Hey, R. D. and Davies, T. D. (eds) *Science Technology and Environmental Management*, Saxon House, Farnborough, pp. 73–88.

Langbein, W. B. and Leopold, L. B. (1966). River meanders and the theory of minimum variance, *United States Geological Survey Professional Paper* 422–H.

Lewin, J. (1983). Changes of channel patterns and floodplains, in K. J. Gregory (ed.), *Background to Palaeohydrology*, Wiley, Chichester, pp. 303–19.

Lewin, J., Bradley, S. B. and Macklin, M. G. (1983). Historical valley alluviation in mid-Wales, *Geological Journal*, **18**, 331–50.

Lewis, G. W. and Lewin, J. (1983). Alluvial cutoffs in Wales and the Borderland, in Collinson, J. D. and Lewin, J. (eds), *Modern and Ancient Fluvial Systems*, International Association of Sedimentologists, Special Publication no. 6, Blackwells, Oxford, 145–54.

Macklin, M. G. and Lewin, J. (1986). Terraced fills of Pleistocene and Holocene age in the Rheidol Valley, Wales, *Journal of Quaternary Science*, **1**, 21–34.

Newson, M. D. (1981). Mountain streams, in Lewin, J. (ed.), *British Rivers*, Allen & Unwin, London, pp. 59–89.

Newson, M. D. (1986). River basin engineering—fluvial geomorphology, *Journal of the Institution of Water Engineers and Scientists*, **40**, 307–24.

Smith, D. G. (1983). Anastomosed fluvial deposits: modern examples from Western Canada, in Collinson, J. D. and Lewin, J. (eds), *Modern and Ancient Fluvial Systems*, International Association of Sedimentologists, Special Publication no. 6, Blackwells, Oxford, 155–68.

Thompson, T. R. E. (1982). Soils in Powys II Sheet SJ21 (Arddleen), *Soil Survey Record*, no. 75.

Thorne, C. R. and Lewin, J. (1979). Bank processes, bed material movement and planform development of a meandering river, in Rhodes, D. D. and Williams, G. P. (eds), *Adjustments of the Fluvial System*, Kendall-Hunt, Dubuque, Iowa, pp. 117–37.

Palaeohydrology in Practice
Edited by K. J. Gregory, J. Lewin and J. B. Thornes
© 1987 John Wiley & Sons Ltd.

# 9

# *Morphological Responses and Sediment Patterns*

## D. J. MITCHELL

*School of Applied Sciences, Wolverhampton Polytechnic*

## A. J. GERRARD

*Department of Geography, Birmingham University*

## INTRODUCTION

Since the establishment of the drainage basin as 'the fundamental geomorphic unit' (Chorley, 1969), the concept of the 'basin sediment system' has become widely accepted in the fields of fluvial geomorphology, sedimentology and hydrology (e.g. Chorley and Kennedy, 1971; Gregory and Walling, 1973; Richards, 1982). Early research was directed to micro- or meso-scale units over short time spans with an emphasis on sediment and solute production, erosion of sediment source areas (Imeson, 1970; Harvey, 1974; Lewin, Cryer and Harrison, 1974; Finlayson, 1977) and catchment yields (Hall, 1964; Douglas, 1967; Fleming, 1969; Walling, 1971). Research in different climatic and topographic environments has allowed the conversion of basin sediment and solute yields into denudation rates on regional (Corbel, 1964; Holeman, 1968; Douglas 1969, 1973), and global scales (Fournier, 1960; Strakhov, 1967; Stoddart, 1969; Walling and Webb, 1983). Some workers have examined the basin sediment system as a whole (e.g. Zaslavsky, 1979; Trimble, 1981; Lehre 1982) but, as much of the research has been conducted over short time spans, few investigations have been made into rates of sediment storage although volumes of sediments stored have been well documented (Beckinsale and Richardson, 1964; Brown, 1983). Research into rates of sediment accumulation, storage times and the reworking of sediments requires long-period studies or new techniques to identify historical changes (Lewin, Davies and Wolfenden, 1977; Oldfield, 1977; Oldfield, Appleby and Thompson, 1980).

Studies of long-term sediment erosion, storage and yield, the palaeohydrology of drainage basins, have usually relied on the interpretation of the stratigraphy and chronology of fluvial and mass wasted deposits. However, erosion rates and basin yields are now available for several decades in the USA and a decade in the UK (Webb and Walling, 1984), and these relatively long-period studies can be used to balance the under- or over-estimations created by one- or two-year studies which may coincide with atypical conditions. Micro-studies over short time periods measure the high degree of landscape sensitivity (Smith, 1958; Fullen, 1984) but, if their results are scaled up, erosion rates become phenomenal with little relevance to the total denudation process (Roels, 1985). Increasing the scale of the 'geomorphic unit' not only extends the complexity of the interactions between physical parameters but reduces the sensitivity of the landscape.

Extensive erosion in one part of a catchment often leads to rapid deposition in another part as shown by examples of catastrophic slope failure resulting in accumulation of footslope debris which can remain *in situ* with minimal modification for thousands of years. Similarly, in mobile floodplains, lateral shifting of rivers causes rapid bank erosion resulting in comparable sediment accumulation on point bars, riffles and overbank areas. Many of these episodes are related to exceptional hydrometric events which have a catastrophic effect on landscape processes, but with limited long-term measurements it is difficult to evaluate the net influence of major events in comparison with the gradual erosion and deposition during the intervening periods. The catastrophic events often receive more attention, but a 1 in 100 year rainfall event may not be more erosive than a 100 years of lesser activity. A spectacular landslip may be less significant than imperceptible soil creep and hill wash acting over a longer period of time. Whether the process is rapid or imperceptible the net result is the progressive denudation of the catchment.

Catchment research has clearly demonstrated the many factors responsible for erosion and sediment yield. These factors can be grouped under two basic headings; active or dynamic factors involving the hydrometric variations which provide the input of energy 'to do work' and the passive or static factors involving the catchment characteristics which provide materials and form for the instigation of the 'work'. Although, by definition, catchment characteristics at a moment in time are static, rates of erosion are controlled by natural and anthropogenic changes. Research has shown that many of these factors cannot be used in a simple way but require 'fine tuning'. Thus, an intense storm following a dry period may result in limited runoff while a moderate storm acting on a catchment with moist antecedent conditions can be more effective.

Therefore, in order to assess the denudational history of a catchment over a long time period, a knowledge of the particular features of that catchment must be possessed together with the critical thresholds that distinguish dynamic and static behaviour. In large river basins, catchment characteristics and hydrometric variables are so complex that each basin can be regarded as unique, with different parts responding in different ways. The Severn and Wye basins have been selected to illustrate these factors, to evaluate the spatial and temporal variations of hydrometric and catchment characteristics and to assess their influence on erosion and deposition of sediments.

## CATCHMENT CHARACTERISTICS

The catchment is composed of two major rivers, the Severn and the Wye, with a common source above 600 m on Plynlimon in Central Wales, and a common outlet to the Bristol Channel. The upper parts of both basins have similar features with tributaries draining east and south-east from an upland area, subjected to heavy orographic rainfall, to a drier lower area, lying in the rain shadow of the Cambrian Mountains. Heavy upland rainfall has a much greater influence on the Wye than on the Severn, which has a longer course in the rain shadow area. Therefore, it is not surprising that the Wye is ranked hydrologically higher than the Severn above the Avon's confluence (Ward, 1981). The Severn is joined by the River Avon 35 km above its tidal limit. The Avon, in contrast to the two other rivers, has a relatively subdued lowland catchment.

The high-energy-efficient system of the upper catchment leads to supply and transportation of coarse material derived from Ordovician and Silurian material and glacial and periglacial debris. In the middle and lower parts of the Severn and Wye catchments, erosion and movement of such material is limited due to a reduction in energy from more subdued relief and gentler slopes. The sediment load of the subcatchments in the Welsh borderland and Avon catchment is dominated by suspended sediment and wash load with minimal bedload transportation. Major rivers are characterized by extensive floodplains which have accumulated over a long period of time but are still subjected to varying rates of modification.

The Severn and Wye exhibit within their catchments contrasting valley types, namely confined valleys, mobile floodplains, stable floodplains and incised channels. In confined valleys, such as in the Upper Elan valley (Lewin and Brindle, 1977), river meanders impinge upon and are restricted by solid rock and superficial deposits. Mobile floodplain areas of the middle zone are dominated by large quantities of coarse bedload derived from either reworked glacial material or from more recent fluvial erosion of the uplands. The continuous accumulation of clastic material leads to rapid channel changes. In a detailed study of the River Severn below Caersws, Thorne and Lewin (1979) measured retreat rates of 0.5 m per year, between 1948 and

1975, and Blacknell (1982), using maps and air photographs, analysed the changes that had taken place on a point bar on the River Lugg at Mousenatch from 1902 to 1979. He found that formation started in 1946 with the greatest rates of channel migration and bar construction having occurred since 1977.

These rapid channel changes have invariably led to the creation of cutoffs which mainly occur on the mobile floodplain zones. In an analysis of 145 cutoffs in Wales and the Welsh Borderland, Lewis and Lewin (1983) found that 45 per cent dated within the last 100 years with a distinct increase in the rate of occurrence per 1000 km of river from one cutoff every two years between 1950 and 1970 to one every five years between 1880 and 1900. For 92 cutoffs, infill sedimentation rates varied from 0.003–0.071 m per year (average 0.015 m per year) with seven out of nine infill rates greater than 0.03 m per year dated since 1946. A sediment core taken from a cutoff formed after 1833 near Glasbury on the Wye floodplain was 2 m long with layers of sediments from different source areas giving an average sedimentation rate of approximately 0.013 m per year (Mitchell, in prep.).

Such channel changes may have been accelerated following increased erosion of the uplands due to land drainage for afforestation and agriculture which have increased significantly since 1948 and accelerated in recent years (Newson and Leeks, 1987). Large quantities of bedload lead to bar formation, channel change and hence rapid bank erosion as calculated by Lewin and Chisholm (Table 9.1). Down-valley, away from the piedmont or 'gravel-bed' zone, more stable floodplains are found on the major rivers, with minimal changes in meander patterns. As a consequence of slow migration, floodplain development is dominated by vertical accretion due to inundation of suspended and wash load sediments (Lewin and Hughes, 1980). The stability of the channel pattern in the Shrewsbury area has been supported by Dury (1983) who has demonstrated that channel adjustments have been made to bed forms such as riffle and pool sequences rather than channel pattern migration. Also, few changes have been found below Hereford on the Wye (Mitchell, in prep.) and between Alberbury and Buildwas on the Severn (Hayward, 1985).

Spectacular channel erosion occurs during major events, for example the

TABLE 9.1   Rates of bank erosion during the last century at selected sites in Mid-Wales (from Lewin and Chisholm 1983)

| | No. of sites | Rate of maximum erosion m yr$^{-1}$ | | | |
| --- | --- | --- | --- | --- | --- |
| | | 1885–1901 | 1901–48 | 1948–70s | 1980s |
| Severn | 12 | 0.86 | 0.63 | 0.97 | 1.97 |
| Wye | 11 | 0.82 | 0.65 | 1.00 | 1.60 |

flood flows resulting from the 6 August 1973 storm destroyed a loop forming at Caersws. But Thorne and Lewin (1979) found that the moderate flow events up to bankfull were largely responsible for channel changes, which agrees with the statements of Wolman and Miller (1960) concerning magnitude, frequency and landform development. Using gravel-bed sections of the Severn and Wye, Hey (1975b) was able to produce a theoretical model to show relationships between discharge, load and maximum work done by rivers. Complications are introduced because both rivers have been subjected to river regulation by dam construction in headwater catchments causing different responses downriver in the alluvial channels (Hey, 1975a, 1976).

Stability of slopes and river banks is a significant factor in the supply of sediments to the river system. Major climatic events often cause spectacular movements (Newson, 1980a; Gerrard, 1980; Gerrard and Morris, 1981) but the translocated material often remains in store for long periods before being removed from the system by rivers. Slope failures occurring within the catchments today are largely associated with relict glacial and periglacial debris which becomes reactivated by climatic conditions or undercutting by the deeply incised stream channels. Potts (1971) has shown that in the uplands many of the fossil periglacial features are associated with river valleys. This extensive mantle of material provides a continuous supply of debris through soil creep but the rate of removal of this material from the catchments is very difficult to estimate because of its slow and intermittent movement, as shown by the exceptionally high sediment yields calculated by systematic sampling techniques in the Afon Cyff catchment (Slaymaker, 1972) compared with subsequent detailed analyses for the same basin (Newson, 1980a, 1985). Some of this variability can be attributed to major climatic events or changes in management, which may reactivate debris and increase supply of sediments to the river systems.

The degree of incision of most of the river channels has had a significant influence on bank erosion. Tributaries of the Farlow catchment, incised as much as 4 m into the Dittonian platform of Shropshire, have a significant influence on the supply of suspended sediment load with bank caving, slumping, mudflows, block slide and headward erosion (Mitchell, 1979). Whereas the movement of bedload and mass movement is highly intermittent, movement of suspended sediment and solute load is more or less continuous. Although some suspended sediments are temporarily redeposited in channels and more permanently on floodplains, the majority of the suspended sediment is carried completely out of the system. Using values of suspended sediment estimated every 15 minutes at Redbrook on the River Wye from 1974 to 1980 it is possible to calculate annual suspended sediment yields (Table 9.2). Although seven years is a small sample it is worthy of note that three of those years were relatively wet (1974, 1977, 1979) and two were very dry (1975, 1976). The average annual suspended sediment yield over these seven years was 125,235 tonnes (31.23 t/km²). Using this figure the total yield in 8000 years would have been in the order of 1000 million tonnes, equivalent to 100 mm lowering over the whole catchment, allowing 2.5 t/m³. If solute yields are included, the average annual sediment yield becomes 94.73 t/km², a threefold increase. Instead of multiplying averages, runoff totals can be reconstructed from climatic data (Lockwood, 1979) or by using river flow reconstruction models (Jones, 1984). Annual rainfall data can be compared by regression analysis with suspended sediment and solute yields for ten contrasting catchments of the Wye (Figure 9.1). Varying slopes of the regression equations indicate different rates of output from the upland and lowland catchments but they are all averaged through the final output by the River Wye into the Severn Estuary (Figure 9.2 and 9.3, Table 9.3). Brookes (1974), using mean daily data of rivers entering the Severn Estuary, estimated larger sediment loads for the River Wye showing greater rates of erosion by the Wye compared with the Severn.

TABLE 9.2   Sediment/solute yields for the Wye catchment at Redbrook (4040 km²) (Mitchell, in prep.)

| Year | Suspended sediment (Tonnes) | Organic (Tonnes) | Inorganic (Tonnes) | Total hardness (Tonnes) | Total suspended (Tonnes) | Hardness + sediment (Tonnes/km²) |
|---|---|---|---|---|---|---|
| 1974 | 197,931 | 36,098 | 161,833 | 290,851 | 488,782 | 121.89 |
| 1975 | 59,006 | 11,371 | 47,635 | 187,786 | 246,792 | 61.54 |
| 1976 | 48,611 | 11,307 | 37,304 | 184,189 | 232,799 | 58.05 |
| 1977 | 162,689 | 30,113 | 132,576 | 299,751 | 462,440 | 115.32 |
| 1978 | 92,480 | 17,540 | 74,940 | 235,256 | 327,737 | 81.73 |
| 1979 | 200,186 | 33,702 | 166,484 | 301,823 | 502,009 | 125.19 |
| 1980 | 115,740 | 23,523 | 92,216 | 282,737 | 398,477 | 99.37 |
| Total | 876,643 | | | | | |
| Average | 125,235 | | | | | |

## GEOLOGIC FRAMEWORK

Erosion and transportation of material are controlled particularly by the nature and supply of sediment, which in turn depends on rock types and superficial deposits. Central Wales and the higher parts of the Welsh Border-land are dominated by Ordovician and Silurian rocks, mainly grits, slates and shales, providing coarse material but with restricted amounts of sand-size and finer fractions. In contrast, the eastern area, which includes the Welsh Borderland, Severn and Avon valleys, is dominated by younger sedimentary rocks of Devonian, Carboniferous, Triassic and Jurassic age. Extensive outcrops of Devonian Old Red Sandstone in the middle and lower catchments are significant in the provision of easily eroded and transported sand-sized sediments. Although more localized, the Carboniferous and

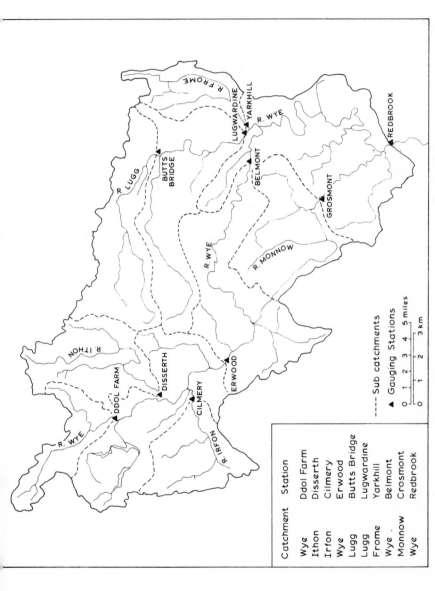

FIGURE 9.1   The ten subcatchments of the River Wye analysed for suspended sediment and solute yields

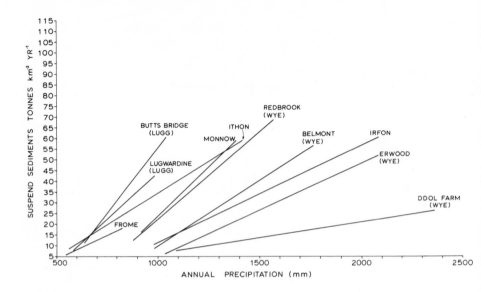

FIGURE 9.2   Relationships between suspended sediment yield and annual precipitation for ten subcatchments of the River Wye

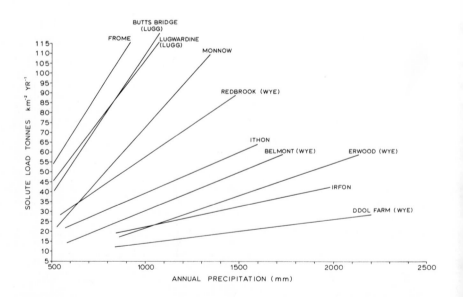

FIGURE 9.3   Relationships between solute yield and annual precipitation for ten subcatchments of the River Wye

TABLE 9.3 Catchment characteristics and sediment yield for selected basins

| Catchment | | Ddol Farm Wye | Cilmery Irfon | Disserth Ithon | Erwood Wye | Belmont Wye | Buttsbridge Lugg | Lugwardine Lugg | Yarkhill Frome | Grosmont Monnow | Redbrook Wye |
|---|---|---|---|---|---|---|---|---|---|---|---|
| Area km² | | 174 | 244 | 358 | 1280 | 1900 | 371 | 886 | 144 | 357 | 4010 |
| Relief ratio | | 0.020 | 0.022 | 0.017 | 0.012 | 0.006 | 0.015 | 0.012 | 0.007 | 0.029 | 0.007 |
| Geology % | Devonian | 0 | 0 | 0 | 3 | 31 | 28 | 57 | 96 | 97 | 55 |
| | Silurian | 90 | 43 | 73 | 69 | 47 | 67 | 38 | 0 | 0 | 31 |
| | Ordovician | 10 | 56 | 22 | 25 | 17 | 0 | 0 | 0 | 0 | 8 |
| | Other | 0 | 1 | 5 | 3 | 5 | 5 | 5 | 4 | 3 | 6 |
| Land use % | Arable | 3 | 2 | 5 | 4 | 7 | 24 | 26 | 31 | 19 | 15 |
| | Pasture | 84 | 67 | 83 | 82 | 80 | 73 | 69 | 61 | 78 | 71 |
| | Woodland | 12 | 31 | 11 | 13 | 12 | 3 | 4 | 7 | 2 | 13 |
| | Orchard | | | | | | | | | | |
| | Other | 1 | 0 | 1 | 1 | 1 | 0 | 1 | 1 | 1 | 1 |
| Arable/Pasture ratio | | 0.038 | 0.030 | 0.060 | 0.044 | 0.092 | 0.327 | 0.378 | 0.509 | 0.248 | 0.210 |
| Cattle and sheep/ha pasture | | 5.5 | 8.8 | 9.1 | 7.8 | 7.0 | 9.1 | 8.5 | 6.0 | 8.3 | 6.8 |
| Av. Annual Precipitation 1974–80, mm | | 1750 | 1507 | 1016 | 1544 | 1374 | 841 | 801 | 693 | 1138 | 1130 |
| Av. Annual Runoff 1974–80, mm | | 1209 | 1199 | 671 | 890 | 719 | 484 | 404 | 262 | 512 | 541 |
| Runoff Coefficient | | 0.69 | 0.80 | 0.66 | 0.58 | 0.52 | 0.58 | 0.50 | 0.38 | 0.45 | 0.47 |
| Av. suspended sediment yield t km⁻²y⁻¹ 1974–80 | | 16.92 | 34.59 | 35.41 | 28.50 | 32.20 | 35.58 | 26.79 | 12.03 | 36.14 | 32.23 |
| Av. solute yield t km⁻²y⁻¹ 1974–80 | | 22.07 | 32.08 | 40.06 | 39.64 | 45.95 | 87.22 | 82.67 | 82.34 | 87.65 | 66.55 |

Jurassic sands and clays tend to dominate the eastern areas, providing suspended and wash load.

Quaternary superficial deposits are important to sediment supply within the basins. The Wolstonian glaciation probably covered the whole area at some time but only remnant lake deposits in the Avon Valley (Shotton, 1953) and isolated deposits of high-level 'old' drift remain. Although not as extensive, the more recent Devensian Glaciation and associated processes have covered much of the solid geology in Wales and the Welsh Borderland with glacial sands and tills, river terrace deposits, solifluction and alluvial material. These superficial deposits have altered the rather simplistic geological divisions described above. A mixture of glacially transported fine and coarse material covers most of the Upper Severn and Upper Avon catchments incorporating material from outside the basin as well as material translocated from within the basin. Further movement and sorting of this material has occurred during the retreat of the Devensian ice forming extensive areas of outwash sands and river terrace deposits.

Upland areas beyond the maximum extent of the Devensian Glaciation, and areas progressively ice-free during the late Devensian period, experienced severe periglacial conditions mantling slopes and valleys with solifluctuated debris and screes. Initially, before vegetation recolonization, these deposits were undoubtedly unstable, leading to landslips and slumping often activated by fluvial erosion associated with increased water availability and changing base levels. Erosion in the late Devension period was probably widespread due to continuing frost action and periodic floods from annual snowmelt in a sparsely vegetated catchment. Subsequent erosion has been dominated by fluvial activity and the deposition of colluvium on slopes and alluvium in the major river valleys.

## SEDIMENT AVAILABILITY

While the geological framework and covering of superficial materials determine the nature of the sediment sources, the availability of sediments for erosion, transport and possible deposition depends on specific catchment and hydrometric variables. Analysis of sedimentation rates has shown that vegetation and land use changes during the last 10,000 years have been important factors in the variation of erosion in parts of the Severn catchment (Bartley, 1960; Jones, Benson-Evans and Chambers, 1985). For example, development of a forest cover during the Holocene reduced erosion, leading to progressive stabilization of the landscape with activity probably limited to river channel incision and headward erosion. Low rates of sediment accumulation of 30 mm/100 years in Llangorse Lake (Jones, Benson-Evans and Chambers, 1985) and evidence of slope stability in the Cotswolds support this premise (Ackerman and Cave, 1968).

But selective clearance of woodland by Mesolithic food gatherers and hunters and progressive forest clearance by Neolithic colonists was significant for the decline in vegetative cover. Neolithic farmers practised a shifting cultivation above 300 m, evidence of which is provided by pollen analysis in the Elan valley (Moore and Chater, 1969; Moore, 1982; Wiltshire and Moore, 1983). Each site was probably cropped for a few years before fertility was exhausted, leaving the area to coppice regeneration which often suffered browsing by animals. This probably resulted in increased runoff and hence erosion as indicated by the increased rates of sedimentation of 132 mm/100 years in Llangorse Lake (Jones, Benson-Evans and Chambers, 1985).

An intensification of forest clearance by Bronze Age farmers coupled with wetter conditions at about 3500 BP, must have resulted in increased rates of erosion and deposition. Following a drier interval, wet conditions returned at approximately 2900 BP with a maximum precipitation around 2700–2800 BP. Cereal pollen in the deposits in Rhosgoch Common indicates increased cultivation in the Late Bronze Age. An accompanying increase in erosion is shown by deposits of stoneless alluvium, devoid of organic matter and buff red in colour, in marked contrast with the earlier greyish alluvial deposits. This abrupt change occurs at a depth of 1.34 m in the Avon alluvial sequence and has been dated as 2650 BC (Shotton and Williams, 1971, 1973; Williams and Johnson, 1976). A similar change has been dated at 2697 ± 85 BP in the Windrush valley where sediment accumulation rates of 39.4 mm/100 years have been estimated (Hazelden and Jarvis, 1979).

As more permanent settlements became established, forest clearance continued in progressively lower zones, as shown by the continual decline of arboreal pollen in Rhosgoch Common (Bartley, 1960). In spite of periods of preservation of vegetal cover, such as the Anglo-Saxon pastoralization and the protection of the Norman hunting forests, other activities have main-tained rates of erosion comparable with those of the Sub-Boreal Period. This is shown by sedimentation of 141 mm/100 years in Llangorse lake between 850 BC and 1840 AD (Jones, Benson-Evans and Chambers, 1985) and increased sedimentation in Rhosgoch Common Lake (Bartley, 1960). Despite some piecemeal afforestation and planting of trees to prevent bank erosion, vegetal cover continued to decline due to increased cultivation and develop-ment of forest-based industries. Land drainage and the ploughing of steep slopes during wartime food shortages also increased the rates of erosion with sedimentation is Llangorse Lake since 1840 amounting to 590 mm/100 years; the maximum rate since the retreat of the Devensian ice.

Using sediment rating curves, the estimated accumulation of suspended sediments on the Wye floodplain between Hereford and Ross-on-Wye, during the December 1960 and December 1979 floods, were 0.134 and 0.010 mm respectively. These estimated rates are considerably less than accumu-lation rates measured using Cs137 dating of cores from an ox-bow lake at

Letton and a core taken from the nearby floodplain. Assuming that the fallout of Cs137 first occurred in 1954, 220 mm and 110 mm have accumulated in the respective locations during which time 14 significant flood events have occurred. The rating curve method assumes an even sedimentation rate over the whole floodplain while the cores may record the accumulation in sediment 'sinks'.

## CONTEMPORARY CATCHMENT CONTROLS

Reafforestation of extensive areas in Central Wales and other land use changes provide a unique opportunity to assess the relative effects of certain catchment characteristics on water yield and erosion. Research catchments established by the Institute of Hydrology to investigate the effects of afforestation took advantage of the contrasting land use management of the grassland Upper Wye and afforested Upper Severn catchments. Results from the Upper Severn catchments (Newson, 1979, 1985) showed the same effects of afforestation as those demonstrated by Bates and Henry (1928) in Colorado, USA. Runoff was significantly reduced from the forest catchment due to interception and evapotranspiration and a reduction in peak flow and rise times (Table 9.4). Decreased runoff would lead normally to a reduction in erosion but for the effect of preparing the land for afforestation. Using results from the Cyff (Upper Wye) and Tanllwyth (Upper Severn) catchments of Plynlimon, Painter *et al.* (1974) constructed a general model to explain the effects of increased drainage density due to forest ditching and the effects of road construction on sediment yield. They suggested that ditching has an immediate effect on sedimentation, giving a 1000-fold increase in total sediment yield.

TABLE 9.4    Effect of afforestation on runoff characteristics, 1970–77 (from Newson, 1979, 1985)

| Land use | Wye | Severn |
|---|---|---|
| Grassland | 98% | 40% |
| Forest | 2% | 60% |
| Precipitation mean mm ($P$) | 2348 | 2213 |
| Runoff ($Q$) | 1944 | 1364 |
| Runoff coefficient | 83% | 62% |
| $P$-$Q$ mm | 405 | 849 |
| Av. rise time, hrs | 2.0 | 2.3 |
| Peak flow | 100% | 66% |
| Av. suspended sediment mg/l | 38 | 69 |
| Bedload $m^3$ $yr^{-1}$ | 2.5 | 8.4 |

As a consequence of afforestation, suspended sediment and bedload are greater in the Upper Severn than in the Upper Wye (Table 9.4). Similar results were found in the Upper Tywi catchment (Jones, 1975). As most of the sediment is composed of bedsize material, movement of material from the Plynlimon catchments to the lower zones is complex (Arkell *et al.*, 1983; Leeks, 1984). Using conventional bedload traps, sediment yields from the forested Tanllwyth catchment were found to be 3.5–20.5 times greater than from the grassed Cyff catchment. This provides clear evidence for small catchment studies but the transport of bedload throughout the larger basin system is more difficult to evaluate. To assess the transport of coarse sediments from drainage ditches in the Plynlimon catchment, use is being made of enhanced magnetic-susceptible material that can be located downriver (Arkell *et al.*, 1983).

There is a need to examine more carefully the hydrological sensitivity of land use changes and land management techniques. Field drainage can have a significant hydrological effect. In a theoretical model Gregory (1971) suggested that an extension of artificial drainage systems should lead to a reduction in drainage density and channel erosion. However, where field drains replace slow infiltrating soils runoff is more rapid, leading to higher discharge and greater erosion of the remaining stream channels (Mitchell, 1979). In addition, localized outfall from field drains in the Farlow catchment leads to extensive erosion by backcutting in a highly entrenched channel network.

In the early 1950s the former Lower Wye Drainage Board carried out an extensive tree-clearance scheme along the Monnow to increase the flow of the river, reducing the risk of flooding. But now erosion has become a serious problem in many middle and lowland sections of rivers due to increased flow rates. Increased velocity of rivers has caused an even greater increase in erosion of the alluvium and Old Red Sandstone banks. In parts of the middle Monnow valley the extent of the erosion can be plotted by lines of alder trees marking the original channel, with unstable banks more than 36 m from the original channel (Fryer, 1960).

Similar intensification of farming techniques has developed in the piedmont and lowland areas. As a result of continuous cultivation and compaction on sandy soils soil erosion in certain areas has become a more significant factor (Reed, 1979, 1983; Fullen 1985a,b). Consequently sediment rating curves, especially of winter data, tend to be much steeper in areas subject to soil erosion and bank caving compared with pastoral catchments (Table 9.5). Although not as critical as in semi-arid areas of Australia and America, stocking ratios and animal husbandry can be influential factors in the acceleration of erosion. Thomas (1964) surveyed an area of 3.64 $km^2$ of Plynlimon and estimated that 5 per cent was affected by sheet erosion induced by sheep with a further 19 per cent likely to be eroded if sheep-grazing remained unrestricted. Lowland areas with a high arable/pasture ratio tend to have lower stocking ratios but in most cases domestic animals are kept on steeper slopes and marshy lands near river courses. Bare soil on steep slopes and bank instability due to animals were found to be a significant factor in channel and slope erosion in the Farlow catchment (Mitchell, 1979).

TABLE 9.5   Sediment rating curve constants and coefficients compared with land use of the Farlow catchment, Shropshire (from Mitchell 1975)

| Subcatchment | Major land use | Sediment rating curves constants and coefficients | | | |
|---|---|---|---|---|---|
| | | Winter | | Summer | |
| | | a | b | a | b |
| Ingardine Brook | Arable/Ley grass | 1.245 | 1.234 | 2.592 | 1.012 |
| Wheathill Brook | Arable/Ley grass | 0.206 | 1.244 | 2.353 | 0.631 |
| Silving Brook | Pasture/Moorland | 9.972 | 0.590 | 3.024 | 0.862 |
| Cleeton Brook | Moorland | 3.459 | 0.757 | 7.124 | 0.601 |
| Shirley Brook | Moorland | 0.528 | 0.548 | 4.041 | 0.132 |

Suspended sediment concentration $= aQ^b$

## CONTRASTING CATCHMENT RESPONSE

Contemporary erosion studies and basin sediment yields can be used to illustrate various causes of erosion and slope failure. Current vegetation/land use patterns can be matched with past vegetation and annual variations in hydrometric data used to simulate former climatic conditions. The generalized observations described so far refer to average hydrometric conditions acting on basin characteristics which change over long periods of time. Although the net erosion and deposition rates over long periods may produce variable trends, the annual variation in climate, acting on contrasting catchment characteristics in different locations within the whole basin, can be used to simulate conditions which have occurred since the Devensian Glaciation. Some conditions were probably more constant than others, which showed either oscillatory patterns or constant changes. In order to assess the past erosion and deposition of the basins, contemporary basin studies can be used, provided that contrasting hydrometric conditions prevail and that catchment characteristics similar to former landscapes are still to be found within the basin (Table 9.3, Figures 9.2 and 9.3). Measurements of erosion and deposition rates within the Severn and Wye basins can be used to illustrate many of the conditions experienced in the catchments' development in the last 12,000 years. Although these hydrological reconstructions are speculative it is possible to make realistic estimates of erosion rates which can be compared with the deposition rates observed in lakes and floodplains.

Denser networks of climatological stations and river recording stations have, in recent years, enabled detailed assessments to be made of hydrometric events. Climatic conditions which result in high discharge and flooding often instigate catastrophic erosion and subsequent deposition. In the Upper Severn and Wye catchments, floods have occurred in the last 50 years due to saturated antecedent conditions in 1946, 1948, 1960, 1964 and 1965, to snowmelt in 1941 and 1947 and to the passage of secondary depressions giving prolonged duration rain in 1929, 1940 and 1964 (Howe, Slaymaker and Harding, 1967). Erosion and sediment yield from major events depends on the distribution of storms as well as their magnitude. Estimates of erosion have been made for relatively small experimental catchments by assessing the localized effects of storms which cannot be increased by areal proportions for the large catchments. Two summer storms in August 1973 and August 1977 have been well documented at the Institute of Hydrology experimental catchments on Plynlimon (Newson, 1975, 1980a). During 5 and 6 August 1973 a well-marked frontal system which gave widespread heavy rainfall, 72.2 mm in 6 hours with a return period of 20–50 years, under highly saturated catchment conditions in the Upper Severn and Wye, produced peak flows of 41.0 and 65.4 m³/sec respectively with a return period of 1000–10,000 years. Supercharged pipes, high pore-water pressure, critical subsurface flow and extended partial areas resulted in peat bursts, debris slides and gullies in unconsolidated material. Although spectacular erosional features occurred, material remained on footslopes, channel changes were minimal and immediate transport of bedload was limited. In contrast a thunderstorm on 15 August 1977 was more intense than the 1973 storm but dry antecedent conditions and the more localized nature of the storm meant that critical slope thresholds were not crossed and slope failure was limited to two small slides. Although peak discharge was only 34 m³/sec, with a return period of less than 100 years, there was extensive channel development with redistribution of gravel and cobble shoals (Newson, 1980a).

This research emphasizes the disjointed nature of sediment erosion and transportation, especially in the headwater catchments. Where the immediate influence of a catchment event can be quantified it is important to assess its magnitude with respect to lesser events acting over longer time periods. In a detailed study of the Farlow catchment, Shropshire, between 1972 and 1975, it was found that erosion and transportation of suspended sediments were related to high rainfall intensity, especially associated with the passage of cold fronts, high antecedent soil moisture conditions and high discharge. During the study period of 2½ years, the most significant factor in the supply of suspended sediment, and to a certain extent solute load, was the effect of major discharge events. Using daily yield as a basic unit, 57 per cent of total suspended sediment was transported on only 3 days (0.3 per cent of the time) and as much of 85 per cent on a total of 20 days (2.2 per cent of the time). The storm of 5 and 6 August 1973, as found on Plynlimon, was the most effective erosion event with an average catchment rainfall of 36.1 mm and a maximum amount of 47.5 mm in 15 hours on Titterstone Clee and intensity of 11.5 mm/hr. This storm, coupled with a comparatively low soil moisture deficit, led to a rapid increase in discharge, peaking at 47.70 m³/ sec. The importance of this major event in contributing to sediment and solute load is shown in Table 9.6. The dramatic increase in suspended sediment is attributed to an increase in available load by an extension of transportable sediment, by an increase in surface contributing area through rill development and overland flow, by increased channel erosion and by flushing out of available sediments (Mitchell, 1979).

TABLE 9.6   The importance of the storm on 5 and 6 August 1973 in the denudation of the Farlow catchment (Mitchell, 1979)

| Time period | Suspended sediment load (tonnes) | t km⁻² | % of 5 and 6 August total | % of August total | % of 1972–73 total |
|---|---|---|---|---|---|
| 1 hour | 1241.0 | 61.6 | 51.7 | 51.5 | 28.4 |
| 2 hours | 2085.7 | 103.6 | 86.9 | 86.7 | 47.8 |
| 4 hours | 2340.9 | 116.2 | 97.5 | 97.2 | 53.6 |
| 5 and 6 August | 2400.2 | 119.2 | — | 99.7 | 55.0 |
| | Dissolved load | | | | |
| 5 and 6 August | 73.49 | 3.7 | — | 41.1 | 3.4 |

The nature of sediment and solute load is governed by rock type, superficial deposits and soil types. Erosion, transportation and deposition are dependent on sediment composition, size and availability. As mentioned earlier, the upper parts of the Severn and Wye catchments are dominated by coarse glacial and periglacial deposits covering Ordovician and Silurian slates, shales and grits. Consequently, bedload size material is abundant compared with finer suspended and wash load. In a first-order tributary of the Afon Cyff, Reynolds (1986) calculated the relative components of total load, indicating the dominance of bedload in the sediment budget for 1980–81 (Table 9.7). In order to measure the sensitive differences between areas, small catchments can be used. Oxley (1974) estimated suspended sediment and dissolved load of two small catchments, Ebyr North and Ebyr South, in Central Wales, both developed on impermeable Silurian greywackes with a thin mantle of locally derived till. Ebyr North was mainly forest with thin soils (<1 m deep) and steep slopes (10–15°), while Ebyr South was mainly pasture developed on deeper acid brown earths (>2 m deep) and gentle slopes (3–5°). Using data for 1971, he concluded that channel material rather than the overall conditions of the catchment was the most important factor affecting suspended sediment load while solute concentrations were controlled by rate and nature of throughflow (Table 9.7).

TABLE 9.7   Sediment transport in contrasting catchments

| Catchment | Area km² | Sediment load tons km² yr⁻¹ | | | Year | Reference |
|---|---|---|---|---|---|---|
| | | Bed | Suspended | Dissolved | | |
| 1st-order tributary of Cyff | 0.04 | 8.0 | 2.8 | 37.6 | 1980–1 | Reynolds |
| Ebyr North | 0.074 | 1.9 | 0.78 | 5.41 | 1971 | Oxley |
| Ebyr South | 0.096 | 9.5 | 1.13 | 4.09 | 1971 | Oxley |
| Preston Montford | 3.15 | 1.05 | 32.8 | 67–100 | 1982 | Job |
| Farlow, Shropshire | 20.14 | 0.002 | 112.03 | 113.02 | 1972–5 | Mitchell |

In contrast to the upland catchments, the Farlow basin is composed of 85 per cent Old Red Sandstone and 15 per cent Carboniferous shales, sandstones, limestones and conglomerates resulting in variable sediment load. Bedload was found to be insignificant in comparison with suspended and solute load, with an estimated 0.0735 tonnes (0.0007 per cent of total load). The low bedload component can be attributed to the lack of bedload-sized material in the catchment's lithology and the nature of the channel bed which is frequently littered with relict bedload requiring traction velocities far in excess of those experienced in the present climatic regime. The sediment yield from the Farlow catchment is dominated by suspended sediment with an estimated 5640.5 tonnes in 2½ years, equivalent to 112.03 t/km²/yr. More than 80 per cent of the suspended sediment is composed of fine sand which requires the lowest erosion velocity of the Hjülstrom (1935) curve. Due to the highly calcareous soils and rocks the solute load was marginally greater than the suspended sediment yield with an average of 113.03 t/km²/yr. A similar pattern of sediments and solutes was estimated for the Preston Montford catchment (3.15 km²) which is composed of glacial and fluvioglacial material of the North Shropshire plain near Shrewsbury (Job, 1982). In 1979–80, total sediment yield was dominated by suspended sediments amounting to 32.8 t/km² while bedload amounted to 1.05 t/km², 3.2 per cent of total sediment yield. Annual dissolved load was estimated to be two to three times greater than suspended and bedload.

## CONCLUSIONS

River basins are complex process-response systems and the larger the basin, the more complex is the system. Concepts such as landscape sensitivity and magnitude and frequency considerations can be applied to small drainage basins and to large drainage basins over short time periods. Increasing the scale of the unit of study increases the complexity and reduces the sensitivity of the system. Thus severe problems are faced when efforts are made to assess morphological responses and changing patterns of erosion, transport

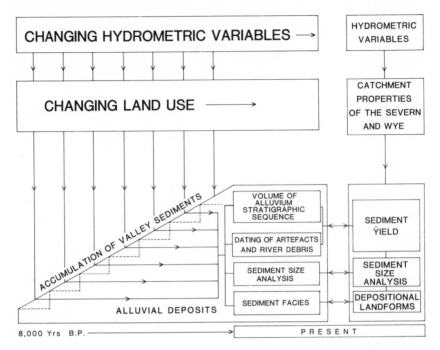

FIGURE 9.4   Potential interaction between catchment characteristics and catchment response (from Mitchell, 1980)

and deposition of sediment. The examples reviewed here have shown that the interplay between catchment controls and characteristics and catchment response is far from straightforward, with many potential linkages (Figure 9.4). It is clear that no one approach will produce a definitive statement of long-term erosion, storage and yield. A stratigraphic analysis of river and slope deposits will enable the broad framework of landscape change to be established. As more efficient models of the relationships between erosion and vegetation type and cover become available, the evidence provided by the stratigraphic approach can be explained in more specific terms. But the 'fine-tuning' can only be achieved by detailed studies of the scale and timing of slope and fluvial activity within basins of differing spatial scales. All three approaches are required for palaeohydrology to become an exact study.

## ACKNOWLEDGEMENTS

The authors are grateful to the Wye Division, Welsh Water Authority and the Institute of Hydrology for providing data and to the Polytechnic, Wolverhampton and the University of Birmingham for financial assistance.

## REFERENCES

Ackerman, K. J. and Cave, R. (1968). Superficial deposits and structures including landslips in the Stroud District, Gloucestershire, *Proceedings Geologists Association*, **78**, 567–86.

Arkell, B., Leeks, G., Newson, M. and Oldfield, F. (1983). Trapping and tracing: some recent observations of supply and transport of coarse sediment from Upland Wales, *Special Publication, International Association of Sedimentology*, **6**, 107–19.

Bartley, D. D. (1960). Rhosgoch Common, Radnorshire; Stratigraphy and pollen analysis, *New Phytologist*, **59**, 238–63.

Bates, G. G. and Henry, A. J. (1928). Forest and stream flow at Wagon Wheel Gap, Colorado, Final Report, *Monthly Weather Review, Supplement*, **30**, 1–79.

Beckinsale, R. P. and Richardson, L. (1964). Recent findings on the physical development of the Lower Severn Valley, *Geographical Journal*, **130**, 87–105.

Blacknell, C. (1982). River migration, channel morphology and channel sedimentation on the River Lugg at Mousenatch (SO469607), *Cambria*, **19**, 16–24.

Brookes, R. E. (1974). Suspended sediment and solute transport for rivers entering the Severn estuary, unpublished Ph.D. thesis, University of Bristol.

Brown, A. G. (1983). Floodplain deposits and accelerated sedimentation in the lower Severn basin, in Gregory, K. J. (ed.), *Background to Palaeohydrology: A Perspective*, Wiley, Chichester, pp. 375–97.

Chorley, R. J. (1969). The drainage basin as the fundamental geomorphic unit, in Chorley, R. J. (ed.), *Water, Earth and Man*, Methuen, London, pp. 77–99.

Chorley, R. J. and Kennedy, B. A. (1971). *Physical Geography: A Systems Approach*, Prentice-Hall, London.

Corbel, J. (1964). L'érosion terrestre étude quantitative (Méthodes-techniques-résultats), *Annales de Géographie*, **73**, 385–412.

Douglas, I. (1967). Natural and manmade erosion in the humid tropics of Australia, Malaysia and Singapore, *International Association for Scientific Hydrology*. Pub. 75, 17–29.

Douglas, I. (1969). The efficiency of humid tropical denudation systems, *Transactions of the Institute of British Geographers*, **46**, 1–16.

Douglas, I. (1973). *Rates of denudation in selected small catchments in Eastern Australia*, University of Hull, Occasional Paper in Geography, **21**, 127 pp.

Dury, G. H. (1983). Osage-type underfitness on the River Severn near Shrewsbury, Shropshire, England, in Gregory, K. J. (ed.), *Background to Palaeohydrology: A Perspective*, Wiley, Chichester, pp. 399–408.

Finlayson, B. L. (1977). *Runoff contributing areas and erosion*, School of Geography, University of Oxford, Research Paper, 18.

Fleming, G. (1969). The Clyde Basin: Hydrology and sediment transport, unpublished Ph.D. thesis, University of Strathclyde, Glasgow.

Fournier, F. (1960). *Climat et érosion*, P.U.F., Paris, 201 pp.

Fryer, N. T. (1960). Farm land lost through river erosion, *Country Life*, 28 July, 176–8.

Fullen, M. A. (1984). An investigation of rainfall, runoff and erosion on fallow arable soils in east Shropshire, unpublished Ph.D. thesis, CNAA.

Fullen, M. A. (1985a). Erosion of arable soils in Britain, *International Journal of Environmental Studies*, **26**, 55–69.

Fullen, M. A. (1985b). Soil compaction, hydrological processes and soil erosion on loamy sands in east Shropshire, England, *Soil and Tillage Research*, **6**, 17–29.

Gerrard, A. J. (1980). Large-scale forms of hillslope failure in the Severn and Wye drainage basins, in Gregory, K. J. (ed.), *Palaeohydrology of the temperate zone, subproject A: fluvial environments, Severn Basin,* unpublished preliminary report.

Gerrard, A. J. and Morris, L. (1981). *Mass movement forms and processes on Bredon Hill, Worcestershire.* University of Birmingham, Department of Geography, Working Paper 10.

Gregory, K. J. (1971). Drainage density changes in South West England, in Gregory, K. J. and Ravenhill, W. D. (eds), *Exeter Essays in Geography,* Exeter, pp. 33–53.

Gregory, K. J. and Walling, D. E. (1973). *Drainage Basin Form and Process: A Geomorphological Approach,* Edward Arnold, London.

Hall, D. G. (1964). The sediment hydraulics of the River Tyne, Unpublished Ph.D. thesis, University of Durham.

Harvey, A. M. (1974). Gully erosion and sediment yield in the Howgill Fells, Westmoreland, *Institute of British Geographers, Special Publication,* **6,** 45–58.

Hayward, M. (1985). Soil development in Flandrian floodplains: River Severn case study, in Boardman, J. (ed.), *Soils and Quaternary Landscape Evolution,* Wiley, Chichester, pp. 281–99.

Hazelden, J. and Jarvis, M. G. (1979). Age and significance of alluvium in the Windrush Valley, Oxfordshire, *Nature,* **282,** 291–2.

Hey, R. D. (1975a). Response of alluvial channels to river regulation, *Water for Human Needs,* Proceedings of the 2nd World Congress on Water Resources, 15, Technology and Ecology, IWRA, New Delhi, India.

Hey, R. D. (1975b). Design discharge for natural channels, in Hey, R. D. and Davies, T. D. (eds), *Science and Technology in Environmental Management,* Saxon House, Farnborough, D.C. Heath.

Hey, R. D. (1976). Impact prediction in the physical environment, in O'Riordan, R. I. and Hey, R. D. (eds), *Environmental Impact Assessment,* Saxon House, Farnborough, pp. 71–81.

Hjulstrom, F. (1935). Studies of the morphological activity of rivers as illustrated by the River Fyris, *Bulletin of the Geological Institute, University of Uppsala,* **25,** 221–527.

Holeman, J. N. (1968). The sediment yield of major rivers of the world, *Water Resources Research,* **4,** 737–47.

Howe, G. M., Slaymaker, H. O. and Harding, D. M. (1967). Some aspects of the flood hydrology of the upper catchment of the Severn and Wye, *Transactions of the Institute of British Geographers,* **41,** 35–58.

Imeson, A. C. (1970). Erosion in three East Yorkshire catchments and variations in dissolved, suspended and bedloads, unpublished Ph.D. thesis, University of Hull.

Job, D. A. (1982). Runoff and sediment output from a small lowland catchment— the example of Preston Montford Brook, Shropshire, *Field Studies,* **5,** 685–729.

Jones, A. (1975). Rainfall, runoff and erosion in the Upper Tywi catchment, unpublished Ph.D. thesis, University College, Swansea.

Jones, P. D. (1984). Rainfall reconstruction from precipitation data, *Journal of Climatology,* **4,** 171–86.

Jones, R., Benson-Evans, K. and Chambers, F. M. (1985). Human influence upon sedimentation in Llangorse Lake, Wales, *Earth Surface Processes and Landforms,* **10,** 227–36.

Leeks, G. J. (1984). Assessment of river erosion and deposition in Mid-Wales, UK, in Burt, T. P. and Walling, D. E. (eds), *Catchment Experiments in Fluvial Geomorphology,* Geobooks, Norwich.

Lehre, A. K. (1982). Sediment budget of a small coast range drainage basin in North-Central California, in Swanson, F. J. *et al.* (eds), *Sediment budgets and routing in forested drainage basins*, US Forest Service General Technical Report, PN-141, 67–77.

Lewin, J. and Brindle, B. J. (1977). Confined meanders, in Gregory, K. J. (ed.), *River Channel Changes*, Wiley, Chichester, pp. 221–33.

Lewin, J. and Chisholm, N. W. T. (1983). *Assessment of the historical rate of erosion and deposition in Mid-Wales*, unpublished report, Institute of Hydrology.

Lewin, J., Cryer, R. and Harrison, D. I. (1974). Sources for sediments and solutes in Mid-Wales, *Institute of British Geographers Special Publication*, 6, 73–85.

Lewin, J., Davies, B. E. and Wolfenden, P. (1977). Interactions between channel change and historic mining sediments, in Gregory, K. J. (ed.), *River Channel Changes*, Wiley, Chichester, pp. 353–67.

Lewin, J. and Hughes, D. (1980). Welsh floodplain studies: application of a qualitative inundation model, *Journal of Hydrology*, 46, 35–49.

Lewis, G. W. and Lewin, J. (1983). Alluvial cutoffs in Wales and the Borderlands, *International Association of Sedimentology, Special Publication*, 6, 145–54.

Lockwood, J. G. (1979). Water balance of Britain 50,000 years BP to the present day, *Quaternary Research*, 12, 297–310.

Mitchell, D. J. (1975). *The relationship between basin characteristics and sediment rating curves*, unpublished paper, British Geomorphological Research group, Basin Sediment System Meeting, Birmingham.

Mitchell, D. J. (1979). Aspects of the hydrology and geomorphology of the Farlow Basin, Shropshire, unpublished M.Sc. thesis, University of Birmingham.

Mitchell, D. J. (1980). Valley sediments and changing sediment yields, in Gregory, K. J. (ed.), *Palaeohydrology of the temperate zone, subproject A: fluvial environments, Severn Basin*, unpublished preliminary report.

Mitchell, D. J. (in prep.). The use of contemporary discharge and suspended sediment concentrations to estimate long-term erosion and deposition in the Wye catchment.

Moore, P. D. (1982). Sub-surface formation of charcoal: an unlikely event in peat, *Quaternary Newsletter*, 38, 13–14.

Moore, P. D. and Chater, E. H. (1969). The changing vegetation of west-central Wales in the light of human history, *Journal of Ecology*, 57, 361–79.

Newson, M. D. (1975). Plynlimon floods of August 5/6 1973, *Institute of Hydrology Report*, 26.

Newson, M. D. (1979). The results of ten years experimental study on Plynlimon, Mid-Wales, and their importance for the water industry, *Journal of the Institute of Water Engineers*, 33, 321–33.

Newson, M. D. (1980a). The geomorphological effectiveness of floods—a contribution stimulated by two recent events in Mid-Wales, *Earth Surface Processes*, 5, 1–16.

Newson, M. D. (1980b). The erosion of drainage ditches and its effect on bedload yields in Mid-Wales: Reconnaissance case studies, *Earth Surface Processes*, 5, 275–90.

Newson, M. D. (1985). Forestry and water on the uplands of Britain—the background of hydrological research and options for harmonious land use, *Journal of Forestry*, 79, 113–20.

Newson, M. D. and Leeks, G. (1987). Transport processes at a catchment scale. A regional study of increasing sediment yield and its effects in Mid-Wales, UK: In Thorne, C., Bathurst, J. and Hey, R. D. (eds.), *Sediment transport in gravel bed rivers*, Wiley, Chichester.

Oldfield, F. (1977). Lakes and their drainage basins as units of sediment-based ecological study, *Progress in Physical Geography*, 1, 460–504.

Oldfield, F., Appleby, P. G. and Thompson, R. (1980). Palaeoecological studies of three lakes in the highlands of Papua New Guinea, *Journal of Ecology*, **68**, 457–77.

Oxley, N. C. (1974). Suspended sediment delivery rates and the solute concentration of stream discharge in two Welsh catchments, *Institute of British Geographers Special Publication*, 6, 141–54.

Painter, R. B., Blyth, K., Mosedale, J. C. and Kelly, M. (1974). The effect of afforestation on erosion processes and sediment yield, *International Association of Hydrological Sciences*, **113**, 62–7.

Potts, A. S. (1971). Fossil cryonival features in Central Wales, *Geografiska Annaler*, **53A**, 39–51.

Reed, A. H. (1979). Accelerated erosion of arable soils in the United Kingdom by rainfall and runoff, *Outlook on Agriculture*, **10**, 41–8.

Reed, A. H. (1983). The erosion risk of compaction, *Soil and Water*, **11**:29, 31–3.

Reynolds, B. (1986). A comparison of element outputs in solution, suspended sediments and bedload for a small upland catchment, *Earth Surface Processes and Landforms*, **11**, 217–22.

Richards, K. S. (1982). *Rivers: Form and Process in Alluvial Channels*, Methuen, London.

Roels, J. M. (1985). Estimation of soil loss at a regional scale based on plot measurements—some critical considerations, *Earth Surface Processes and Landforms*, **10**, 587–95.

Shotton, F. W. (1953). Glaciation and sediment supply: Pleistocene deposits of the area between Coventry, Rugby and Leamington and their bearing on the topographic development of the Midlands, *Philosophical Transactions of the Royal Society*, **B237**, 209–60.

Shotton, F. W. and Williams, R. E. G. (1971). Birmingham University radiocarbon dates V, *Radiocarbon*, **13**, 141–56.

Shotton, F. W. and Williams, R. E. G. (1973). Birmingham University radiocarbon dates VI, *Radiocarbon*, **15**, 1–12.

Slaymaker, H. O. (1972). Patterns of present sub-aerial erosion and landforms in Mid-Wales, *Transactions of the Institute of British Geographers*, **55**, 47–68.

Smith, D. D. (1958). Factors affecting rainfall erosion and their evaluation, *International Association for Scientific Hydrology*, **43**, 97–107.

Stoddart, D. R. (1969). World erosion and sedimentation, in Chorley, R. J. (ed.), *Water, Earth and Man*, Methuen, London, pp. 43–64.

Strakhov, N. M. (1967). *Principles of Lithogenesis*. Oliver and Boyd, Edinburgh.

Thomas, T. M. (1964). Sheet erosion induced by sheep in the Pumlumon (Plynlimon) area, Mid-Wales, *British Geomorphological Research Group, Occasional Publication*, 2, 11–14.

Thorne, C. R. and Lewin, J. (1979). Bank processes, bed material movement and planform development in a meandering river, in Rhodes, D. D. and Williams, G. P. (eds), *Adjustments of the Fluvial System*, Kendall-Hunt Ubuque, pp. 117–37.

Trimble, S. W. (1981). Changes in sediment storage in the Coon Creek Basin, Driftless area, Wisconsin, 1853–1975, *Science*, **214**, 181–3.

Walling, D. E. (1971). Instrumented catchments in S.E. Devon: Some relationships between drainage basin characteristics and catchment responses, unpublished Ph.D. thesis, University of Exeter.

Walling, D. E. and Webb, B. W. (1983). Patterns of sediment yield, in Gregory, K. J. (ed.), *Background to Palaeohydrology: A Perspective*, Wiley, Chichester, pp. 69–100.

Ward, R. C. (1981). River systems and river regimes, in Lewin, J. (ed.), *British Rivers*, George Allen & Unwin, London, pp. 1–33.

Webb, B. W. and Walling, D. E. (1984). Magnitude and frequency characteristics of suspended sediment transport in Devon rivers, in Burt, T. and Walling, D. E. (eds), *Catchment Experiments in Fluvial Geomorphology*, Geobooks, Norwich.

Williams, R. E. G. and Johnson, A. S. (1976). Birmingham University radiocarbon dates VII, *Radiocarbon*, **18**, 249–67.

Wiltshire, P. E. J. and Moore, P. D. (1983). Palaeovegetation and palaeohydrology in upland Britain, in Gregory, K. J. (ed.), *Background to Palaeohydrology: A Perspective*, Wiley, Chichester, pp. 433–51.

Wolman, M. G. and Miller, J. P. (1960). Magnitude and frequency of forces in geomorphic processes, *Journal of Geology*, **68**, 54–74.

Zaslavsky, M. N. (1979). *Erozia Pochv (Soil erosion)*. Mysl Publishing House, Moscow.

Palaeohydrology in Practice
Edited by K. J. Gregory, J. Lewin and J. B. Thornes
© 1987 John Wiley & Sons Ltd.

# 10

# Climatic History of the Severn Valley during the last 18,000 Years

## K. E. Barber

*Palaeoecology Laboratory, Department of Geography, University of Southampton*

## G. R. Coope

*Department of Geology, University of Birmingham*

### THE LATE DEVENSIAN

The period chosen here is intended to include the time from the retreat of the lowland ice from its maximum extent to the Wolverhampton line until the sudden climatic amelioration that ushered in the post-glacial period. The climate of the Severn valley must be interpreted in a broad geographical context because critical sites lie outside the strict limits of the watershed and the gross climatic changes mentioned here are no doubt applicable to the whole area of the Severn Basin and its immediate surroundings.

From the palaeohydrological point of view these climatic changes have significant implications. During the period when ice sheets occupied the northern parts of the Severn valley the supply of water to the main valley would have had an important component of meltwater unrelated to the catchment area of the present-day valley system. At this stage the hydrology of the main Severn valley would thus have been quite different from that in the Avon valley whose upper reaches were not invaded by the ice sheets. After all the glacier ice had gone from the total valley system, the climatic changes outlined below should have had drastic effects on the palaeo-hydrology. For example during episodes of climatic continentality when winters were well below 0°C, all the annual runoff would have been concentrated during the spring thaw and the short summer. The sedimentological and geomorphological consequences of such a climatic regime would differ greatly from anything that we can measure today in Britain. When the annual temperatures were, on average, below 0°C the further complication of continuous and discontinuous permafrost would have altered the runoff and recharge properties of the rocks within the basin. Cryoturbation and solifluction of valley sides, coupled with permafrost and a fairly vigorous active layer, must have had drastic effects upon the supply of sediment to the main rivers and their tributaries. During temperate episodes when climatic conditions were similar to those of the present day, the palaeohydrology of the valley system could still have been very different from present-day hydrological conditions. In particular the temperate episode between 13,000 BP and 12,000 BP was accompanied by a landscape almost entirely devoid of trees with a thin and probably patchy vegetation cover in spite of seasonal temperatures not much different from those of the present day. It is likely that this vegetational impoverishment was due to nutritional problems associated with the raw soils (van Geel, de Lange and Wiegers, 1984; Pennington 1986). The runoff regime at this stage would not have a close analogue in modern Britain.

These comments are intended to highlight some of the palaeohydrological problems that arise from inferred climatic changes between 18,000 BP and 9500 BP. It now remains to put figures on the inferred climatic regime of this period. The palaeoclimatic interpretation for this time is derived from assemblages of Coleoptera because: (a) they are abundant and diverse; (b) they can frequently be identified to the species level; (c) they readily change their geographical ranges in such a manner as to leave little doubt that they are responding to changes in the thermal environment; (d) temporal sequences of faunas are now available for much of this period; and (e) a computer program is now available that relates species range to the geographical distribution of various climatic parameters (Atkinson *et al.*, 1986). Coleoptera may be preferred to palynological data for the earlier part of the period because of the evident lag in the response of the flora to changes of climate compared with the response time of the Coleoptera. As yet they provide information on changes in the thermal regime but no direct evidence of precipitation levels between 18,000 and 13,000 BP.

## 18,000–13,000 BP

Evidence from this period is restricted to Glanllynnau (Coope and Brophy, 1972) and Glen Ballyre (Coope and Joachim, 1980). Average July temperatures were close to 10°C ± 2°C. Average February temperatures were probably close to −20°C ± 10°C. The large error in the estimation of winter temperatures is probably due to uncertainties introduced by the fact that Coleoptera hibernate and are therefore insulated to some extent from the controls exerted by winter cold. With a thermal regime of this type, average annual temperatures may well have been about −5°C and permafrost would be expected on low ground at least discontinuously. Insolation-favoured sites may have thawed out completely each year, however. The river would have flowed freely for about four months of the year at which time the accumulated snows of winter would have almost entirely melted (the fact of glacial retreat shows that at this time the net ice budget was in chronic deficit) and for the Severn, and probably the Wye also, there must have been a significant contribution from the melting glaciers during the summer only. The water in the Avon at this time must have been derived largely from spring-thaw runoff. The hydrology of the Severn–Wye system must, at this time, have been rather different from that of the Avon. It is not certain when glacier ice eventually went from the Severn drainage but it must surely have done so by 14,000 BP if radiocarbon dates from Scotland suggesting that ice retreat was extensive there by that date can be taken at face value. By this later date, therefore, the gross hydrology of all the rivers of the Severn Basin was uncomplicated by the presence of glacier ice in some of their head waters.

The pronounced continentality of the thermal regime outlined above may imply a dimunition in the frequency of depressions with their associated frontal systems crossing central Britain, with the consequent reduction in precipitation levels. Lower precipitation is not at all inconsistent with the obvious vigour of the rivers' flow indicated by the coarse gravels transported in the broad low-gradient lower reaches of both Severn and Avon at this time. It does, however, imply catastrophic spring floods probably well above the range of historic extremes.

### 13,000–12,000 BP

The evidence from this period is chiefly from four sites, Glanllynnau (Coope and Brophy, 1972), Glen Ballyre (Coope, 1971; Joachim, 1978), Church Stretton (Osborn, 1972) and St Bees (Coope and Joachim, 1980). It is clear from the Coleopteran record that there was a sudden amelioration in the climate close to 13,000 BP. Average July temperatures were 17°C ± 2°C. Average February temperatures were near to 0°C ± 5°C. This marked change in the thermal conditions was achieved very suddenly, in a matter of a century or so, and coincides in both timing and intensity with the northward passage of the polar front in the Atlantic Ocean off the west coast of Britain (Ruddiman and McKintyre, 1973).

The great diminution in climatic continentality may carry the implication of increase in the number of depressions crossing central Britain with their associated increased levels of precipitation. There would have been decreased seasonality of runoff and, in the absence of active solifluction, a diminution in the supply of sediment to the river. An episode of riverine incision probably characterized this period, especially where the rivers crossed unconsolidated sediments laid down during the previous climatic regime. This episode of incision may explain why organic sediments of this age are almost unknown in riverine contexts in central Britain.

### 12,000–11,000 BP

The evidence from this period is chiefly from the four sites previously mentioned but numerous other localities yield insect faunas of this general age. At or shortly before 12,000 BP there was a moderate deterioration in the thermal environment. Average July temperatures were close to 15°C ± 2°C and were maintained at this level for the following thousand years (NB this time interval is largely equivalent to the Allerød period of other authors). Average February temperatures were about −5°C ± 4°C. It was during this period that birch trees spread widely in England, though there is evidence of the presence of tree birch much earlier than this particularly in north-west England (Pennington, 1977). Organic deposits of this age occur in cutoff channel deposits in the valley bottom at Wilden near Stourport (Shotton and Coope, 1983) suggesting that at this time the rivers adopted a meandering mode with lateral cut and fill.

### 11,000–10,000 BP

Numerous sites are available that have yielded insect faunas dating from this period. They indicate that at or close to 11,000 BP there was a further deterioration in the thermal environment and average July temperatures fell to near 10°C ± 2°C. Winter temperatures were depressed still further so that they probably had average February temperatures near −20°C ± 10°C. In other words the average annual temperature was now below 0°C and may have been well below that figure. This thermal regime is thus very similar to that of the earlier fully glacial period prior to 13,000 BP. Ice-wedge casts are known from this episode from as far south as the Isle of Man though we are unaware of any from within the Severn valley system. Extensive gravel deposition even in mature reaches of central British rivers also occurred at this time. Cryoturbation and extensive solifluction probably contributed significantly to the seasonal loading of the rivers.

It is interesting to note that northern insect species, absent from Britain today, occurred at this time as far south as Bodmin Moor (Coope, 1977) and at Folkstone (Coope in Kerney, Preece and Turner, 1980).

### 10,000–9500 BP

A very sudden climatic amelioration must have taken place at about 10,000 BP when all the exclusively high northern species of Coleoptera became extinct in lowland Britain and were replaced by relatively southern species in central England (Ashworth, 1973; Osborne, 1974), south-west Scotland (Bishop and Coope, 1977) and even as far north as Rannoch Moor (Coope,

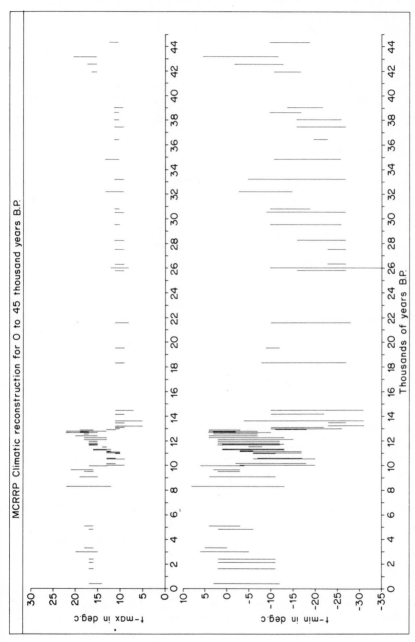

FIGURE 10.1 Estimates for the mean temperatures of the warmest month (*t*-max) and the coldest month (*t*-min) for central Britain for the last 45,000 years, based on Mutual Climatic Range analysis of coleopteran assemblages (Coope, 1987)

1977). There is ample evidence of temperate conditions at this early post-glacial phase as warm or even warmer than those of the present day. There is no evidence of any greater climatic continentality at this time.

## THE HOLOCENE

A number of compilations of Holocene climatic change have been published recently (Lamb, 1977, 1982; Simmons and Tooley, 1981; Harding, 1982; Wigley, Ingram and Farmer 1981; Flohn and Fantechi, 1984; Morner and Karlen, 1984; Tooley and Sheail, 1985; and Tyson, 1986, amongst others). There is still a need for further studies, however, particularly in the early Holocene period, and for more studies which, using transfer functions derived from macrofossil analysis of peats or microfossil pollen data (Prentice, 1983), cover a span of time with a reliable reconstructed climatic sequence rather than isolated data points. The Mutual Climatic Range computer program used earlier for the quantification of the climatic inferences from Coleoptera (Atkinson *et al.*, 1987) should be equally applicable to plant fossils, especially those that can be identified to the species level.

### 10,000–7500 BP

As noted above there is good Coleopteran evidence of a very sudden rise in temperatures at the beginning of the Holocene, with temperatures reaching today's summer levels or above by 9500 BP, as a result of the retreat of polar waters to the north and the concomitant movement of the polar front in the north Atlantic (Ruddiman and McIntyre, 1973). Large-scale events such as these would have had an effect which was evident over the whole of the British Isles and much of north-west Europe, and to a large extent we may apply the results of studies of climatic change from such an area to the Severn Basin. To what degree the basin reacted in any kind of individual manner is difficult to discern, especially in the absence of much detailed data from the Severn Basin itself, but the peat stratigraphic evidence (Barber and Twigger, Chapter 11 this volume) indicates a degree of parallelism in the later Holocene period. In the early Holocene global climatic parallelism was complicated by the continued existence of the North American ice-cap after the disappearance of the Fennoscandian ice. But within the north-west European area the record can be reconstructed on a continental and regional basis from the following evidence.

The water levels of closed-basin lakes have been shown to fluctuate in a manner suggestive of climatic control (Digerfeldt, 1972, 1986; Gaillard, 1984; Harrison and Metcalfe, 1985; Street-Perrott and Harrison, 1985), though the problems associated with this kind of proxy data must be recognized and have recently been critically reviewed by Dearing and Foster (1986). In a European context the work undertaken at Lund on Swedish lakes provides good evidence for a period of water-level lowering in Preboreal–early Boreal times, about 9500 BP. This is consistently shown in seven sites, followed by a second phase of lowering in late Boreal times, *c.* 8000 BP at three sites.

In northern Germany and in the Netherlands, Behre (1978) and Van Geel and Kolstrup (1978) have identified climatic oscillations from pollen-analytical evidence in the Preboreal; first a temperature rise from *c.* 10,150 to 9850 BP, the Friesland phase, followed by a drier and perhaps cooler period, the Rammelbeek phase from *c.* 9850 to 9700 BP, after which westerly winds penetrated into the area and temperatures and rainfall rose rapidly to full postglacial levels. Van Geel, Bohncke and Dee (1981), in a very detailed analysis of a 3 m-deep section of Holocene sediments at De Borchert near Denekamp, the Netherlands, confirmed this oscillation and discuss the reasons for it, as well as identifying a number of relatively wet and relatively dry phases (10 in all) between 9700 and 6200 BP, the length of each phase varying between 100 and 900 years. While such changes remain to be confirmed by further work—which must, to be comparable, be in similar detail (microfossil samples every 0.8 cm)—it seems reasonable to assume that the recovery from lateglacial conditions may not have been in the form of smoothly rising temperature and precipitation curves. Van Geel, Bohncke and Dee (1981) make the comparison with Barber's (1981) curve of surface wetness on an ombrotrophic bog over the last 2000 years, and it is clear that *detailed* analysis may resolve palaeoclimatic variations of less than 100 years—there is no reason why the climate of the period around 8000 years ago should be less variable than that of around 800 or 80 years ago.

Numerical estimates of temperature and rainfall for the Holocene are rare and their reliability is difficult to judge. Lamb (1977, p. 404) estimates Preboreal summer temperatures to be slightly higher than today's at 16.3°C (against 15.8 for AD 1900–50 over England and Wales), with winter mean monthly temperatures of 3.2 (against 4.2)—but see Atkinson *et al.* (1986) for details of a possible range of values. Rainfall figures are even more problematical but Lamb (1977) is more conservative than others in calculating a range of 92–95 per cent of the AD 1916–50 England and Wales average.

The good local record from Crose Mere (Beales, 1980, and see Barber and Twigger, Chapter 11 this volume), has about 1.5 m of sediment from this period (10,000–7500 BP) but the coarseness of the sampling interval compared to that of the Dutch work (Van Geel, Bohncke and Dee, 1981) precludes any derivations of climatic parameters. However, there is no reason to doubt that the above figures from Lamb (1977) would apply to the Severn Basin area, and that this level of temperatures continued in the Boreal period, *c.* 9100–7500 BP. Rainfall figures, or at least groundwater conditions dependent upon the precipitation/evapotranspiration index, are more difficult to assess. Van Geel, Bohncke and Dee (1981) give an interpretation which includes three drier episodes culminating at 8900, 8200 and 7950 BP, with a further dry spell in early Atlantic times *c.* 7500 BP, the Boreal being drier overall than the Atlantic period. This of course fits all the orthodox views of the relative dampness of the Boreal and Atlantic periods, following from Blytt and Sernander's subdivision of the postglacial in the late nineteenth century, and while such generalizations should not be taken too far (Magny, 1982), a number of lines of evidence do back up this general view (Simmons and Tooley, 1981). There is most certainly a need for more studies along the lines of Gaillard (1984) and Van Geel, Bohncke and Dee (1981) described above, which can give proxy data for climatic change, and a further extension of isotopic and chemical analyses (Pennington and Lishman, 1971), but there is nevertheless 'substantial evidence for a period of dryness' covering a wide area of the British Isles in late Boreal times (Smith, 1984).

### 7500–5000 BP

This time period coincides with the Atlantic period, generally acknowledged to be the time of greatest postglacial warmth. Lamb (1977) gives summer mean monthly temperatures (°C) of 17.8 (cf. 15.8 for AD 1900–50) and 5.2 (against 4.2) for winter, 10.7 for the annual average (against 9.4). Rainfall was estimated at between 110–115 per cent of the AD 1916–50 average, but due to the forest cover runoff to rivers may have been as much as 50 per cent below present-day values (Lockwood, 1979). This 'warm and wet' view of the Atlantic is supported by many different types of proxy data including lake-levels (Gaillard, 1974), latitudinal and altitudinal tree-lines (Godwin, 1975), more northerly faunal limits of species such as *Emys orbicularis* (Stuart, 1979) and the accumulation of ombrotrophic peat—though the date at which this began is also dependent upon a developmental time-lag in that hydroseral change had to progress to the point when *Sphagnum* species could invade the habitat in question.

While there is evidence for some fluctuations in the climate of Atlantic times the overall picture is one of stability. This is well shown in the oxygen isotope curve from Camp Century (Dansgaard *et al.* 1971, Figure 6) where for about 2000 years the values fluctuate very little. There is less direct evidence of rainfall values but the almost uniform dark-brown peat stratigraphy of ombrotropic mires at this time (Godwin, 1981) is evidence that climatic changes were probably slight. Van Geel (1978) records some changes in the largely non-sphagnaceous Atlantic peats (6250–4850 BP) of Engbertsdijksveen I, but notes that these may be only of local significance and that more correlative work is needed. Reconstructed atmospheric circulation charts (Lamb, 1977) show a much slacker meridional circulation for 6500 BP than for the present Subatlantic period, with depression tracks steered far to the north of Scotland by an extended Azores anticyclone.

Pollen diagrams covering the Atlantic period give a similar impression of stability (Godwin, 1975; Pennington, 1974) as the so-called climax vegetation stabilizes, and the pollen percentage values of the major trees fluctuate little from level to level. Those few diagrams (e.g. Van Geel, Bohncke and Dee, 1981) which cover this period in detail, however—with close counts spaced at 1–2 cm—do show fluctuations of the individual taxon curves but within an overall level of tree pollen of *c*. 95 per cent of that pollen. Clearly there was a dense forest cover over most of the European lowlands, and the Severn Basin was no exception (Barber and Twigger, Chapter 11 this volume).

### 5000–2500 BP

This time period is equivalent to the Subboreal zone of Blytt and Sernander and was traditionally viewed as one of lower temperatures but also lower rainfall than the preceding Atlantic period. Lamb (1977) estimates falls of 1°C in both mean summer and mean annual temperatures to 16.8 and 9.7 respectively, with a fall of 1.5 in winter means to 3.7 (0.5 of a degree cooler than AD 1900–50), and with rainfall decreasing about 10 per cent to levels equivalent to today's—resulting in a slightly more 'continental' regime than at present. Recent research has tended to show greater variability within Subboreal climates than was formerly assumed (see Smith's 1981 review for example) but the evidence for it being warmer than the present Subatlantic period has been maintained and strengthened.

Frenzel (1966), in a thorough review of the botanical and other evidence then extant for changes at the Atlantic/Subboreal transition (which in fact takes in evidence for the whole postglacial), concludes that the Atlantic ended with a cold spell between about 5400–5000 BP, followed by increased continentality interrupted by minor oscillations. From about 3500 BP, but especially since 2900 BP, temperature fell, culminating in the Subatlantic period about 2500 BP. This analysis has not been overturned since and research in the last 20 years has rather added strength to Frenzel's conclusions, which were elaborated in Frenzel (1977). One apparent inconsistency in the evidence—that of a damper climate during the Subboreal of central Europe—is explained by Magny (1982) as being due to the westward penetration of Polar Continental air masses to the north of the Polar Front, which then lay across central France.

The evidence for climatic change becomes richer and more diverse from 5000 BP on. One of the oldest types of data, used by Blytt and Sernander a century ago, is that of the stratigraphy of ombrotrophic bogs. The well-known Recurrence Surfaces of Granlund (1932) from Swedish raised bogs (wherein a change in stratigraphy from darker humified peat to lighter brown unhumified peat, often with pool layers, can only result from changes in the precipitation/evaporation ratio) were dated on archaeological grounds as:

RY I 1200 AD; RY II 400 AD; RY III 600 BC (2550 BP); RY IV 1200 BC (3150 BP); RY V 2300 BC (4250 BP)

This topic has been extensively researched and reviewed (Godwin, 1954, 1975; Aaby; Tinsley, 1981; Turner, 1981; Barber, 1981, 1982, 1985 for example), and it is now fairly clear that different bogs reacted in a gross enough way to produce the major changes which we term Recurrence Surfaces at different times, but with some clustering around the original Granlund dates. More or less obvious shifts from humified to unhumified peat have been recorded from the Subboreal at about 4500 BP and 4105 BP in Northern Ireland (Smith, 1981); 4700 BP, 4500 BP, 4100 BP, 3600 BP and 3200 BP in Denmark (Aaby, 1976); and at 3900 BP, 3500 BP and 2900 BP in Holland (Van Geel, 1978). Clearly these and other bog stratigraphic data (see Barber, 1982, for references) are evidence of recurring cooler and/ or wetter phases of climate in the Subboreal. Evidence for drier phases comes again from lake-level studies such as that of Gaillard (1984) with a major lowering at around 3500 bp. Coleopteran evidence is also useful during this period. Osborne (1982) reporting on three beetle faunas from Bronze Age and earlier contexts points to 'optimum temperatures for this particular group of beetles having been reached between 4000 and 3000 years ago and not having been attained since.'

There is no evidence directly concerned with climate at this period within the Severn Basin, lake sediment changes such as that at Crose Mere (Beales, 1980; Barber and Twigger, Chapter 11 this volume) being more definitely influenced by human land clearance activity.

**2500–0 BP**

The deterioration in climate at the opening of the Subatlantic shows itself most dramatically in the formation of the major Recurrence Surface known as the *Grenzhorizont*, after Weber, a German bog scientist writing in 1900 (Barber, 1982). This major humification change varies in date from bog to bog because of local climatological factors, altitude and latitude, antecedent conditions and the local hydrology and vegetation of each bog (Barber, 1985).

Lamb (1977) estimates that at the onset of the Subatlantic, which he dates to 2850–2400 BP, temperatures fell to 15.1 in summer (compared with 17.8 for the Atlantic, 16.8 for the Subboreal, 15.8 for AD 1900–50), but that winters became milder (4.7 against 3.7 for the preceding Subboreal) and rainfall stayed at about 100–105 per cent of today's level. The lower summer temperatures would of course have affected evapotranspiration to quite a marked degree, having a direct effect on ombrotrophic mire water levels, which in turn directly effect the composition of the mire vegetation and the humification and accumulation rate of the peat.

The evidence from proxy data sources and later on from documentary sources is comprehensively summarized in Lamb (1977, 1982 and the texts mentioned earlier (p. 208) and will not be reviewed again here. There appears to have been something of a less stable climate during the Subatlantic than earlier in the Holocene, though of course this may also reflect more available data. Aaby's (1976) demonstration of a 260-year periodicity of wetter surface conditions on Danish bogs—shifts *c*. 600 BC, 300 BC, 0 BC/AD, AD 250, AD 500, AD 1000, AD 1250 and AD 1500—is also interesting in that not all shifts were shown in all of the five bogs he studied. There is room therefore for variation in the registration of climatic changes by raised bogs, a theme explored by Conway in a seminal paper published in 1948, which was itself echoing much earlier ideas of Von Post in the early 1900s. More recently Barber (1981), in a monograph on a single site in Cumbria, and Turner (1981), in a cogent review of Bronze and Iron Age climates, have shown how impressively and sensitively raised bog stratigraphy reflects climatic change. Indeed for the time period under review here it forms a unique archive by virtue of its accessibility and temporal continuity as well as the spatial spread of suitable sites from Ireland across to eastern and central Europe.

In summary Turner (1981) sees clear evidence for a climatic deterioration from about 1200 BC down to 600 BC, with then a warmer climate from 400 BC to AD 450. One of the latter pieces of evidence called upon by Turner (p. 261) is that of the pine stump layer at Whixall Moss, Shropshire, within the Severn Basin, which Godwin and Willis (1960) dated to 357 BC (Q-383: 2307 ± 110 BP). As part of a study of the main humification change (MHC) in bogs on a transect from Ireland to Poland, three further radiocarbon dates have been obtained from Whixall Moss (Barber and Haslam, unpublished data). The zone of tree remains extends from about 30–60 cm below the present cut surface of the peat and the tree stumps are encased in humified peat (H6–7 von Post scale) with a maximum of 25 per cent identifiable *Sphagnum* remains, mainly *S. imbricatum* and Acutifolia with increasing proportions in Sphagna Cuspidata. Peat from 40–44 cm of this tree-remains stratum has been dated to 2180 ± 50 BP (SRR 3074) or *c*. 230 BC. The marked stratigraphic break is at 18 cm, the peat above this level being almost 100 per cent *Sphagnum imbricatum*. The date for peat below this level (20–26 cm) is 1930 ± 50 BP (SRR 3036) or *c*. 20 BC; for a sample from 11–18 cm the date is 1750 ± 60 BP (SRR 3035) or *c*. AD 200, very similar to the date for Fenn's Moss nearby (Barber and Twigger, Chapter 11 this volume). Further analyses are underway on this and other sites by C. J. Haslam and it would be premature to speculate on the outcome of this investigation but clearly there is evidence for phaes of moister and drier conditions in these data. Barber (1982) summarized the peat stratigraphic evidence of climatic change over the Subboreal–Subatlantic time period as follows:

1. There was probably a climatic deterioration $c$. 3950–3850 BP (2000–1900 BC) even if it was only of a minor nature.
2. A more pronounced decline followed $c$. 3450–3350 BP (1500–1400 BC).
3. There was a catastrophic decline to a cooler and/or wetter climate around 2850–2550 BP (900–600 BC); taking into account the 'sensitivity' of individual bogs the earlier date may be more generally applicable.
4. Some evidence for further decline around 2050 BP (100 BC).

Further confirmation of the unstable climate of the Subboreal–Subatlantic transition comes from detailed palaeoecological and isotopic analyses on Dutch ombrotrophic peat (Van Geel, 1978; Brenninkmeijer, Van Geel and Mook, 1982; Dupont, 1985; Dupont and Brenninkmeijer, 1984). These studies again show a fluctuating climate and local differences in the response of the bog vegetation (Brenninkmeijer, 1983) with a number of wet phase-shifts apparent between 3000–2000 BP and particularly wet conditions around 2600 BP (650 BC).

During the historic period we are increasingly able to correlate proxy data and documentary evidence for climatic change (Lamb, 1982). Lamb's well-known curves of temperature from AD 900 to the present (Lamb, 1984) show a 'Little Optimum' of the period AD 1150–1300, the first decline into the Little Ice Age between AD 1300–1500, and amelioration AD 1500–50, followed by the culmination of the Little Ice Age around 1700 but continuing down to the 1850s. The data derived from a mixture of proxy sources and fragmentary written records are eventually replaced by instrumental records such as Manley's monthly means from 1659 to 1973 (Manley, 1974) and the general rainfall figures over England and Wales since 1727 (Nicholas and Glasspoole, 1931; Lamb, 1977, pp. 621–5). The combining of these and other records into a consolidated 'surface wetness' record of direct palaeohydrological usefulness has not yet been attempted—though it should be reasonably easy to compute.

We may turn again, however, to the record of bog stratigraphy, in that this record integrates physically the effects of precipitation, snowmelt, temperatures and evapotranspiration into one parameter—the bog water table. In turn this controls the species present and the rate of decomposition and therefore by macrofossil and stratigraphic analysis one can reconstruct in a semi-quantitative but sequential manner the past surface wetness of the bog. Clearly this must be related to groundwater tables and thus to river flow. Figure 10.2, from Barber (1981) is such a reconstructed sequence for the last 2000 years and in view of its close relation to independently known

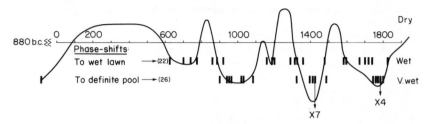

FIGURE 10.2 Surface wetness curve for Botton Fell Moss, Cumbria for the last 2000 years (from Barber, 1981)

climatic trends from about AD 900 it can be claimed to have wider relevance than the north Cumbrian area. Making the connection between this kind of proxy data and, for example, the 30-year running averages of runoff of the Rhine at Andernach (Flohn, 1984, pp. 110–16), would be a transfer function worth having.

## REFERENCES

Aaby, B. (1976). Cyclic climatic variations in climate over the last 5500 years reflected in raised bogs, *Nature*, **263**, 281–4.

Ashworth, A. C. (1973). The climatic significance of a Late Quaternary insect fauna from Rodbaston Hall, Staffordshire, England, *Entomol. Scand.*, **4**, 191–205.

Atkinson, T. C., Briffa, K. R., Coope, G. R., Joachim, M. J. and Perry, D. W. (1986). Climatic calibration of coleopteran data, in Berglund, B. E. (ed.), *Handbook of Holocene Palaeoecology and Palaeohydrology*, Wiley, Chichester, pp. 851–8.

Atkinson, T. C., Briffa, K. R. and Coope, G. R. (1987). Seasonal temperatures in Britain during the last 22,000 years reconstructed using beetle remains, *Nature*, **325**, 587–592.

Barber, K. E. (1981). *Peat Stratigraphy and Climatic Change: a Palaeoecological Test of the Theory of Cyclic Peat Bog Regeneration*, Balkema, Rotterdam.

Barber, K. E. (1982). Peat-bog stratigraphy as a proxy climate record, in Harding, A. F. (ed.), *Climatic Change in Later Prehistory*, Edinburgh University Press, pp. 103–13.

Barber, K. E. (1985). Peat stratigraphy and climatic change: some speculations, in Tooley, M. J. and Sheail, G. M. (eds), *The Climatic Scene: essays in honour of Gordon Manley*, Allen & Unwin, London, pp. 175–85.

Beales, P. W. (1980). The Late Devensian and Flandrian vegetational history of Crose Mere, Shropshire, *New Phytologist*, **85**, 133–61.

Behre, K.-E. (1978). The climatic oscillations during preboreal time in Europe, *Petermanns Geographische Mitteilungen*, **2**, 97–103.

Bishop, W. W. and Coope, G. R. (1977). Stratigraphic and faunal evidence for late glacial and early Flandrian environments in South West Scotland, in Gray, J. M. and Lowe, J. J. (eds), *Studies in the Scottish Late Glacial Environments*, Pergamon, Oxford, pp. 61–88.

Brenninkmeijer, C. A. M. (1983). Deuterium, Oxygen-18 and Carbon-13 in tree rings and peat deposits in relation to climate, unpublished Ph.D. thesis, University of Groningen.

Brenninkmeijer, C. A. M., Van Geel, B. and Mook, W. G. (1982). Variations in the D/H and $^{18}O/^{16}O$ ratios in cellulose extracted from a peat bog core, *Earth and Planetary Science Letters*, **61**, 283–90.

Conway, V. M. (1948). Von Post's work on climatic rhythms, *New Phytologist*, **47**, 220–37.

Coope, G. R. (1971). The fossil Coleoptera from Glen Ballyre and their bearing upon the interpretation of Late Glacial environments, in Thomas, G. S. P. (ed.), *Field Guide to Isle of Man*, Quaternary Research Association, Liverpool.

Coope, G. R. (1977). Fossil coleopteran assemblages as sensitive indicators of climatic changes during the Devensian (Last) cold stage, *Philosophical Transactions of the Royal Society of London*, **B280**, 313–40.

Coope, G. R. (1987). Fossil beetle assemblages as evidence for sudden and intense climatic changes in the British Isles during the last 45,000 years, in *Abrupt Climatic Change—A NATO/NSF Workshop*, Reidel, *Biviers (Grenoble)*.

Coope, G. R. and Brophy, J. A. (1972). Late glacial environmental changes indicated by a coleopteran succession from North Wales, *Boreas*, **1**, 97–142.

Coope, G. R. and Joachim, M. T. (1980). Lateglacial environmental changes interpreted from fossil Coleoptera from St. Bees, Cumbria, NW England, in Lowe, J. J., Gray, J. M. and Robinson, J. E. (eds), *Studies in the Lateglacial of North-West Europe*, Pergamon, Oxford, pp. 55–68.

Dansgaard, W., Johnson, S. J., Clausen, H. B. and Langway, C. C. (1971). Climatic record revealed by the Camp Century Ice Core, in Turekian, K. K. (ed.), *The Late Cenozoic Glacial Ages*, Yale University Press, New Haven and London, pp. 37–56.

Dearing, J. A. and Foster, I. D. L. (1986). Lake sediments and palaeohydrological studies, in Berglund, B. E. (ed.), *Handbook of Palaeoecology and Palaeohydrology*, Wiley, Chichester, pp. 67–89.

Digerfeldt, G. (1972). The Post-Glacial development of Lake Trummen, *Folia Limnologica Scandinavica*, **16**, 104 pp.

Digerfeldt, G. (1986). Studies on past lake-level fluctuations, In Berglund, B. E. (ed.), *Handbook of Holocene Palaeoecology and Palaeohydrology*, Wiley, Chichester, pp. 127–43.

Dupont, L. M. (1985). Temperature and rainfall variation in a raised bog ecosystem— a palaeoecological and isotope—geological study, unpublished Ph.D. thesis, University of Amsterdam.

Dupont, L. M. and Brenninkmeijer, C. A. M. (1984). Palaeobotanic and isotopic analysis of late Subboreal and early Subatlantic peat from Engbertsdijksveen VII, The Netherlands, *Review of Palaeobotany and Palynology*, **41**, 241–71.

Flohn, H. (1984). Climate variability and its time changes in European countries, based on instrumental observations, in Flohn, H. and Fantechi, R. (eds), *The Climate of Europe: past, present and future*, Reidel, Dordrecht, pp. 102–17.

Flohn, H. and Fantechi, R. (eds) (1984). *The Climate of Europe: past, present and future*, Reidel, Dordrecht.

Frenzel, B. (1966). Climatic change in the Atlantic/sub-Boreal transition on the northern Hemisphere: botanical evidence, in *World Climate from 8000–0 BC*, London Royal Meteorological Society.

Frenzel, B. (1977). *Dendrochronologie und postglaziale Klimaschwankungen in Europa*, Steiner, Wiesbaden.

Gaillard, M.-J. (1984). A palaeohydrological study of Krageholmssjon (Scania, South Sweden)—regional vegetation history and water-level changes, *Lundqua Report* No 25.

Van Geel, B., Bohncke, S. J. P. and Dee, H. (1981). A palaeoecological study of an Upper Late Glacial and Holocene sequence from 'De Borchert', The Netherlands, *Review of Palaeobotany and Palynology*, **31**, 367–448.

Van Geel, B. and Kolstrup, E. (1978). Tentative explanation of the late glacial and early Holocene climatic changes in N-W Europe, *Geologie en Mijnbouw*, **57**(1), 87–9.

Van Geel, B., de Lange, L. and Wiegers, J. (1984). Reconstruction and interpretation of the local vegetation succession of a late glacial deposit from Usselo (The Netherlands), based on the analysis of micro and macro fossils, *Acta Botanica Neerlanica*, **3**(4), 535–46.

Godwin, H. (1954). Recurrence Surfaces, *Danm. Geol. Unders.* IIR, **80**, 22–30.

Godwin, H. (1975). *History of the British Flora*, 2nd edn, Cambridge University Press.

Godwin, H. (1981). *The Archives of the Peat Bogs*. Cambridge University Press.

Godwin, H. and Willis, E. H. (1960). Cambridge University Natural Radiocarbon Measurements II, *Am. J. Sci. Radiocarbon Suppl.*, **2**, 62–72.

Granlund, E. (1932). De svenska hogmossarnas geologi, *Sver. geol. Unders.*, **C26**, 1–193.

Harding, A. F. (ed.) (1982). *Climatic Change in Later Prehistory*, Edinburgh University Press.

Harrison, S. P. and Metcalfe, S. E. (1985). Spatial variations in lake levels since the last glacial maximum in the Americas north of the Equator, *Zeitschrift fur Gletscherkunde und Glazialgeologie*, **21**, 1–15.

Joachim, M. J. (1978). Late glacial coleopteran assemblages from the west coast of the Isle of Man, Ph.D. thesis, University of Birmingham.

Kerney, M. P., Preece, R. C. and Turner, C. (1980). Molluscan and plant biostratigraphy of some Late Devensian and Flandrian deposits in Kent, *Philosophical Transactions of the Royal Society of London*, **B290**, 1–43.

Lamb, H. H. (1977). *Climate: Present, Past and Future*, vol. 2, *Climatic History and the Future*, Methuen, London.

Lamb, H. H. (1982). *Climate, History and the Modern World*, Methuen, London.

Lockwood, J. G. (1979). Water balance of Britain, 50,000 yr BP to the present day, *Quaternary Research*, **12**, 297–310.

Magny, M. (1982). Atlantic and Sub-boreal: dampness and dryness? in A. F. Harding (ed.), *Climatic Change in Later Prehistory*, Edinburgh University Press, pp. 33–43.

Manley, G. (1974). Central England temperatures: monthly means 1659–1973, *Quarterly Journal Royal Meteorological Society*, **100**, 389–405.

Morner, N.-A. and Karlen, W. (eds) (1984). *Climatic Changes on a Yearly to Millenial Basis*, Reidel, Dordrecht.

Nicholas, F. J. and Glasspoole, H. (1931). General monthly rainfall over England and Wales 1727–1931, *British Rainfall*, **1931**, 299–306.

Osborne, P. J. (1972). Insect faunas of Lat Devensian and Flandrian age from Church Stretton, Shropshire, *Philosophical Transactions of the Royal Society of London*, **B263**, 327–67.

Osborne, P. J. (1974). An insect assemblage of early Flandrian age from Lea Marston, Warwickshire and its bearing on contemporary climate and ecology, *Quaternary Research*, **4**, 471–86.

Osborne, P. J. (1982). Some British later prehistoric insect faunas and their climatic implications, in Harding, A. F. (ed.), *Climatic Change in Later Prehistory*, Edinburgh University Press, pp. 68–74.

Pennington, W. (1974). *History of British Vegetation*, 2nd edn, English University Press, London.

Pennington, W. (1977). The Late Devensian flora and vegetation of Britain, *Philosophical Transactions of the Royal Society of London*, **B280**, 247–71.

Pennington, W. (1986). Lags in adjustment of vegetation to climate caused by the pace of soil development: evidence from Britain, *Vegetatio*, **67**, 105–18.

Pennington, W. and Lishman, J. P. (1971). Iodine in lake sediments in Northern England and Scotland, *Biological Review*, **46**, 279–313.

Prentice, I. C. (1983). Postglacial climatic change: vegetation dynamics and the pollen record, *Progress in Physical Geography*, **7**, 273–86.

Ruddiman, W. F. and McIntyre, A. (1973). Time-transgressive deglacial retreat of polar waters from the North Atlantic, *Quaternary Research*, **3**, 117–30.

Shotton, F. W. and Coope, G. R. (1983). Exposures in the Power House Terrace of the river Stour at Wilden, Worcestershire, England, *Proc. Geol. Ass.*, **94**, 34–44.

Simmons, I. G. and Tooley, M. J. (eds) (1981). *The Environment in British Prehistory*, Duckworth, London.

Smith, A. G. (1981). The Neolithic, in Simmons, I. G. and Tooley, M. J. (eds), *The Environment in British Prehistory*, Duckworth, London, pp. 125–209.

Smith, A. G. (1984). Newferry and the Boreal-Atlantic transition, *New Phytologist*, **98**, 35–55.

Street-Perrott, F. A. and Harrison, S. P. (1985). Lake levels and climate reconstruction, in A. D. Hecht (ed.), *Quantitative Palaeoclimate Analysis and Modelling*, Wiley, New York.

Stuart, A. J. (1979). Pleistocene occurrences of the European pond tortoise (*Emys orbicularis L.*) in Britain, *Boreas*, **8**, 359–71.

Tinsley, H. M. (1981). The Bronze Age, in Simmons, I. G. and Tooley, M. J. (eds), *The Environment in British Prehistory*, Duckworth, London, pp. 210–49.

Tooley, M. J. and Sheail, G. M. (1985). *The Climatic Scene: essays in honour of Gordon Manley*, Allen & Unwin, London.

Turner, J. (1981). The Iron Age, in Simmons, I. G. and Tooley, M. J. (eds), *The Environment in British Prehistory*, Duckworth, London, pp. 250–81.

Tyson, P. D. (1986). *Climatic Change and Variability in Southern Africa*, Oxford University Press, Cape Town.

Wigley, T. M. L., Ingram, M. J. and Farmer, G. (eds) (1981). *Climate and History*, Cambridge University Press.

Palaeohydrology in Practice
Edited by K. J. Gregory, J. Lewin and J. B. Thornes
© 1987 John Wiley & Sons Ltd.

# 11

# Late Quaternary Palaeoecology of the Severn Basin

## K. E. BARBER AND S. N. TWIGGER

*Palaeoecology Laboratory, Department of Geography, University of Southampton*

The hydrological history of any river or basin is intimately connected to its ecological history, and of that ecology the most important part is the vegetation. So it is that palynology or pollen analysis of the organic sediments within a catchment provide us with the best evidence of the connection between the ecosystem and the fluvial system in the times before written or mapped historical records, but the palynological record is neither easily read, nor is it free from bias or error. The aim of this contribution is to review and synthesize the palaeoecology of the Severn Basin, including the River Wye, placing it in the context of changes in the British Isles generally, and to supplement this regional view with a detailed case-study from the Baschurch area, some 6 km north of the Severn near Shrewsbury. The contemporary vegetation of the region has recently been analysed in an Ecological Flora of Shropshire (Sinker *et al.*, 1985).

### 11.1 SITE DISTRIBUTION AND BIAS

The number of pollen-analysed sites within and just outside the Severn Basin has increased appreciably in recent years. The watershed of the catchment is not taken as a cut-off point in this review since it does not in general mark any kind of vegetational boundary, particularly in the lowlands. In particular the results of four research projects have been of great importance—Moore (1966), Beales (1976), Brown (1983a) and Twigger (1987). Allied to the palaeoecological work of Shotton (e.g. 1978), Turner (1964, 1965), Rowlands and Shotton (1971) Pannet and Morey (1976), Wiltshire and Moore (1983), Jones, Benson-Evans and Chambers (1985) and studies of environmental archaeology, (e.g. Limbrey, 1983), these projects allow us to have a reasonable overview of vegetational change at the basin scale.

We may then set that overview into the framework of what is known about the Devensian Lateglacial (Pennington, 1977; Lowe, Gray and Robinson, 1980; Lowe and Walker, 1984) and Flandrian postglacial environments (Simmons and Tooley, 1981; Bell and Limbrey, 1982) more widely in the British Isles. However, there are first some important *caveats* to be made regarding site distribution, site type and the relevance of the palaeoecological record to palaeohydrological change.

Ideally the pollen-analysed sites should conform in spacing and type to the guidelines laid down by Berglund (1979, 1983, 1986) and a countryside dotted with medium-size closed lakes interspersed with ombrotrophic bogs would serve very well. The Severn Basin is, however, very diverse, encompassing the Welsh mountains, the Shropshire plain and the Lower Severn Valley grading into its large estuary. Furthermore, the British approach to palaeoecology has not in general been a classificatory/survey type of approach but rather via a series of individual problems, often tackled through Ph.D. research, or studies such as that of Greig (1982) and Chambers and Price (1985) on a single species. In some areas where a single researcher or research group has worked over a number of years, and where there is an abundance of suitable sites, then regional syntheses have been produced, such as Walker (1966) on the Cumbrian Lowland, Pennington (1970) on the English Lake District, Pennington *et al.* (1972) on north-west Scotland, and Simmons and various co-workers on the North York Moors (e.g. Simmons and Cundill, 1974). Birks (1977) gives a preliminary synthesis for the whole of Scotland but this has not as yet been attempted for the rather more diverse and anthropogenically disturbed landscapes of England and Wales, though the environment of the various archaeologically-defined periods is well reviewed in Simmons and Tooley (1981).

In the Severn Basin there is only one well-dated 'reference site' covering the Devensian Lateglacial and Flandrian at Crose Mere (Beales, 1980) located just within the lowland catchment area, although there is another well-dated standard pollen diagram from a site just over the Welsh mountain divide at Tregaron (Hibbert and Switsur, 1976). Both of these sites figure prominently in Huntley and Birks' (1983) isopoll maps as having the highest of their four levels of 'dating quality' throughout the Flandrian. Because of its importance as a reference site Beales' main pollen diagram is reproduced here as Figure 11.1, and his pollen concentration diagram for the Devensian Lateglacial as Figure 11.2. In addition we have a scatter of diagrams from lakes and peats of various sorts with a spectrum of chronological control running from vague correlations with postglacial climatic conditions to a sequence of radiocarbon datings. Some of these diagrams cover quite short spans of time but because of their nearness to the River Severn, and because

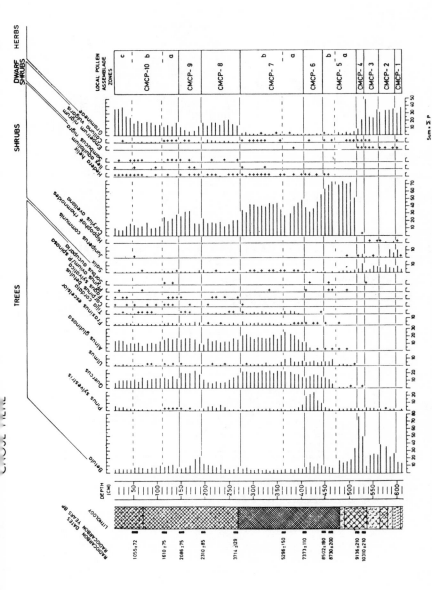

FIGURE 11.1   Percentage pollen and spore diagram from Crose Mere (includes all taxa occurring at five or more levels). Reproduced by permission of The New Phytologist Trust (From Beales, 1980)

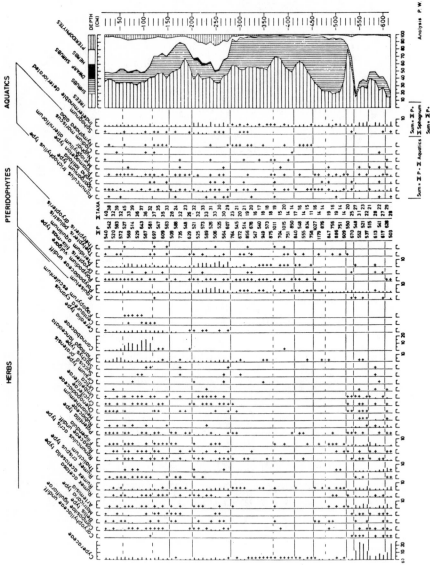

Analysis P. W. BEALES 1973

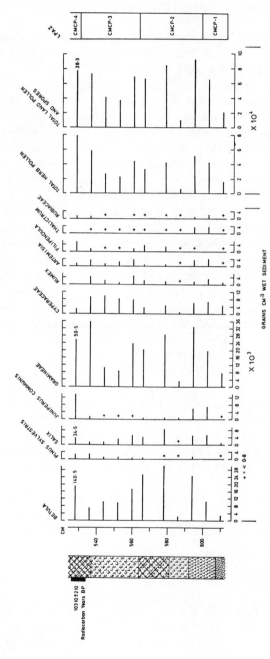

FIGURE 11.2 Late Devensian pollen and spore concentrations of the most abundant taxa from Crose Mere. Reproduced by permission of The New Phytologist Trust (From Beales, 1980)

of the events they encompass, they are of primary relevance to palaeohydrological studies (e.g. Brown and Barber, 1985).

Brown (1983b, p. 380) published a map showing the distribution of most of the important sites in the basin. These fall into two main types: fairly large open lake basins and mires such as Crose Mere, Tregaron bog and Rhosgoch Common (Bartley, 1960), and smaller sites covering a shorter temporal range such as the Old River Bed at Shrewsbury (Pannet and Morey, 1976), Moreton Morell (Shotton, 1967) and Brown's (1983a) palaeochannel sites such as Ripple and Callow End. All the lowland valley sites are, as would be expected from their position within the Holocene alluvial fill, mid to late postglacial in age; for the long uninterrupted sequences we have to turn to the lake and bog sites. Unfortunately those so far worked upon are at some distance from the main Severn channel, e.g. Crose Mere, though work is in progress at the Palaeoecology Laboratory, Southampton University, on lakes nearer to the Severn with promising lateglacial and postglacial sediments, some of which show stratigraphic evidence of lake-level changes— e.g. Alkmund Park Pool in the Lowlands near Shrewsbury (grid reference SJ 480160) and Llyn Mawr in the uplands near Newtown (SO 008972). Unfortunately Isle Pool, within the great meander loop north-west of Shrewsbury (SJ 461170) has been subject to Victorian engineering works and the stratigraphy appears disturbed, but Shrawardine Pool (SJ 398160) now largely infilled with a diverse rich fen community, has an interesting stratigraphy and will be the subject of further work.

Nevertheless Crose Mere (Beales, 1980) is clearly the outstanding site of regional importance and will serve as the standard sequence for the purposes of this discussion. As a lake of 15.2 ha with a smallest diameter of about 280 m and 9.2 m maximum depth (certainly a little larger before drainage operations in 1864) it may be expected, on the basis of the basin size to pollen source area statistics worked out by, amongst others, Jacobson and Bradshaw (1981) and Prentice (1985) to receive pollen inputs as follows:

50 per cent from 'regional' vegetation more than several hundred metres away;
35 per cent from 'extra-local' vegetation between 20 m and several hundred metres away;
15 per cent from 'local' vegetation up to 20 m from the edge of the sampling basin.

This contrasts strongly with the site at Ripple (Brown and Barber, 1985) which, being only 46 m across, would on the above basis be reflecting pollen inputs classed as 10 per cent regional, 25 per cent extra-local, and 65 per cent local.

These figures are certainly oversimplifications of a complex relationship between basin size, pollen source area and pollen representation (Oldfield, 1970), but while they could be refined with further work and analysis of individual taxa, taking into account the differential settling velocities of heavier and lighter pollen grains (e.g. Prentice, 1985, 1986), they do give us an idea of the 'bias' of individual sites.

Clearly a combination of the two sorts of site is required to give insights into the general vegetation pattern of the basin and into the vegetation of more local areas such as the Lower Severn terraces (Brown, 1983a) and changes *within* small areas of a few square kilometres, as in the case-study on the Baschurch Pools reported here.

The greatest gap in the distribution of sites lies between the Welsh upland sites and the lowland Crose Mere–Shrewsbury–Church Stretton area. Work in progress at Southampton on Llyn Mawr will help rectify the gap in our records of this transitional area. A good site between Shrewsbury and Brown's (1984) site at Hartlebury would also be worth researching.

The relevance of vegetational history to the fluvial system is well illustrated by Lockwood's (1983, p. 36) simulated figures for runoff, following heavy rain after a dry period, from a landscape covered with oak forest—15.8 mm rainfall equivalent—against a grass-covered landscape with 68 mm. Even at the simplest level, therefore, and without involving any climatic changes, the gross vegetational changes of the last 15,000 years, from tundra to broadleaf high forest to open farmland, would be expected to produce gross changes in runoff. Similarly, vegetation cover and type are very significant influences on sediment yield (Walling and Webb, 1983), the amount of bare ground due to cold conditions or arable agriculture (Fullen and Reed, 1986) being particularly important as shown by Brown and Barber (1985) and Brown (Chapter 14 this volume). The impact of man in prehistory is therefore integral to any palaeohydrological analysis and is shown particularly well in some of the pollen diagrams discussed here, as well as by Limbrey (Chapter 12 this volume).

## 11.2   LATEGLACIAL VEGETATIONAL HISTORY

The environmental conditions of the Devensian Lateglacial have been well reviewed recently by Gray and Lowe (1977), by the suite of papers published in 1977 (Pennington, 1977), by Lowe, Gray and Robinson (1980) and by Lowe and Walker (1984). Beginning at some time between 14,000 and 13,000 years BP a dramatic warming of Britain's glacial climate followed a shift in the distribution of polar and subpolar water further to the north in the Atlantic. The exact sequence of climatic and general environmental changes from about 13,000 to the beginning of the Flandrian/Holocene at around 10,000 years ago is still a matter for debate. Watts (1980) and Lowe and Gray (1980), in very thorough revisions of both the evidence and the thinking behind various stratigraphic subdivisions, stress the spatial differences, and the time-transgressive nature of environmental changes in the Lateglacial period. Lowe and Gray's scheme for a climatostratigraphical subdivision involving unnamed transition of between 500 and 1000 years' duration (1980, p. 163, Table 2) is probably the best we have at the moment but for simplicity of discussion here the transitions are omitted so that the Lateglacial Interstadial is taken to cover 13,000–11,000 BP and the Younger Dryas Stadial covers 11,000–10,000 BP.

The continuing debate (Pennington, 1975; Watts, 1980; Lowe and Gray, 1980) over the significance of any climatic deterioration between about 12,000 and 11,800 BP, giving the sequence Bolling/Older Dryas/Allerod/Younger Dryas, gains only a little support from the Severn Basin. At Crose Mere, Beales (1980) points to a drastic fall in pollen concentration within a sand and silt layer between 581–592 cm (Figure 11.2). Beales is reluctant, on the evidence of a single pollen spectrum and the lithostratigraphic change, to claim this as an Older Dryas type regression and further work on sites with more amenable lithostratigraphy than Crose Mere is needed to clarify the issue; Crose Mere has steep slopes immediately to the north which would readily produce a largely inorganic lithostratigraphy (loss on ignition was only *c.* 10 per cent).

As can be seen from Beales's main pollen diagram (Figure 11.1) the Devensian Lateglacial vegetation of the area, represented by pollen zones CMCP1–3, ending at 10,310 BP, was characterized by large amounts of grass (Gramineae) and sedge (Cyperaceae) pollen, and a great variety of herb pollens, indicating a largely treeless environment at first (zone CMCP 1), into which birch trees (*Betula*) immigrated to form open woodland (zone CMCP-2)—grasses and herbs are still over 50 per cent of the total—which then reverted to a herb-rich grassland (Zone CMCP3). This is a typical Devensian Lateglacial Interstadial sequence, equivalent to the classical Older Dryas cold period–Allerod warmer period–Younger Dryas cold period sequence in north-west Europe and to Zones I, II and III in Godwin's scheme (Godwin, 1975). As Beales (1980) notes, the amount of birch pollen in CMCP-2, up to 40 per cent of all dry land pollen, is in marked contrast to the lowland sites to the north such as Moss Lake, Liverpool (Godwin, 1959), and Bagmere, Cheshire and Chat Moss, Lancashire (Birks, 1965) where percentages are so low (below 20 per cent) as to indicate very sparse amounts of woodland, if any, during the interstadial. According to Huntley and Birks (1983) values of more than 25 per cent 'can indicate local birch-dominated woodland and values in excess of 50% normally reflect areas where birch-dominated woodland covers the landscape' (p. 125). Further north still, however, the crucial site of Blelham Bog, with its excellent radiocarbon-dated pollen diagrams, both percentage and influx (absolute), shows clearly the establishment of birch woodland with percentages of over 40 per cent of total pollen and influx figures of over 1000 grains per $cm^{-2}$ $yr^{-1}$ (Pennington and Bonny, 1970; Pennington, 1975). Coope and Pennington (1977), in proposing the Windermere Interstadial as a standard reference section, have clear evidence in terms of tree birch macroscopic remains, abundant birch pollen and woodland beetle species, for birch woods being present between 12,500 and 11,000 BP. Both of these lowland sites have plenty of south-facing slopes where insolation values may rise to 50–60 per cent above the values for a horizontal surface (Geiger, 1965), and similarly there are south-facing slopes on the northern shore of Crose Mere. The angle of incidence of the solar beam would, of course, be the same as now, leading to some distinctly warm summer days and this important difference from present-day high-latitude tundra areas, evidenced clearly by the differences between beetle and floral assemblages (Coope and Pennington, 1977), must always be borne in mind. In the extreme south of Britain, at Church Moor in the New Forest, there is new evidence of plentiful birch pollen by 12,400 BP with a peak of over 70 per cent of dryland pollen. Following Prentice (1981) these birch pollens were separated into three classes with a clear abundance of *Betula pendula*, the main tree birch, indicating more or less closed birch woodland (Barber, 1981b, and in prep.). Conversely in upland Wales, near the head of the Wye, Moore (1970) found *Betula* percentages which exceeded 10 per cent with only one value of 20 per cent for a short period which he equated with Godwin Zone II. Birch pollen values are also lower at Church Stretton (Rowlands and Shotton, 1971).

The picture that emerges in western Britain is therefore one of fairly dense birchwoods in the south with local birchwoods further north, especially on south-facing slopes, but with only scattered birch copses set in herb-rich grassland in upland Wales and on the Lancashire–Cheshire plain. The amount of bare ground and of inorganic sediment yield varied according to local topographic factors as well as vegetation cover but was often (Elan Valley, Crose Mere) a silty detritus mud with distinctly more allochthonous input than, for example, the closed broadleaf high forest of the mid-Holocene.

The Younger Dryas chronozone (*c.* 11–10,000 years BP) was, in the Severn Basin as elsewhere in Britain, a period of intense cold, of periglacial activity and lower vegetation cover. In upland areas which escaped glaciation short turf plant communities with disturbed soil persisted (Moore, 1970) but at Crose Mere 'the suggestions of soil disturbance . . . are not as pronounced as in many lowland sites' especially those in Cumbria (Beales, 1980, p. 146). However, there is evidence of severe conditions in southern England from the soliflucted chalk debris in sites investigated by Kerney (1963) and at Church Moor (Barber, 1981b and in prep.) where a black clay seals the interstadial deposits. Even in the most favoured environment of the Isle of Wight, Scaife (1982) sees a much more open herbaceous vegetation, with only perhaps a few birch stands, than was inferred by Godwin (1975), and the presence of pine pollen in Younger Dryas deposits is probably as a result of long-distance transport from the pinewoods of northern Iberia and the Alps (Huntley and Birks, 1983).

This vegetational picture of a grass- and sedge-dominated landscape with many 'weeds' now typical of bare and disturbed soil (Lowe and Walker, 1984) is paralleled by instability in the fluvial system and it is probable that conditions in the Severn Basin were similar to those discussed by Rose *et al* (1980) with the added input of high snowmelt discharge from the Welsh mountains.

## 11.3  EARLY HOLOCENE VEGETATIONAL HISTORY

Though conventionally taken as 10,000 years BP the opening of the Holocene (Flandrian Stage in Britain) occurred somewhat earlier on the evidence of radiocarbon datings from various sites. At Bleham Bog pollen concentration rises between 10,650 and 10,490 BP (Pennington and Bonny, 1970); further north in Perthshire two sites at the Loch Lomond Stadial ice limit show a similar sequence of juniper and then of birch rises between 10,670 and 10,420 BP (Lowe, 1978), and at Crose Mere the event is dated at 10,310 ± 210 years BP. Considering the problems inherent in radiocarbon dating lateglacial/early postglacial sediments (e.g. Sutherland, 1980) it is probably best to retain the conventional date of 10,000 BP. Whether one then divides Holocene time into relatively short or relatively long periods depends upon one's purpose. In terms of catchment vegetation/hydrology relationships the most significant Holocene event is taken to be the introduction of agriculture at around 5000 BP and the sensible split therefore appears to be 10,000–5000 BP the period of natural vegetational change, and 5000–0 BP, the period of culturally-forced vegetational change.

A more or less complete vegetation cover, with soil stability and a change to mainly autochthonous organic accumulation in lakes, followed rapidly upon the climatic amelioration. Between 10,310 and 5296 BP the pollen curves from Crose Mere show a typical pattern of successive immigrations and establishment of the main arboreal taxa, following the usual succession of western Britain where pine expands after the arrival of oak and elm, the dominant species of the mid-Holocene mixed broadleaf forest, in contrast to the pronounced early pine period seen in eastern and south-eastern Britain (Godwin, 1975; Bennett, 1984). With values just touching 25 per cent of total pollen, pine can never have been a major forest component around Crose Mere, although the diagram from Tregaron (Hibbert and Switsur, 1976) and the data reviewed by Moore (1972) suggest greater amounts in the western upland area of the Severn/Wye catchment. Brown (1984) records large amounts of pine pollen (>40 per cent total land pollen), together with hazel and birch, in the sandy area of Hartlebury Common, together with a very early date on pine wood—9710 ± 130 BP. This suggests local edaphically-induced variation in the woodland of the Severn Basin, overlain by the climatic and altitudinal variation from the lowlands of Shropshire–Worcestershire to the Welsh uplands.

Crose Mere (Figure 11.1) may be taken to represent the general trends in the area, with the important *caveat* tnat it almost certainly underrepresents the importance of lime in the woodlands (Greig, 1982; Brown, 1983a), as brought out by the Baschurch Pools study (see p. 240) reported later. After the initial expansion of birch (zone CMCP-4), hazel scrub woodland dominates for about 400 years from around 9100 BP, taking the arboreal pollen percentage to in excess of 90 per cent, the only other major component being birch. There are indications, in the continuing inorganic inputs during zone CMCP-5a, of areas of open ground, and the pollen deposition rate diagram shows more clearly than the percentage diagram that grass and herb pollens were still contributing to the pollen assemblage at around 3–4000 grains $cm^{-2}$ $yr^{-1}$ out of a total deposition rate of 80,000 grains $cm^{-2}$ $yr^{-1}$. However as Dearing and Foster (1986) point out, bank erosion may produce significant allochthonous inputs.

Hazel continues to play an important part in the pollen stratigraphy, continuing into the establishment of a mixed oak forest (zones CMCP 5b, 6 and 7a) which, from the very low herb pollen presence must have been quite closed with hazel probably flowering freely only on the lake shore.

The colonization of valley bottoms by alder was a significant event in the early Flandrian, both pollen-analytically and hydrologically. New evidence for its early expansion in north-west Wales, *c.* 8465 BP, is given in Chambers and Price (1985), who also review other sites, as does Smith (1984). Taken together they emphasize the time-transgressive nature of the alder rise, from *c.* 8500–6200 BP in England and Wales and, discounting somewhat the older climatic argument that the expansion of the hygrophilous tree was due to the Holocene rise in sea level establishing a more oceanic climate (Godwin, 1975, p. 464), they conclude that 'human influence in the expansion of alder may have been considerable' (Smith, 1984, p. 50), linking this effect with the secondary maximum of hazel representation seen at the Boreal–Atlantic Transition at many sites (Smith, 1970), including Crose Mere (zones 6–7a boundary). The alternative view, of virtually no human influence on vegetation until Neolithic farming begins *c* 5000 BP, has most recently been expounded by Birks (1986) who makes no reference to the now considerable body of evidence for pre-elm decline interference with vegetation in the British Isles (see for example Edwards and Hirons, 1984; Smith, 1984; Simmons, Rand and Crabtree, 1983; and Simmons, Dimbleby and Grigson, 1981, for earlier references). Within the Severn/Wye Basin early human influence is not pronounced and in the highlands especially evidence is sparse (Limbrey, Chapter 12 this volume).

The Crose Mere diagram shows some evidence of openings in the forest, or of a more open lake shore, in the continuous presence of ash, *Fraxinus excelsior*, in zone 7a but the almost complete absence of herbs and grasses and, using Smith's (1984) arguments, the very slow expansion of alder over some 1500 years, point to a more or less complete forest cover, with all that implies hydrologically in terms of precipitation interception, runoff and soil (and river-bank) stability. It is likely that by 5300 BP at the latest, and considerably earlier in some places, almost all river banks would have had a dense alder/willow vegetation, binding the bank material with their root networks and creating a considerable amount of organic debris. Hayward and Fenwick (1983) point to the greater amount of silt and clays in earlier Flandrian floodplain fills than today's rather coarser debris. Pre-clearance alder pollen values of 55 per cent or so are common in Brown's (1983a) floodplain sites at Ripple Brook, Ashmoor Common and Callow End, and Brown and Barber (1985) calculate erosion estimates of around 20 tonnes $km^{-2}$ $year^{-1}$ for the early and mid-Holocene period due to high forest cover, increasing to late-Holocene figures of around 120–140 tonnes $km^{-2}$ $year^{-1}$ once agriculture began on the floodplain.

In complete contrast to the lateglacial picture therefore, we now have, between 10,000 and 5000 years BP, a picture of a forest-dominated landscape, even into the uplands on the western side of the catchment (Wiltshire and Moore 1983), yielding low amounts of sediment to a stable fluvial system.

## 11.4  LATE HOLOCENE VEGETATIONAL HISTORY I

The period after around 5000 BP was increasingly dominated by man's activity but there were natural changes taking place as well, and changes in which man played only a minor part. Amongst the former we may note the spread of peatlands in both lowlands and uplands (perhaps aided by man in the uplands at least); the abandonment and infill of channel cutoffs; the toll of tree disease and the arrival of late immigrants to the forest flora; the impoverishment of soils due to leaching, and the rise of water-tables generally due to climatic deterioration (especially after about 3000 BP). As to the man-induced changes we are dealing mainly with the clearance of woodland for pasture and for arable crops. The temporal pattern of *clearance-farming–abandonment–woodland regeneration* was repeated many times in parts of the Severn Basin, as in the rest of north-west Europe, though recent work is tending to emphasize the longevity of clearance episodes (Smith, 1975; Buckland and Edwards, 1984). In some favoured areas, however, as in the case of the area near Berth Pool reported in this chapter, clearance in the Bronze Age led to a permanently 'open' landscape.

The area of peatland within the Severn Basin probably reached its zenith around AD 800, before the population increase of the Middle Ages triggered attempts at drainage for fuel and cultivation. As Walker (1970) demonstrated, the more or less stable end-point of the hydroseral succession in lakes in the British Isles is raised bog. Besides such large ombrotrophic mire systems, best exemplified in the Severn area today by the Whixall-Fenn's Moss complex, there were once widespread areas of valley and floodplain mire (such as the Baggy Moor area of the Perry catchment and Ashmoor Common (Brown, 1983b), and in the uplands large areas of spreading blanket mire (Moore, 1973; Moore, Merryfield and Price, 1984).

It is difficult, looking at today's landscape, to appreciate just how widespread mires once were in lowland England. Enormous tracts of Lincolnshire and Cambridgeshire are now reclaimed fen (Wheeler, 1984) and in the New Forest, probably the largest undrained area of southern England, virtually every valley contains oligotrophic fen communities, locally known as 'valley bogs' (Rose, 1953; Barber, 1981b). In the Severn Basin, Brown (1983a) investigated a number of riverine peat deposits, including an extensive palaeochannel at Ashmoor Common, a meander cutoff at Callow End and peats at Ripple Brook (Brown and Barber, 1985) which are part of the infill of a former valley. Pollen analysis of these deposits, and of the Old River Bed at Shrewsbury (Pannet and Morey, 1976), all indicate a floodplain dominated by wet alder carr from before 5000 BP until extensive clearance during the Iron Age, and from that time the areas were used primarily for grazing animals when ground conditions were favourable. Extensive clearance of terrace woodlands, many dominated by lime (Greig, 1982) also took place during the Iron Age, somewhat earlier than the alderwood clearances to judge from Brown's evidence (1983a), and these changes have had dramatic palaeohydrological consequences (Brown, Chapter 14 this volume), with floodplain accretion and slopewash burying or modifying the riverine peatlands.

In the uplands, on the other hand, man's activities from quite an early date may have encouraged the initiation and spread of peatlands, specifically blanket bogs, through the removal of tree cover leading to reduced interception and evapotranspiration and hence waterlogged conditions suitable for the growth of *Sphagna*. Obviously, for the relative importance of the human factor to be assessed, accurate radiocarbon dates are essential, as are pollen diagrams. Work by Moore (1975), Smith (1975), Chambers (1981) and others showed that the date of initiation of peat at sites in Wales and Northern Ireland could vary by thousands of years within the same upland region, and there is useful discussion of the topic in Edwards and Hirons (1982), Chambers (1982) and Moore, Merryfield and Price (1984). As the latter point out, blanket peat development started in early Holocene times in Norway and Scotland before man's impact could have been very extensive or intensive, and may therefore be more or less wholly attributed to climatic deterioration acting together with topographic and pedogenic factors, but that 'as the climatic limits of blanket mires are approached, their initiation and continued development became more dependent on human activity' (p. 232). In the Welsh uplands radiocarbon dates for peat initiation range from 4830 ± 55 BP (Moore, Merryfield and Price, 1984), to 4380 ± 70 BP (Chambers, 1981) and may therefore be associated with Neolithic activity, down to the Dark Ages with a date of 1310 ± 70 BP (Chambers, 1981). Wiltshire and Moore (1983), in a study of two sites in the Wye catchment, concluded that hydrological changes leading to peat initiation and a speeding up of peat accumulation could be dated to two periods—Neolithic/elm decline times, and late Bronze Age/Iron Age times, both periods of coincident climatic change and intensified human pressure. Clearly more work is needed on this problem, with as many radiocarbon dates as possible.

From purely ombrotrophic mires in the late Holocene we have an excellent potential palaeoclimatic record, one capable of further development (Barber, 1985; Moore, 1986). Phase-shifts to wetter bog surfaces have been dated from various parts of Europe (Barber, 1982) and detailed records of mire palaeohydrology have been constructed by Aaby (1976) for Denmark since 5500 years BP, by Van Geel (1978) for the Netherlands from 4300 BC to 600 BC and from 1400 BC to AD 300, and by Barber (1981a) for north-west England since 2000 BP. In the later period these records agree in showing wetter periods *c.* AD 1000 and AD 1450–1500. Seen in the light of the accumulated results of bog stratigraphic work since Granlund's proposal (1932) for five climatically-conditioned renewals of bog growth (his Recurrence Surfaces, RYs, of RYI: AD 1200; RYII: AD 400; RYIII: 600 BC; RYIV: 1200 BC; and RYV 2300 BC), it is clear that a series of shifts to a wetter, cooler climate affected north-west Europe during the late Holocene. Exactly how each raised bog responded may have depended on local or regional factors such as rainshadow effects or coastal influences but detailed studies, backed by careful stratigraphic work and plentiful radiocarbon dates, such as the recent work by Smith (1985) and Wimble (1986) confirm that a major climatic deterioration occurred in the late Bronze Age/early Iron Age. This is backed up by isotopic and other analyses from Dutch peat (Brenninkmeijer, Van Geel and Mook, 1982; Dupont, 1985; Dupont and Brenninkmeijer, 1984) and by various other indicators (see papers in Flohn and Fantechi, 1984; and Morner and Karlen, 1984). While it is arguable that some of the earlier changes in climate during the period 5000–0 BP were too minor in nature to effect hydrological changes on a large scale, the evidence is now growing that the deterioration at the Subboreal–Subatlantic transition was marked enough to increase discharges and lead to higher lake-levels and more rapid peat accumulation, and the dates on peat for this event now cluster around 900–600 BC and possibly a century or two earlier in western Britain.

Within the Severn Basin we have so far only a few radiocarbon dates, from ombrotrophic peat, referring to this period. They are, however, of more than ordinary interest since they relate to an unusual phenomenon in the stratigraphy of British raised mires—a well-developed layer of treestumps of pine, *Pinus sylvestris*, first described by Hardy (1939) and dated as part of Turner's (1964) research on the site. Radiocarbon dates (Godwin and Willis, 1960) from the site, Fenn's Moss and Whixall Moss, which are continuous parts on the same large raised bog, give us three periods of phase-shifts to a wetter bog surface. Firstly, pine wood was dated to 2307 ± 100 BP, 357 BC; presumably the pines were killed by rising water-tables soon after this date, but further work is being carried out at this site by C. J. Haslam (Southampton) involving dendrochronology, peat macrofossil analysis and more radiocarbon dates. Secondly, at Site B, Fenn's Moss, dates of 1842 ± 100 BP (AD 108) and 1670 ± 110 BP (AD 280), from respectively below and above a recurrence surface denote another climatic deterioration in early Roman times (comparable with the flooding of the lake village at Meare in the Somerset Levels—Coles and Orme, 1980). Thirdly, an upper recurrence surface was dated as 746 ± 90 BP (AD 1204) below and 655 ± 90 BP (AD 1295) above the humification change, very close to the dates for a similar feature at Tregaron (dates of 760 ± 70 BP and 768 ± 90 BP, Godwin and Willis, 1960) and to the climatic deterioration evident further north at Bolton Fell Moss, beginning around AD 1300 (Barber, 1981a). The palaeohydrological record of the Fenn's Moss/Whixall Moss complex is therefore of great value in giving us three distinct shifts to wetter/colder conditions at about 300 BC, AD 200 and AD 1300. The 300 BC deterioration is paralleled by the date of 310 BC for the pool peat associated with Lindow Man, the bog body found in Lindow Moss, some 55 km north of Whixall Moss (Barber, 1986).

The history of dry-land vegetation since about 5000 BP may be summarized as being a story of very little impact in Neolithic times, a major impact shortly after the opening of the Bronze Age at about 3900 BP which persisted into the Iron Age in some areas, some evidence of woodland regeneration in early Roman times, but renewed clearance thereafter with the landscape taking on its modern appearance in terms of for example the percentage of woodland by medieval times.

The evidence for the lack of any real Neolithic impact comes not only from the lack of archaeological evidence (Rowley, 1972; Smith, 1974, Figures 12, 14, 15 and 16), but from pollen diagrams such as that of Beales (1980) shown in Figure 11.1. This is supported by diagrams from Tregaron (Hibbert and Switsur, 1976); Llyn Mire (Moore, 1978—there are some cereals recorded here but very little other evidence of large openings in the forest);

Borth Bog and Plynlimmon (Moore, 1968; Moore and Chater, 1969); Rhosgoch Common (Bartley, 1960), and at Ashmoor Common (Brown, 1983a). This is in contrast to the evidence for early, and in some cases permanent, clearance in south-east England (Waton, 1982a; Waton, Barber and Fasham, in prep.) and in the north-west (Pennington, 1970), and in some contrast to the views of Taylor (1984) who sees a 'general pattern . . . of total occupation by a large number of Neolithic people in all parts of England' (p. 43), and 'from 4000 BC onwards (approximately 3180 BC uncalibrated) there is evidence of widespread forest removal on a huge scale' (p. 41). The detailed review of the pollen-analytical evidence from the Neolithic period by Smith (1981) does not support this generalization. While acknowledging the problems of detection of small clearances in pollen diagrams, and the degree of bias in the regional spread of sites (see also Limbrey, Chapter 12 this volume), Smith concludes that 'We can hardly doubt, therefore, that the countryside remained in general quite densely forested' (1981, p. 201), as indeed have some local areas throughout the Holocene (Barber, 1981b).

The well-known elm decline around 5000 BP, over the cause of which opinion has swung from climatic-edaphic factors, to human impact and now to a disease hypothesis (perhaps aided by man's activities), has most recently been reviewed by Sturludottir and Turner (1985) and Birks (1986) and will not be dealt with here, except to note that the decline at Crose Mere, from 9 to 1 per cent of total pollen over an estimated 450 years does not coincide with the appearance of cereal pollen grains nor with any significant increase of herbs. Not until *c.* 3200 BP is there any marked woodland clearance, at the opening of zone CMCP-8 (Figure 11.1). The date of *c.* 3900 BP given by Beales (1980) is thought to have been affected by the inwash of old carbon due to soil erosion. The proportion of tree and shrub pollen falls dramatically from *c.* 90 to 50 per cent of total pollen, similar to the levels of today. Caution must be applied to any simplistic analogy though; trees in today's parkland settings, flowering freely from top to bottom and in a wind field quite different from a forest microclimate, must produce far more better-dispersed pollen than a single tree in high forest conditions. Nevertheless this first major impact of man in the early Bronze Age, which also reduced the sediment organic content to 50 per cent, indicating considerable erosion, was clearly an event of regional importance and so is discussed in detail in the following case-study.

## 11.5 LATE HOLOCENE CASE-STUDY: THE BASCHURCH POOLS

The four small lakes, or meres, collectively referred to here as the Baschurch Pools are: Berth Pool, which is 2.9 ha in extent; Birchgrove Pool, 1.7 ha; Fenemere, 9.4 ha; and Marton Pool, 6.8 ha (Figure 11.3).

These meres occupy discrete hollows, probably kettle holes, within fluviog-

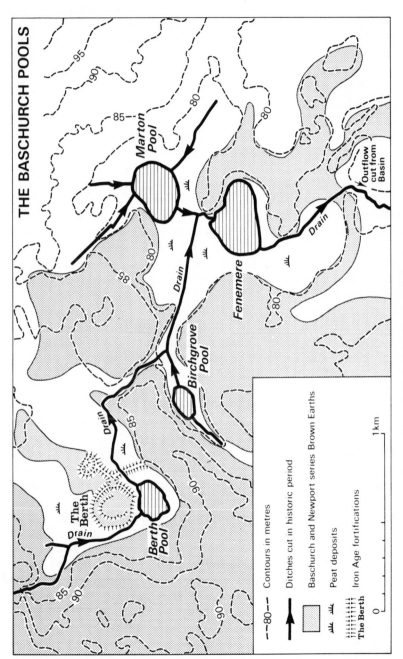

# THE BASCHURCH POOLS

**Marton Pool**

**Fenemere**

**Birchgrove Pool**

**The Berth**

Drain

**Berth Pool**

Outflow cut from Basin

Drain

Drain

Drain

Drain

95

90

85

80

80

80

80

80

85

90

90

85

98

95

90

80

- - 80 - -    Contours in metres

   Ditches cut in historic period

☐    Baschurch and Newport series Brown Earths

   Peat deposits

**The Berth**    Iron Age fortifications

0         1 km

FIGURE 11.3   The Baschurch Pools

lacial sands and gravels deposited by the wasting Devensian ice-masses of the area (Shaw, 1972). Although glacial till forms the dominant surface drift type of the Shropshire–Cheshire Plain, most of the meres and peat mosses lie within belts of sand and gravel which stretch out across the till (Reynolds, 1979). Three of the Baschurch Pools, Fenemere, Marton Pool and Birchgrove Pool, are located within a low-lying basin surrounded by sand and gravel ridges. This basin has no natural outflow and peat deposits have built up in between the pools. A terrestrial core taken in between Fenemere and Marton Pool revealed 1.5 m of peat overlying a yellowish-brown shelly mud. Pollen analysis of the peat, which contained mesotrophic *Sphagnum*, suggest that it began accumulating in Boreal times *c.* 9000 BP. The meres are therefore believed to have existed as separate entities during most of the postglacial. Local reports of water surface coalescence between Fenemere and Marton Pool almost certainly reflect the effect of the drainage ditches which have been cut into the basin, and which have artificially increased the catchment area of the meres. Berth Pool occupies a similar basin at a slightly higher elevation. All the inflow and outflow streams linking the four meres have been cut during drainage operations which probably date back to the fifteenth century (Hey, 1977). The four Baschurch Pools are no more than 2 km apart, and therefore provide an excellent opportunity to test pollen profile replicability, allowing a multiple profile study to be made in a relatively restricted locality (Figure 11.3). Since the sites had no inflow streams during prehistoric and early historic times, they will tend to represent a more local-ized vegetational record during these times (Pennington, 1979). Detailed surface sample studies of pollen frequency variation have been produced for many sites, for example Davis (1967), Berglund (1973), Bonny (1980) and Bonny and Allen (1984), but fewer studies relate directly to fossil pollen data, as Edwards (1983, p. 587) points out. Four lakes within 2 km of one another could theoretically have at least some extra-local and regional pollen source areas in common (Janssen, 1966, 1973; Tauber, 1965, 1977; Jacobson and Bradshaw, 1981; Prentice, 1985). Quantitative differences in the contri-bution of common pollen source areas could be due, at least in part, to intrinsic site factors such as size, morphology and sedimentary type (Jacobson and Bradshaw, 1981; Prentice, 1985). Pollen diagrams have been prepared from each of the Baschurch Pools and three are presented here (Figures 11.4 and 11.5). Sediment cores were taken in the deep-water zones of the pools. Single deep-water cores have been shown to be representative of the whole deep water zone (Bonny, 1976; Davis, Brewster and Sutherland, 1969). Deep-water cores have also been shown to be closely comparable with shallow water samples (Edwards and Thompson, 1984), and several multiple profile studies from the same site have shown no substantial variability in relative pollen frequencies, for example Davis, Moeller and Ford (1984) and Tolonen (1984). Some uncertainty does, however, persist on the subject of inter- and intra-site pollen frequency variation (Huntley and Birks, 1983). Recently, however, more rapid methods of core correlation have fostered the evolution of more comprehensive multiple coring strategies (Oldfield *et al.*, 1983; Dearing, 1983). An additional pollen profile has been prepared from Berth Pool, and two profiles prepared from nearby Boreatton Moss, SJ 417225, in order to examine intra-site pollen frequency variation (Twigger, 1987). Profiles were correlated by means of identifying a distinctive phase of pollen frequency change which has already been outlined elsewhere in the region (Turner, 1964; Beales and Birks, 1973; Beales, 1980). The inter-site correlation and comparison of this phase provides the basis for an examination of the degree of replicability of palynological data, and indicates the possible extent of the dominant pollen source areas at the individual sites.

**Vegetational changes**

Secondary, post-*Ulmus* decline woodland

| | |
|---|---|
| Berth Pool: | below 202 cm |
| Birchgrove Pool: | below 308 cm |
| Fenemere: | below 324 cm |

Comparatively little inter-site pollen frequency variation is observed at these levels. *Quercus* (oak) and *Alnus* (alder) pollen frequencies are high, *Betula* (birch), *Ulmus* (elm), *Tilia* (lime) and *Fraxinus* (ash) pollen frequencies are much lower, but exhibit relatively little fluctuation. Gramineae (grasses) pollen is present at low frequencies, while the open habitat indicators *Plantago lanceolata* (ribwort plantain) and *Rumex acet.* type (including docks and sorrels) and Liguliflorae types (e.g. dandelion) are either absent or, again, present at very low frequencies.

The possible forest composition can be inferred from these pollen frequencies, although the uncritical transfer of pollen representation factors, or R values (Andersen, 1973) from one area, or time, to another has been shown to be inadvisable (Oldfield, 1970). Ideally only pollen representation factors derived from contemporary analyses of pollen–vegetation relationships at the site or sites under study should be used, since it has been shown that site diameter can give rise to quantitative changes in pollen representation (Bradshaw and Webb, 1985). While a homogenous vegetation can be assumed, in order to facilitate theoretical formulations (Prentice, 1985), the natural heterogeneity of vegetation communities will inevitably complicate even site-specific derivations of representation factors (Birks and Gordon, 1985). High *Alnus* pollen frequencies at the Baschurch Pools probably represent locally dominant alder stands on fen peat around the pool margins. *Quercus*, *Ulmus* and *Tilia* pollen percentages indicate a broad-leaved high forest bordering

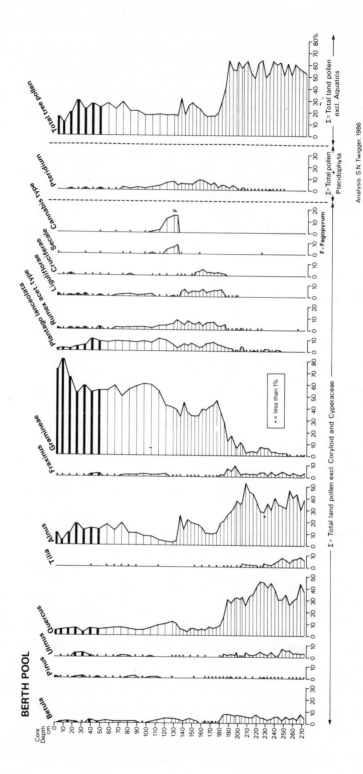

FIGURE 11.4  Berth Pool: selected pollen taxa. Samples from 0–48 cm are 2 cm thick, from a Mackereth mini-core

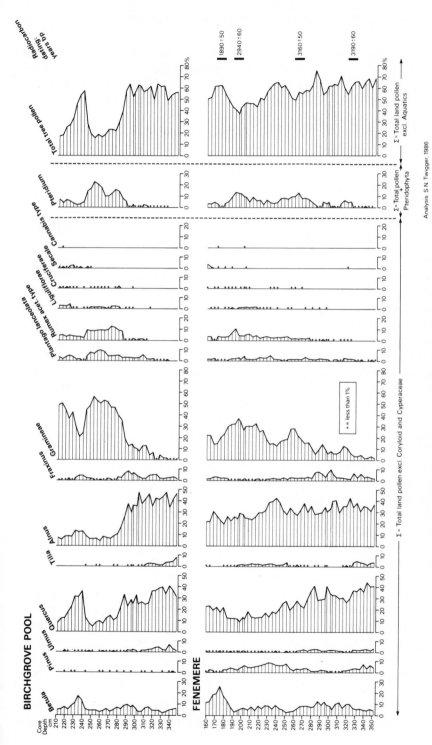

FIGURE 11.5  Birchgrove Pool and Fenemere: selected pollen types

the sites. The continuous representation of *Betula* and *Fraxinus* pollen is probably due to the fact that a degree of woodland disturbance has already occurred by this time (Beales and Birks, 1976; Beales, 1980); these two arboreal taxa tend to expand in the vegetation as a result of developing breaks in the high broadleaf canopy.

### The *Tilia* decline

| | |
|---|---|
| Berth Pool: | 202 cm |
| Birchgrove Pool: | 308 cm |
| Fenemere: | 324 cm |

A marked post-*Ulmus* decline reduction in *Tilia* pollen frequencies, accompanied by an increase in pollen taxa indicating land clearance has been observed in many British pollen diagrams, and has been shown to be diachronous, and to be anthropogenic in origin (Turner, 1962, 1965). Beales (1980) gives a date of 3714 ± 129 BP for the *Tilia* decline at the onset of zone CMCP-8 at Crose Mere (Figure 11.1), which is only 7 km from the Baschurch Pools. The *Tilia* decline at Whixall Moss, 8 km north-east of Crose Mere and 16 km north of Baschurch is dated by Turner (1964) to 3238 ± 115 BP. Beales believed that inwashing of old carbon, as a result of vegetation and soil disturbance, had caused a reversal of the radiocarbon date at the *Tilia* decline horizon at Crose Mere, and that *c.* 3700 BP was too early a date for the *Tilia* decline in this region. Radiocarbon dates obtained for this study from Boreatton Moss, 2 km west of the Baschurch Pools, and from New Pool, 5 km to the south, place an earlier *Tilia* decline at 3660 ± 50 BP (SRR 2831) at Boreatton Moss, and 3550 ± 50 bp (SRR 2833) at New Pool. A radiocarbon date from Fenemere places a later *Tilia* decline at 3190 ± 60 BP (SRR 2923) or *c.* 1240 bc (Figure 11.5). It is believed that this later decline correlates to the *Tilia* declines at 202 cm at Berth Pool (Figure 11.4) and 308 cm at Birchgrove Pool (Figure 11.5) and also to the *Tilia* decline at the onset of CMCP-8 at Crose Mere (Figure 11.1) and to the *Tilia* decline dated by Turner (1964) at Whixhall Moss to 3238 ± 115 BP. The dates from Boreatton Moss and New Pool represent a much earlier phase of vegetation change which probably correlates to the earlier *Tilia* decline at 224 cm at Berth Pool (Figure 11.4). In the three diagrams shown (Figures 11. 4 and 11.5) *Quercus* and *Alnus* pollen frequencies do not decline markedly at the time of the 3227 BP *Tilia* decline, whereas *Betula* and *Fraxinus* frequencies increase. At the Baschurch Pools it is evident that the 3227 BP *Tilia* decline is paralleled by a short-lived peak in Gramineae pollen frequencies, reaching *c.* 10 per cent. Immediately above this peak, a clear peak in *Fraxinus* pollen is then seen at each site. This *Fraxinus* pollen peak then declines and is succeeded by a second small Gramineae peak. A second *Graxinus* peak is then discernible. In vegetational terms, these pollen frequency variations indicate two distinctive cycles of woodland interference and regeneration. Increases in the frequencies of open habitat indicator taxa are associated with the Gramineae peaks. The lack of any clear reduction in total tree pollen in the phase immediately succeeding the *Tilia* decline points to a level of human activity which altered forest structure without substantially reducing total tree cover. A phase of forest exploitation, with small-scale clearances, is therefore indicated, similar to the type of interference reported for post-*Tilia* decline levels at Whixall Moss, Shropshire (Turner, 1964). Following these small-scale clearances, the frequency of Gramineae pollen rises to over 30 per cent at Fenemere, and 40 per cent at Berth Pool and Birchgrove Pool. While Gramineae pollen is not, in isolation, of great indicator value, it can be clearly seen that other open-habitat indicators in the pollen spectra, such as *Plantago*, *Rumex*, Cruciferae undiff., Liguliflorae undiff. and *Pteridium aquilinum* (bracken) spores all increase considerably in close association with the Gramineae increases. A considerable expansion in the cleared area around the pool is indicated. The radiocarbon dates for the 280 cm clearance expansion at Fenemere, 3160 ± 50 BP (SRR 2922) and the 190 cm clearance maximum, 2940 ± 60 BP (SRR 2921) are believed to be too old; inwashing of older carbon during clearance activity is thought to be responsible (Pennington *et al.*,1976). The clearance expansion at 280 cm at Fenemere (Figure 11.5) is estimated, by extrapolation, to have occurred at *c*. 2750 BP palaeoenvironmental data from a number of sites in the Severn Basin have shown that vegetational clearances in the Mid First Millennium bc were of sufficient magnitude to cause extensive erosion and a consequent increase in floodplain alluviation (Brown, 1983; Brown and Barber, 1985). The exact interpretation of anthropogenic indicator pollen can be problematic, due to the lack of modern analogues of prehistoric vegetation clearances (Behre, 1981). A valuable starting point in the clearer understanding of prehistoric vegetation change can be made by at least confirming, with a multiple site study, that observed patterns of pollen frequency change amongst the clearance indicators can be replicated from site to site (Simmons and Innes, 1985; Edwards and Ralston, 1984). The spectra from Fenemere and Birchgrove Pool follow the same sequences of vegetational change but, in detail, different pollen frequencies can be clearly seen, particularly the contrast in *Alnus* frequencies.

Regeneration or continued clearance?

Birchgrove Pool: 232–244 cm
Fenemere:        184 cm and above
Berth Pool:      *c*. 150 cm

Increases in the frequencies of *Betula*, *Quercus* and *Fraxinus* pollen at the above levels at Fenemere and Birchgrove Pool indicate a phase of woodland regeneration. Clear reductions are seen in the pollen frequencies of non-arboreal open-habitat indicator taxa. A degree of land abandonment is indicated; a similar phase to this is also observed at Crose Mere (Figure 11.1) and also following post-*Tilia* decline clearances at Whixall Moss.

A regeneration phase is less easy to discern at Berth Pool. A slight decrease in Gramineae and *Plantago lanceolata* pollen, and a decline in the pollen of Cruciferae can be seen between 160 and 150 cm at Berth Pool, with an associated increase in *Betula* and *Alnus* pollen. It is thought that this is the approximate point of correlation with the regeneration phase at Fenemere and Birchgrove Pool. A second pollen profile prepared from the deep-water zone of Berth Pool has produced an identical series of pollen frequency changes (Twigger, 1987). A change in the pattern of land use, from a phase of extensive clearance activity, to a phase of more intensive and more localized land use is indicated. The pollen spectra from Birchgrove Pool show that the regeneration phase is a discrete feature and is succeeded by a re-expansion of cleared land.

### Baschurch Pools conclusions

The contrast in pollen frequencies between Fenemere and Birchgrove Pool is more marked than the contrast between Berth and Birchgrove, particularly after the decline in arboreal pollen at 280 cm at Fenemere and 282 cm at Birchgrove Pool (Figure 11.5). The well-drained, brown earth soils of the Baschurch series (Crompton and Osmond, 1954) occupy more extensive tracts closer to Berth Pool and Birchgrove Pool, whereas more extensive areas of low-lying poorly drained peat soils exist around the margins of Fenemere and Marton Pool (Figure 11.3). The result of this difference in the proximity of soil types most suitable for clearance and farming could be a difference in the proximity of clearance activity to each of the meres, clearance activity being locally more intensive around Berth Pool and Birchgrove Pool. The radiocarbon date for the woodland regeneration phase at Fenemere, 1890 ± 50 BP (SRR 2920) or *c*. AD 60, suggests that regeneration occurred in late Iron Age or early Roman times. The fact that a regeneration phase cannot be clearly discerned at Berth Pool suggests that land occupation continued in the vicinity of this site while woodland regenerated around Birchgrove Pool and Fenemere. It is evident that pollen from the regenerated woodland around Birchgrove Pool and Fenemere did not reach Berth Pool in sufficient quantity to make the precise point of correlation between Berth Pool and Birchgrove Pool discernible. Although *c*. 3 ha in area, and therefore theoretically large enough to receive much of its pollen from a kilometre or more away (Jacobson and Bradshaw, 1981), the continued presence of open land close to Berth Pool appears to have provided a dominant non-arboreal pollen input to the site. The only arboreal pollen taxon to increase in frequency above the initial clearance at Berth Pool is *Alnus*, probably reflecting an expansion of alder carr on the fen peat or wetter brown earths around the mere and its enclosing basin. The clear potential for highly localized contrasts in fossil pollen spectra must therefore be borne in mind when vegetational inferences are attempted at the subregional scale on the basis of data from only one site. Beales (1980) acknowledges the fact that the evidence for woodland regeneration at 2310 ± 85 BP at Crose Mere (Figure 11.1) contrasts with evidence for a country-wide expansion of clearances in the Iron Age (Turner, 1981). The Welsh Marches have a high concentration of fortified Iron Age sites (Stanford, 1972a; 1972b); Beales suggests that the hillforts, many of which are on higher sandy ridges, had only a localized influence. It is believed, however, that the regeneration phase CMCP-9 at Crose Mere (Beales, 1980, Figure 11.1) is the same phase as seen at 180 cm at Fenemere and 240 cm at Birchgrove Pool; obviously there is a disparity in the radiocarbon dates. Many of the hillforts of the Welsh Marches were occupied at the suggested time of the regeneration at Crose Mere (Stanford, cited above). It is known that there was a cessation of activity at many hillforts during the mid first century AD (Varley, 1948). Cultural disruption associated with Roman occupation could have resulted in some land abandonment, and for this reason a late Iron Age or early Roman date for the regeneration is thought to be more likely than a mid Iron Age date. The results from the Baschurch Pools show that a single pollen profile cannot be satisfactorily representative in respect of the probable spatial variation in the foci of human activity at a time of cultural and economic change. The derivation of subregional pollen zones from one or two sites dominated by 'regional' pollen, especially during periods of cultural change, can only be very generalized, and cannot take account of local variation. The inter-site contrasts in the pollen spectra between Berth Pool and Birchgrove Pool suggest that the pollen source areas of sites of this size are very restricted (cf. Jacobson and Bradshaw, 1981). The results of this study indicate that the sampling strategy adopted in any palaeoecological study should always be designed to anticipate questions about single site representativity and profile replicability (Edwards, 1983); only then can detailed inferences of even localized environmental change be made with confidence.

## 11.6 LATE HOLOCENE VEGETATIONAL HISTORY II

Returning to the more generalized picture afforded by the Crose Mere diagram we see an increasing tempo of human activity throughout the uppermost zone, CMCP-10, which Beales (1980) splits into subzones 10a, b, and c. Beales argues (1980, pp. 156–7) that all three upper radiocarbon dates are too old due to the inwashing of old carbon in the soil and/or a hard-water effect, and proposes that zone 10a should be correlated, on pollen-analytical/agricultural history grounds, with late Romano-British times, 10b with Anglo-Saxon/medieval times and 10c with the period since AD 1600. This seems very reasonable, especially in view of the peak of Cannabiaceae pollen, thought to be due to hemp cultivation; the appearance of *Fagopyrum esculentum*, buckwheat, grown on poor sandy soils in medieval times, and the rise of pine, and the appearance of exotic spruce and hemlock pollen (not shown on Figure 11.1), during zone 10c. What is striking about this diagram, and the one from Berth Pool (Figure 11.4) is the way in which the pastoral nature of agriculture in Shropshire is reflected, compared to other detailed diagrams which extend to the present day from areas with more arable activity. For example the diagrams of Godwin (1962) from Kent; Oldfield (1969) from southern Cumbria; Godwin (1968) from East Anglia; Barber (1981a) from northern Cumbria and Waton (1982a,b) from Dorset and Hampshire, all show a much more varied herb flora with continuous curves of taxa such as *Artemisia*, and with cereal pollen extending at higher percentages up to the present day. Cross Mere and Berth Pool are in good agreement in showing an earlier phase of more arable activity, with curves of cereal-type/*Secale*, and Cannabiaceae probably referable to Anglo-Saxon times, followed by continuous high grass pollen levels, averaging 55 per cent of total land pollen at Berth Pool, and peaking at 80 per cent TLP in very recent times. Across the watershed in Wales, Turner (1964) considers that the beginning of really extensive deforestation and the spread of grassland around Tregaron bog dates from Iron Age times onward, and Turner (1981) details the extension of clearance in northern Britain at this time, concluding that by Roman times most of the landscape had taken on an open, modern appearance. In the lowlands of the Lower Severn valley Brown (1983a) and Brown and Barber (1985) document massive Iron Age deforestation with tree and shrub pollen at Ripple Brook falling from *c.* 90 per cent to *c.* 10 per cent TLP. Although many of these riverine peats are capped with an inorganic unit due probably to slope wash from arable fields (Brown, 1982), so that the pollen evidence does not continue up to the present day, there is no evidence to suggest a resurgence of woodland in the historical record (Rowley, 1972; Taylor, 1984), though of course human populations and farming practices varied enormously throughout the historic period. A difficulty which can only be solved by further work is that pollen evidence becomes sparser due to site destruction and is to a degree replaced by documentary evidence from which it is difficult to extract the kind of 'vegetational summary' that a pollen diagram provides. We know from work elsewhere that there were high tidemarks of agricultural expansion in the Middle Ages prior to the Black Death and during the Napoleonic Wars, and this is shown to some extent by the pollen diagrams from Hartlebury Common (Brown, 1984), and Ashmoor Common (Brown, 1983a), but

Rowley (1972, page 165) notes that in 'north-west Shropshire . . . despite the medieval colonisation, there was dense forest until the early sixteenth century'. Work in progress at Southampton on the recent ecological history of the area should show more exactly the history of this woodland and its clearance, but there is a need for detailed pollen diagrams linked to the documentary record, as pioneered by Oldfield (1969), before we can be more precise about the small-scale differentiation of vegetation during the historic period.

The linkages between vegetational history and fluvial change may not yet be known in sufficient detail but the outlines are clear, and perhaps no more striking example of change can be seen than the massively expanding grass pollen curve from Berth Pool, from virtually zero 4000 years ago to 75 per cent today, eloquent testimony of man's impact on the landscape.

## ACKNOWLEDGEMENTS

The work reported on the Baschurch Pools was supported by a NERC studentship held by Simon Twigger. We are grateful to Tony Brown, Chris Haslam and Colin Prentice for helpful discussions, and to Adrian Bayley and the staff of the Preston Montford Field Studies Centre for practical assistance. The radiocarbon dates from the Baschurch sites were kindly provided by the NERC Radiocarbon Laboratory, East Kilbride.

## REFERENCES

Aaby, B. (1976). Cyclic climatic variations in climate over the last 5,500 years reflected in raised bogs, *Nature*, **263**, 281–4.
Andersen, S. Th. (1973). The differential pollen productivity of trees and its significance for the interpretation of a pollen diagram from a forested region, in Birks, H. J. B. and West, R. G. (eds), *Quaternary Plant Ecology*, Blackwell Scientific, Oxford, pp. 109–15.
Barber, K. E. (1981a). *Peat Stratigraphy and Climatic Change: a palaeoecological test of the theory of cyclic peat bog regeneration*, Balkema, Rotterdam.
Barber, K. E. (1981b). Pollen-analytical palaeoecology in Hampshire: problems and potential, in Shennan, S. J. and Schadla-Hall, R. T. (eds), *The Archaeology of Hampshire: From the Palaeolithic to the Industrial Revolution*, Monograph No 1, Hampshire Field Club and Archaeological Society, pp. 91–4.
Barber, K. E. (1982). Peat-bog stratigraphy as a proxy climate record, in Harding, A. F. (ed.), *Climatic Change in Later Prehistory*, Edinburgh University Press, pp. 103–13.
Barber, K. E. (1985). Peat stratigraphy and climatic changes: some speculations, in Tooley, M. J. and Sheail, G. M. (eds), *The Climatic Scene: essays in honour of Gordon Manley*, Allen & Unwin, London, pp. 175–85.
Barber, K. E. (1986). Peat macrofossil analyses as indicators of the bog palaeoenvironment and climatic change, in Stead, I. M., Bourke, J. M. and Brothwell, D. R. (eds), *Lindow Man: the Body in the Bog*, British Museum Publications, London, pp. 86–89.

244 Palaeohydrology in Practice

Bartley, D. D. (1960). Rhosgoch Common, Radnorshire: stratigraphy and pollen analysis, *New phytol.*, **59**, 238–62.

Beales, P. W. (1976). Palaeolimnological studies of a Shropshire Mere, unpublished Ph.D. thesis, University of Cambridge.

Beales, P. W. (1980). The Late Devensian and Flandrian vegetational history of Crose Mere, Shropshire. *New Phytol.*, **85**, 133–61.

Beales, P. W. and Birks, H. J. B. (1973). Palaeolimnological studies in North Shropshire and Cheshire, *Shropshire Conservation Trust Bulletin*, No. 28, 12–15.

Behre, K. E. (1981). The interpretation of anthropogenic indicators in pollen diagrams, *Pollen et Spores*, **23**, 225–45.

Bell, M. and Limbrey, S. (eds) (1982). *Archaeological Aspects of Woodland Ecology*, BAR International Series 146, British Archaeological Reports, Oxford.

Bennett, K. D. (1984). The post-glacial history of *Pinus sylvestris* in the British Isles, *Quaternary Science Reviews*, **3**, 133–55.

Berglund, B. E. (1973). Pollen dispersal and deposition in an area of South-Eastern Sweden—some preliminary results, in Birks, H. J. B. and West, R. G. (eds), *Quaternary Plant Ecology*, Blackwell Scientific, Oxford, pp. 117–29.

Berglund, B. E. (ed.) (1979). *Palaeohydrological Changes in the Temperate Zone in the Last 15,000 Years: Subproject B, Lake and Mire Environments*. Project Guide, Vol. 1, Dep. Quat. Geol., Lund University.

Berglund, B. E. (1983). Palaeohydrological studies in lakes and mires—a palaeoecological research strategy, in Gregory, K. J. (ed.), *Background to Palaeohydrology*, Wiley, Chichester, pp. 237–54.

Berglund, B, E. (ed.) (1986). *Handbook of Holocene Palaeoecology and Palaeohydrology*, John Wiley, Chichester.

Birks, H. J. B. (1965). Late Glacial deposits at Bagmere, Cheshire, and Chat Moss, Lancashire, *New Phytol.*, **64**, 270–85.

Birks, H. J. B. (1977). The Flandrian forest history of Scotland: a preliminary synthesis, in Shotton, F. W. (ed.), *British Quaternary Studies*, Oxford University Press, pp. 119–35.

Birks, H. J. B. (1986). Late Quaternary biotic changes in terrestrial and lacustrine environments, with particular reference to north-west Europe, in Berglund, B. E. (ed.), *Handbook of Holocene palaeoecology and palaeohydrology*, Wiley, Chichester, pp. 3–66.

Birks, H. J. B. and Gordon, A. D. (1985). *Numerical Methods in Quaternary Pollen Analysis*, Academic Press, London.

Bonny, A. P. (1976). Recruitment of pollen to the seston and sediment of some Lake District lakes, *Journal of Ecology*, **64**, 859–87.

Bonny, A. P. (1980). Seasonal and annual variation over 5 years in contemporary airborne pollen trapped at a Cumbrian lake, *Journal of Ecology*, **69**, 45–70.

Bonny, A. P. and Allen, P. V. (1984). Pollen recruitment to the sediments of an enclosed lake in Shropshire, England, in Haworth, E. Y. and Lund, J. W. G. (eds), *Lake Sediments and Environmental History*, Leicester University Press, Chapter 9.

Bradshaw, R. H. W. and Webb, T. (1985). Relationships between contemporary pollen and vegetation data from Wisconsin and Michigan, USA, *Ecology*, **66**(3), 721–37.

Brenninkmeijer, C. A. M., Van Geel, B. and Mook, W. G. (1982). Variations in the D/H and $^{18}O/^{16}O$ ratios in cellulose extracted from a peat bog core, *Earth and Planetary Science Letters*, **61**, 283–90.

Brown, A. G. (1983a). Late Quaternary palaeohydrology, palaeoecology and floodplain development of the Lower River Severn, unpublished Ph.D. thesis, University of Southampton.

Brown, A. G. (1983b). Floodplain deposits and accelerated sedimentation in the lower Severn basin, in Gregory, K. J. (ed.), *Background to Palaeohydrology*, Wiley, Chichester, pp. 375–97.

Brown, A. G. (1984). The Flandrian vegetational history of Hartlebury Common, Worcestershire, *Proceedings of the Birmingham Natural History Society*, **25**, 89–98.

Brown, A. G. and Barber, K. E. (1985). Late Holocene palaeoecology and sedimentary history of a small lowland catchment in Central England, *Quaternary Research*, **24**, 87–102.

Buckland, P. C. and Edwards, K. J. (1984). The longevity of pastoral episodes of clearance activity in pollen diagrams: the role of post-occupation grazing, *Journal of Biogeography*, **11**, 243–9.

Chambers, F. M. (1981). Date of blanket peat initiation in upland South Wales, *Quaternary Newsletter*, **25**, 24–9.

Chambers, F. M. (1982). Two radiocarbon-dated pollen diagrams from high-altitude blanket peats in South Wales, *Journal of Ecology*, **70**, 445–59.

Chambers, F. M. and Price, S. M. (1985). Palaeoecology of *Alnus* (Alder): early post-glacial rise in a valley mire, north-west Wales, *New Phytol.*, **102**, 333–44.

Coles, J. M. and Orme, B. J. (1980). *Prehistory of the Somerset Levels*, Somerset Levels Project, Cambridge.

Coope, G. R. and Pennington, W. (1977). The Windermere Interstadial of the Late Devensian, *Philosophical Transactions of the Royal Society of London*, **B280**, 337–9.

Crompton, E. and Osmond, D. A. (1954). The soils of the Wem District of Shropshire, *Mem. Soil Survey G.B.*, **138**.

Davis, M. B. (1967). Pollen accumulation rates at Rogers Lake, Connecticut, during late-and post-glacial time, *Review of Palaeobotany and Palynology*, **2**, 219–30.

Davis, M. B., Moeller, R. E. and Ford, J. (1984). Sediment focussing and pollen influx, in Haworth, E. Y. and Lund, J. W. G. (eds), *Lake Sediments and Environmental History*, Leicester University Press, Chapter 10.

Davis, R. B., Brewster, L. A. and Sutherland, J. (1969). Variation in pollen spectra within lakes (1), *Pollen et Spores*, **11**, 557–71.

Dearing, J. A. (1983). Changing patterns of sediment accumulation in a small lake in Scania, southern Sweden, *Hydrobiologia*, **103**, 59–64.

Dearing, J. A. and Foster, I. D. L. (1986). Lake sediments and palaeohydrological studies, in Berglund, B. E. (ed.), *Handbook of Holocene Palaeoecology*, Wiley, Chichester, pp. 67–89.

Dupont, L. M. (1985). Temperature and rainfall variation in a raised bog ecosystem— a palaeoecological and isotope-geological study, thesis, University of Amsterdam.

Dupont, L. M. and Brenninkmeijer, C. A. M. (1984). Palaeobotanic and isotopic analysis of late Subboreal and early subatlantic peat from Engbertsdijkveen VII, The Netherlands, *Review of Palaeobotany and Palynology*, **41**, 241–71.

Edwards, K. J. (1983). Quaternary palynology: multiple profile studies and pollen variability, *Progs. in Phys. Geog.*, **7**, 587–609.

Edwards, K. J. and Hirons, R. K. (1982). Date of blanket peat initiation and rates of spread—a problem in research design, *Quaternary Newsletter*, **36**, 32–7.

Edwards, K. J. and Hirons, K. R. (1984). Cereal pollen grains in pre-Elm Decline deposits: implications for the earliest agriculture in Britain and Ireland, *Journal of Archaeological Science*, **11**, 71–80.

Edwards, K. J. and Ralston, I. (1984). Postglacial hunter-gatherers and vegetational history in Scotland, *Proc. Soc. Antiq. Scotland*, **114**, 15–34.

Edwards, K. J. and Thompson, R. (1984). Magnetic, palynological and radiocarbon correlation and dating comparisons with long cores from a Northern Irish lake, *Catena*, **11**, 83–9.

Flohn, H. and Fantechi, R. (eds) (1984). *The Climate of Europe: past, present and future*, Reidel, Dordrecht.

Fullen, M. A. and Reed, A. H. (1986). Rainfall, runoff and erosion on bare arable soils in East Shropshire, England, *Earth Surface and Processes and Landforms*, **11**, 413–25.

Van Geel, B. (1978). A palaeoecological study of Holocene peat bog sections in Germany and The Netherlands, based on the analysis of pollen, spores and macro- and microscropic remains of fungi, algae, cormophytes and animals, *Review of Palaeobotany and Palynology*, **25**, 1–120.

Geiger, R. (1965). *The Climate near the Ground*, Harvard University Press.

Godwin, H. (1959). Studies in the post-glacial history of British vegetation: XIV Late-glacial deposits at Moss Lake, Liverpool, *Philosophical Transactions of the Royal Society of London*, **B242**, 127.

Godwin, H. (1962). Vegetational history of the Kentish Chalk Downs as seen as Wingham and Frogholt, *Veroff. Geobotan. Inst. Rubel*, **37**, 83–99.

Godwin, H. (1968). Studies of the Post-Glacial History of British Vegetation. XV: Organic deposits of Old Buckenham Mere, Norfolk, *New Phytol.*, **67**, 95–107.

Godwin, H. (1975). *History of the British Flora*, 2nd edn, Cambridge University Press.

Godwin, H. and Willis, E. H. (1960). Cambridge University Natural Radiocarbon Measurements II, *Am. J. Sci. Radiocarbon Suppl.*, **2**, 62–72.

Granlund, E. (1932). De Svenska Hogmossarnas geologi, *Sver. geol. Unders.*, **C 26**, 1–193.

Gray, J. M. and Lowe, J. J. (eds) (1977). *Studies in the Scottish Late Glacial Environment*, Pergamon, Oxford.

Greig, J. (1982). Past and present lime woods of Europe, in Bell, M. and Limbrey, S. (eds), *Archaeological Aspects of Woodland Ecology*, BAR International Series 146.

Hardy, E. M. (1939). Studies in the post-glacial history of British Vegetation: V: The Shropshire and Flint Maelor mosses, *New Phytol.*, **38**, 364–96.

Hayward, M. and Fenwick, I. (1983). Soils and hydrological change, in Gregory, K. J. (ed.), *Background to Palaeohydrology*, Wiley, Chichester, pp. 167–87.

Hey, D. G. (1977). *An English Rural Community. Myddle under the Tudors and Stuarts*, Leicester University Press.

Hibbert, F. A. and Switsur, V. R. (1976). Radiocarbon dating of Flandrian pollen zones in Wales and Northern England, *New Phytol.*, **77**, 793–807.

Huntley, B. and Birks, H. J. B. (1983). *An Atlas of Past and Present Pollen Maps for Europe 0–13000 Years Ago*, Cambridge University Press.

Jacobson, G. L. Jr. and Bradshaw, R. H. W. (1981). The selection of sites for palaeovegetational studies, *Quaternary Research*, **16**, 80–96.

Janssen, C. R. (1966). Recent pollen spectra from the deciduous and coniferous-deciduous forests of northeastern Minnesota: a study in pollen dispersal, *Ecology*, **47**, 804–25.

Janssen, C. R. (1973). Local and Regional pollen deposition, in Birks, H. J. B. and West, R. G. (eds), *Quaternary Plant Ecology*, Blackwell Scientific, Oxford, pp. 31–42.

Jones, R., Benson-Evans, K. and Chambers, F. M. (1985). Human influence upon sedimentation in Llangorse Lake, Wales, *Earth Surface Processes and Landforms*, **10**, 227–35.

Kerney, M. P. (1963). Late-glacial deposits on the chalk of south-east England, *Philosophical Transactions of the Royal Society of London*, **B246**, 203–54.

Limbrey, S. (1983). Archaeology and palaeohydrology, in Gregory, K. J. (ed.), *Background to Palaeohydrology*, Wiley, Chichester, pp. 189–212.

Lockwood, J. G. (1983). Modelling climatic change, in Gregory, K. J. (ed.), *Background to Palaeohydrology*, Wiley, Chichester, pp. 25–50.

Lowe, J. J. (1978). Radiocarbon-dated Lateglacial and early Flandrian pollen profiles from the Teith Valley, Perthshire, Scotland, *Pollen et Spores*, **20**, 367–97.

Lowe, J. J. and Gray, J. M. (1980). The stratigraphic subdivision of the Lateglacial of North-west Europe, in Lowe, J. J., Gray, J. M. and Robinson, J. E. (eds), *Studies in the Lateglacial of North-west Europe*, Pergamon, Oxford, pp. 157–75.

Lowe, J. J., Gray, J. M. and Robinson, J. E. (eds) (1980). *Studies in the Lateglacial of North-West Europe*, Pergamon, Oxford.

Lowe, J. J. and Walker, M. J. C. (1984). *Reconstructing Quaternary Environments*, Longman, London.

Moore, P. D. (1966). Stratigraphical and Palynological investigations of upland peats in central Wales, unpublished Ph.D. thesis, University of Wales, Aberystwyth.

Moore, P. D. (1968). Human influence upon vegetational history in North Cardiganshire, *Nature*, **217**, 1006–9.

Moore, P. D. (1970). Studies in the vegetational history of Mid-Wales II: The Late-Glacial Period in Cardiganshire, *New Phytol.*, **69**, 363–75.

Moore, P. D. (1972). Studies in the vegetational history of Mid-Wales III: Early Flandrian pollen data from West Cardiganshire, *New Phytol.*, **71**, 947–59.

Moore, P. D. (1973). The influence of prehistoric cultures upon the initiation and spread of blanket bog in upland Wales, *Nature*, **241**, 350–3.

Moore, P. D. (1975). Origin of blanket mires, *Nature*, **256**, 267–9.

Moore, P. D. (1978). Studies in the vegetational history of Mid-Wales V: Stratigraphy and pollen analysis of Llyn Mire in the Wye Valley, *New Phytol.*, **80**, 281–302.

Moore, P. D. (1986). Hydrological changes in mires, in Berglund, B. (ed.), *Handbook of Holocene Palaeoecology and Palaeohydrology*, Wiley, Chichester, pp. 91–107.

Moore, P. D. and Chater, E. H. (1969). Pollen analysis and the changing vegetation of west-central Wales in the light of human history, *Journal of Ecology*, **57**, 361–79.

Moore, P. D., Merryfield, D. L. and Price, M. D. R. (1984). The vegetation and development of blanket mires, in Moore, P. D. (ed.), *European Mires*, Academic Press, London, pp. 203–35.

Morner, N. A. and Karlen, W. (eds) (1984). *Climatic Changes on a Yearly to Millenial Basis*, Reidel, Dordrecht.

Oldfield, F. (1969). Pollen analysis and the history of land-use, *Advancement of Science, London*, **25**, 298–311.

Oldfield, F. (1970). Some aspects of scale and complexity in pollen-analytically based palaeoecology, *Pollen et Spores*, **12**, 163–72.

Oldfield, F. (1983). Man's impact on the environment: some recent perspectives, *Geography*, **68**, 245–56.

Pannett, D. and Morey, C. (1976). The origin of the Old River Bed at Shrewsbury, *Shropshire Conservation Trust Bulletin*, No. 35, 7–12.

Pennington, W. (1970). Vegetation history in the north-west of England: a regional synthesis, in Walker, D. and West, R. G. (eds), *Studies in the Vegetational History of the British Isles*, Cambridge University Press, pp. 41–80.

## 248  Palaeohydrology in Practice

Pennington, W. (1975). A chronostratigraphic comparison of Late-Weichselian and Late-Devensian sub-divisions, illustrated by two radiocarbon-dated profiles from western Britain, *Boreas*, **4**, 157–71.

Pennington, W. (1977). The Late Devensian flora and vegetation of Britain, *Philosophical Transactions of the Royal Society of London*, **B280**, 247–71.

Pennington, W. (1979). The origins of pollen in lake sediments: an enclosed lake compared with one receiving inflow streams, *New Phytol.*, **83**, 189–213.

Pennington, W. and Bonny, A. P. (1970). An absolute pollen diagram from the British late-glacial, *Nature*, **226**, 871–3.

Pennington, W., Haworth, E. Y., Bonny, A. P. and Lishman, J. P. (1972). Lake sediments in Northern Scotland, *Philosophical Transactions of the Royal Society of London*, **B264**, 193–294.

Pennington, W., Cambray, R. S., Eakins, J. D. and Harkness, D. D. (eds) (1976), Radionuclide dating of the recent sediments of Blelham Tarn, *Freshwater Biology*, **6**, 317–31.

Prentice, I. C. (1981). Quantitative birch (*Betula* L.) pollen separation by analysis of size frequency data, *New Phytol.*, **89**, 145–57.

Prentice, I. C. (1985). Pollen representation, source area, and basin size: toward a unified theory of pollen analysis, *Quaternary Research*, **23**, 76–86.

Prentice, I. C. (1986). Forest-composition calibration of pollen data, in Berglund, B. (ed.), *Handbook of Holocene Palaeoecology and Palaeohydrology*, Wiley, Chichester, pp. 799–816.

Reynolds, C. S. (1979). The Limnology of the Eutrophic Meres of the Shropshire—Cheshire Plain, *Field Studies*, **5**, 93–173.

Rose, F. (1953). A survey of the Ecology of the British Lowland bogs, *Proc. Linn. Soc. Lond.*, **164**, 186–211.

Rose, J., Turner, C., Coope, G. R. and Bryan, M. D. (1980). Channel changes in a lowland river catchment over the last 13,000 years, in Cullingford, R. A., Davidson, D. A. and Lewin, J. (eds), *Timescales in Geomorphology*, Wiley, Chichester, pp. 159–175.

Rowlands, P. H. and Shotton, F. W. (1971). Pleistocene deposits of Church Stretton (Shropshire) and its neighbourhood, *Journal of the Geological Society*, **127**, 599–622.

Rowley, T. (1972). *The Shropshire Landscape*, Hodder and Stoughton, London.

Scaife, R. G. (1982). Late-Devensian and early Flandrian vegetation changes in southern England, in Bell, M. and Limbrey, S. (eds), *Archaeological Aspects of Woodland Ecology*, BAR International Series 146, pp. 57–74.

Shaw, J. (1972). Irish sea glaciation of north Shropshire—some environmental reconstructions, *Field Studies*, **3**, 603–31.

Shotton, F. W. (1967). Investigation of an old peat moor at Moreton Morrell, *Proc. Coventry and District Nat. Hist. & Scientific Soc.*, **4**, 13–16.

Shotton, F. W. (1978). Archaeological inferences from the study of alluvium in the lower Severn-Avon valleys, in Limbrey, S. and Evans, J. G. (eds), *The Effect of Man on the Landscape: the Lowland Zone*, Research Report No 21, Council for British Archaeology, London, pp. 27–32.

Simmons, I. G. and Cundill, P. R. (1974). Late Quaternary vegetational history of the N. York Moors, *Journal of Biogeography*, **1**, 253–61.

Simmons, I. G., Dimbleby, G. W. and Grigson, C. (1981). The Mesolithic, in Simmons, I. G. and Tooley, M. J. (eds), *The Environment in British Prehistory*, Duckworth, London, pp. 82–124.

Simmons, I. G. and Innes, J. B. (1985). Late Mesolithic land-use and its impact in

*Late Quaternary Palaeoecology of the Severn Basin* 249

the English Uplands, in Smith, R. T. (ed.), *The Biogeographical Impact of Land Use Change*, Biogeographical Monographs 2, Biogeography Study Group, Leeds, pp. 7–17.

Simmons, I. G., Rand, J. I. and Crabtree, K. (1983). A further pollen analytical study of the Blacklane peat section on Dartmoor, England, *New Phytol.*, **94**, 655–67.

Simmons, I. G. and Tooley, M. J. (eds) (1981). *The Environment in British Prehistory*, Duckworths, London.

Sinker, C. A. (1962). The North Shropshire meres and mosses: a background for ecologists, *Field Studies*, **1**, 1–38.

Sinker, C. A., Packham, J. R., Trueman, I. C., Oswald, P. H., Perring, F. H. and Prestwood, W. V. (1985). *Ecological Flora of the Shropshire Region*, Shropshire Trust for Nature Conservation, Shrewsbury.

Smith, A. G. (1970). The influence of Mesolithic and Neolithic man on British vegetation: a discussion, in Walker, D. and West, R. G. (eds), *Studies in the Vegetational History of the British Isles*, Cambridge University Press, pp. 81–96.

Smith, A. G. (1975). Neolithic and Bronze Age landscape changes in Northern Ireland, in Evans, J. G., Limbrey, S. and Cleere, H. (eds), *The Effect of Man on the Landscape: the Highland Zone*, CBA Research Report No 11, Council for British Archaeology, London, pp. 64–74.

Smith, A. G. (1981). The Neolithic, in Simmons, I. G. and Tooley, M. J. (eds), *The Environment in British Prehistory*, Duckworth, London, pp. 125–209.

Smith, A. G. (1984). Newferry and the Boreal-Atlantic transition, *New Phytol.*, **98**, 35–55.

Smith, B. M. (1985). A palaeoecological study of raised mires in the Humberhead Levels, unpublished Ph.D. thesis, University of Wales (Cardiff).

Smith, I. F. (1974). The Neolithic, in Renfrew, C. (ed.), *British Prehistory: a New Outline*, Duckworth, London.

Stanford, S. C. (1972a). The function and population of hill-forts in the Central Marches, in Lynch, F. and Burgess, C. (eds), *Prehistoric Man in Wales and the West*, Adams and Dart, Bath.

Stanford, S. C. (1972b). Welsh border hill forts, in Thomas, A. C. (ed.), *The Iron Age in the Irish Sea Province*, CBA Research Report 9, Council for British Archaeology, London, pp. 25–36.

Sturludottir, S. A. and Turner, J. (1985). The Elm Decline at Pawlaw Mire: an anthropogenic interpretation, *New Phytol.*, **99**, 323–9.

Sutherland, D. G. (1980). Problems of radiocarbon dating deposits from newly deglaciated terrain: examples from the Scottish Late-glacial, in Lowe, J. J., Gray, J. M. and Robinson, J. E. (eds), *Studies in the Late Glacial of North-West Europe*, Pergamon, Oxford, pp. 139–49.

Tauber, H. (1965). Differential pollen dispersion and the interpretation of pollen diagrams, *Danmarks Geol. Unders.*, Ser. II, 89.

Tauber, H. (1977). Investigations of aerial pollen transport in a forested area, *Dansk Botanisk Arkiv*, **32**(1), 1–121.

Taylor, C. (1984). *Village and Farmstead: a History of Rural Settlement in England*, George Philip, London.

Tolonen, M. (1984). Differences in pollen and macrophytic remains in sediments from various depths in a small kettle-hole lake in southern Finland, *Boreas*, **13**, 403–12.

Turner, J. (1962). The Tilia decline: an anthropogenic interpretation, *New Phytol.*, **61**, 328–41.

Turner, J. (1964). The anthropogenic factor in vegetational history. I. Tregaron and Whixall Mosses, *New Phytol.*, **63**, 73–90.

Turner, J. (1965). A contribution to the history of forest clearance, *Proc. R. Soc. London*, **B161**, 343–52.

Turner, J. (1981). The Iron Age, in Simmons, I. and Tooley, M. (eds), *The Environment in British Prehistory*, Duckworth, London.

Twigger, S. N. (1987). *Pollen diagram variability and prehistoric land-use change: A case study from the North Shropshire Meres and Mosses.* Manuscript in preparation.

Varley, W. J. (1948). Hillforts of the Welsh Marches, *Archaeological Journal*, **105**, 41–66.

Walker, D. (1966). The Late Quaternary history of the Cumberland Lowland, *Philosophical Transactions of the Royal Society of London*, **B251**, 1–210.

Walker, D. (1970). Direction and rate in some British Post-glacial hydroseres, in Walker, D. and West, R. G. (eds), *Studies in the Vegetational History of the British Isles*, Cambridge University Press, pp. 117–139.

Walling, D. E. and Webb, B. W. (1983). Patterns of sediment yield, in Gregory, K. J. (ed.), *Background to Palaeohydrology*, Wiley, Chichester, pp. 69–100.

Waton, P. V. (1982a). A palynological study of the impact of man on the landscape of Central Southern England and special reference to the chalklands, unpublished Ph.D. thesis, University of Southampton.

Waton, P. V. (1982b). Man's impact on the chalklands: some new pollen evidence, in Bell, M. and Limbrey, S. (eds), *Archaeological Aspects of Woodland Ecology*, BAR International Series 146, British Archaeological Reports, Oxford, pp. 75–91.

Waton, P. V., Barber, K. E. and Fasham, P. J. (in prep.). Palynological evidence for the early and permanent clearance of the chalkland near Winchester, Hampshire.

Watts, W. A. (1980). Regional variation in the response of vegetation to lateglacial climatic events in Europe, in Lowe, J. J., Gray, J. M. and Robinson, J. E. (eds), *Studies in the Lateglacial of North-West Europe*, Pergamon, Oxford, pp. 1–21.

Wheeler, B. D. (1984). British Fens—a review, in Moore, P. D. (ed.), *European Mires*, Academic Press, London, pp. 237–282.

Wiltshire, P. E. J. and Moore, P. D. (1983). Palaeovegetation and palaeohydrology in upland Britain, in Gregory, K. J. (ed.), *Background to Palaeohydrology*, Wiley, Chichester, pp. 433–451.

Wimble, G. T. (1986). The palaeoecology of the lowland coastal raised mires of South Cumbria, unpublished Ph.D. Thesis, University of Wales (Cardiff).

Palaeohydrology in Practice
Edited by K. J. Gregory, J. Lewin and J. B. Thornes
© 1987 John Wiley & Sons Ltd.

# 12

# Farmers and Farmland: Aspects of Prehistoric Land Use in the Severn Basin

S. LIMBREY

*Department of Ancient History and Archaeology, University of Birmingham*

People began to turn the forests of the Severn Basin into farmland in the later part of the sixth millenium BP, and by the time of the Roman conquest much of the lowlands had experienced a long history of farming, while in the uplands the mantle of moorland had been draped over the shoulders of the hills. Archaeological and palynological studies can be used to obtain a general impression of the temporal and spatial patterns involved, but palaeohydrology really needs a more sharply focused series of images: a series of well-dated land use maps. To appreciate why it may never be possible to obtain this, we need to consider the nature of the archaeological record.

## SOIL, SETTLEMENT AND SURVEY

Soil characteristics influence choice of locality for agricultural settlement, the use to which the land will be put, the subsequent fate of the remains of such settlement and land use, and the chance of their discovery. The weighting of soil factors in settlement depends on the level of agricultural technology: power available for cultivation, knowledge of and capability for drainage practices, range of crops available, manuring practices, and the relationship between crop-growing and animal husbandry. The survival of remains associated with a particular phase of land use depends on the subsequent history of soil conditions and soil disturbance, and these in turn are affected by each phase of agricultural activity. Some soils have sustained continuous or repeated arable use which disturbs and damages earlier archaeological remains. Others, because of the early onset of degradation under agricultural exploitation or because of location in climatically marginal areas, retain the evidence of one or two episodes of exploitation, no subsequent activity having significantly disturbed them. In the former case, remaining traces can be found by field walking, that is by systematic search of the surfaces of ploughed fields for the potsherds, flints and fragments of building materials brought to the surface as the plough cuts across truncated remains, and by aerial photography, which reveals the differences in moisture retention of the truncated pits and ditches and the bases of walls, floors or other stony remains, in the form of crop marks. The very fact that these remains are, usually, detectable only in arable land emphasizes that they are surviving traces caught in passing and that in these areas of frequent or continuous arable land use the greater part of the archaeological record has probably been lost or has been buried beneath soil accumulations in hollows and lower slopes beyond the reach of surface or aerial detection methods. In the case of areas which have experienced few and short phases of ploughing, on the other hand, there is a better chance of survival of archaeological remains. Those remains originally built of durable materials which formed upstanding features may remain as detectable surface relief, even if reduced to faint humps and bumps visible only as shadow marks or when emphasized by light snow or blown leaves. Anything having no surface visibility, such as the remains of settlements composed of wooden buildings, and all the ground level and subsurface traces of occupation, may be well preserved but undetectable under permanent pasture, woodland or moorland.

As in other field sciences, distribution maps reflect the distribution of field workers, whose density is correlated with present kind and intensity of land use—there are simply more people around in areas where arable land predominates. Moreover, archaeological survey is perhaps a more strongly self-reinforcing activity than equivalent procedures in other subjects, so dense concentrations and vacant areas can reflect real patterns. The motivation and the status of all survey activity in an area needs to be known before a pattern can be assessed. A student may carry out survey as part of a project for no other reason than location of home base; an amateur archaeologist might carry out survey regularly within a practicable range of home or where there is agricultural or forestry work being done; school teachers might work with pupils. This is particularly likely to be self-reinforcing since while an individual enthusiast may be conscientious in checking negative results or persisting with areas of low productivity, a teacher is more likely to return to localities in which the boredom threshold will not be crossed. Professional survey is usually a response to a specific threat, such as gravel-winning, road-building or other forms of development, and so is very selective, as well as being concentrated in areas of higher population density.

Success in survey is a function of soil type and timing of survey. The light soils which would have been the first choice for cultivation by farmers having a low tractive power resource and lightweight tools are also those in which archaeological remains are easiest to detect, whether by field walking or by aerial photography. Heavy soils, which require greater power for cultivation and may need draining, will have been left under forest or used for grazing until the iron-shod plough and its draught team were developed, and so will have a lower density of settlement until then. Such soils, however, are reluctant to yield up any evidence they do contain when they are under arable use today, since they can be physically difficult to walk, farmers are reluctant to give access to them for fear of structural damage, rain does not wash clay off the flints and potsherds and so render them visible, and their water deficit does not often reach the levels at which crop marks develop. These heavy soils also conceal their remains most effectively when under grass. Contrasts in settlement evidence are therefore exaggerated by, and in some cases entirely produced by, contrast in soil type.

We can now begin to appreciate that the palimpsest of successive agricultural landscapes as presented by the archaeological record offers opportunities for understanding different periods according to the sequence of more and less drastic soil disturbances. In the lowlands, on those soils which have always supported the bulk of the population, much of the evidence for earlier periods has been destroyed or obscured by later ones. In the uplands, at altitudes at which arable land use has been limited to short periods of favourable climate or high economic stimulus, and also above this where land has only ever been used for grazing, there is a good chance of preservation of remains. Many of those remains are upstanding earthworks and stone structures which are readily detected. In the lowlands too there are areas of fragile soils where early use has been followed by marginality. It is one of the paradoxes of archaeological research that a disproportionate amount of our understanding of prehistoric societies is derived from these areas of low intensity, and little continuity, of settlement. For palaeohydrology, however, the impact of a few people in agriculturally marginal areas may well be at least as significant as that of their more numerous contemporaries living in the richer lands.

## DATING AND THE CHRONOLOGICAL FRAMEWORK

For the prehistoric period, the radiocarbon timescale diverges from solar years in a way which causes quite considerable, and very uneven, contraction of the actual perods concerned. While it is simpler, until an agreed calibration is in general use, to stick to radiocarbon dates, it has to be remembered that apparent rates of change and lengths of periods can be misleading. In Figure 12.1 an outline of prehistoric chronology is given, and the approximate real timescale is indicated.

The chronological framework indicated in Figure 12.1 is based on large numbers of dates from excavated sites throughout Britain. Relatively few

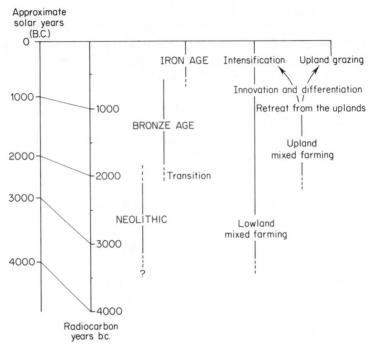

FIGURE 12.1   Archaeological chronology and changing land use in the
Severn Basin

radiocarbon dates for sites in the Severn Basin have been published. Surface
finds and unexcavated sites are dated by typological comparison, and can
only be set in broad periods. When using archaeological evidence for palaeohydrological purposes, we need to know not just when a particular site was
occupied, but for how long its occupation continued. Given the uncertainty
inherent in radiocarbon dating, demonstrating contemporaneity of occupation is difficult even when sites have been excavated and it is impossible
for unexcavated ones. This is a major handicap in assessing settlement density
and in testing hypotheses about the relationship between, for example,
upland and lowland farmsteads, and it is probably the biggest limitation on
the use of archaeological evidence for palaeoecological purposes.

## THE AGRICULTURAL RESOURCES

### Soils

The range of soil types in the Severn Basin is as great as that in the whole
of Britain, and in both uplands and lowlands there are soils contrasting

strongly in their potential for use by prehistoric farmers. In the lowlands of the east, the extremes of texture are represented by soils on the coarse sandstones and the marls of the Triassic period, together with the drifts derived from them, from blown sands to boulder clays. The history of these soils has been one of convergence as the forest soils developed, followed by divergence and exaggeration of their adverse characteristics under the impact of agriculture, some of the coarse siliceous soils podzolizing, and the clays becoming heavily gleyed. The podzolization certainly happened in prehistoric times, and many of these soils have been subsequently reclaimed for agriculture after a period as heathland, while others were always maintained in a more fertile condition by careful management. The history of the gleyed soils is more difficult to determine, but forest clearance would have exposed them to more water at the same time as making their drainage worse, and depletion of organic matter and mechanical damage under agriculture would have caused loss of structure, so they could only be farmed with careful management, and the use of drainage techniques.

In the uplands the soils on the Devonian sandstones of the Black Mountains and the clay-rich soils formed on Ordovician and Silurian shales and mudstones, have textural extremes which are less strongly marked. Again, the early history of these soils would have been convergent, but it remains to be determined to what extent adverse climatic conditions had already produced divergence under forested conditions. The extent to which human impact was responsible for eventual convergence on the complex of variably gleyed and podzolized soils we see today, and the timing of the critical phases of that impact, is probably very variable. The soils themselves are locally variable, and the emphasis given to the question of the causes and the consequences of the development of blanket peat by both archaeologists and palaeobotanists (see Barber and Twigger, Chapter 11 this volume, for references) is, for the Severn Basin, and indeed for the whole of upland Wales, somewhat exaggerated in relation to the actual area of peat. The western uplands of the Severn Basin are above all the province of brown podzolic soils, with the plateaux mostly occupied by stagnopodzols whose organic horizons only become deep enough to be classified as peat in relatively restricted areas (Soil Survey of England and Wales 1983, Rudeforth *et al.*, 1984). While the significance of the development of these waterlogged and acid soils for the farming communities having access to them is undoubtedly great, the resource they offer is not so poor as that of peat, since although in the edaphically wettest areas they carry the same flora as blanket peat, elsewhere plants are still rooting into the mineral soil, and they are dry enough in summer to provide some grazing.

As important in terms of agricultural history are the soils which lie between the ultimately geologically and climatically determined extremes. A very large proportion of the basin, and particularly of the broad hilly zone where upland and lowland meet, where climatic factors too lie between the extremes, has soils formed on limestone, calcareous and non-calcareous fine sandstones and siltstones, and shales. Some of these soils, which are predominantly of the loamy textural classes and when well-structured will produce a light tilth, are probably little changed from their condition under forest. However, there are substantial areas where gleying has become serious enough, whether as a result of change in groundwater level which may be due to human or climatic factors, or due to development of argillic horizons, again of uncertain timing and causation, to render them far less flexible in agricultural use than they were when they were first farmed. The fine sand and silt content of many of these soils, and especially of the upper horizons if clay migration has occurred, renders them particularly susceptible to erosion. Such erosion has strong implications for the archaeological record, in addition to its hydrological and agricultural significance.

**Domestic animals**

From the beginnings of agriculture in Britain sheep, cattle, pigs and a few goats were kept; domestic dogs had been kept by hunter-gatherers long before this. Because of the factors affecting the acquisition and utilization of animal products and the fate of discarded bone, it is difficult to interpret assemblages of bones in terms of ratios of the different species with any degree of certainty. In spite of strenuous efforts by archaeologists on the basis of very inadequate information, there is no real evidence that a high degree of specialization within this quartet occurred, though it is reasonable to expect that there were regional variations related to climatic factors, and changes with time as availability of pasture increased and agriculture spread. The status of the horse in Britain in the early stages of agricultural settlement is uncertain, but they were not important farm animals until the later Bronze Age and thereafter they were both a meat source and a working animal.

Because preservation of bone is poor in acid soil, there is little direct evidence for the farm animals for the uplands and for many areas of gravels and sands in the lowlands of the Severn Basin.

**Crop plants**

The early farmers in Britain brought with them wheat and barley. They could have brought beans and peas, which were known in northern Europe by that time, but we have no evidence for them here until the late Bronze Age. Pulses are much less likely to get preserved by charring than are the cereals, and we have no evidence from waterlogged deposits for the early stages of farming. This is an important point so far as land use and agricultural resources are concerned, relating to the protein economy of the people and the nitrogen economy of the soil, and it seems unlikely that pulses were not grown and some rotation of crops practised.

The status of rye as a weed introduced with wheat or as a crop in its own right is discussed by Chambers and Jones (1984) in the context of the identification of rye pollen in deposits of Llangorse Lake, on the southern watershed of the Severn Basin, at the remarkably early date of mid-fourth millenium BP. In the case of oats, charred remains do not usually allow identification of the domestic form, and since there are indigenous wild oats the occasional grains which are found could well be weeds. It is not until the mid-third millenium BP that we can be sure that oats were deliberately grown as a crop. The possession of a range of crops, including both emmer and bread wheat, naked and hulled barley, would have allowed the farmers to use the full range of soils, apart from those which were waterlogged or markedly acid. The addition of oats and rye, and spelt wheat, again probably a later Bronze Age introduction, would have extended the climatic tolerance of the arable system.

Although much has been written about the changing proportions of different crops through the prehistoric period and the significance of an apparent predominance of wheat or barley for the exploitation of different soil types or for farming under different climatic conditions, the record is biased by the accidents of preservation. It is only when chaff fragments and weed seeds are studied that crops can be identified as having been grown in the area around a site rather than acquired by trade and the condition of the soil in which they were grown can be established. This kind of study is of fairly recent origin and so far few sites have provided suitable samples for a comprehensive study of crop remains from the earliest stages of agriculture in Britain.

Jones (1981) suggests that the great period of agricultural innovation was the later Bronze Age, the first half of the third millenium BP, when diversification of crop types occurred and allowed the fullest range of soil types to be exploited.

## THE ARRIVAL OF FARMERS: THE NEOLITHIC PERIOD

The earliest dating for farming communities in the Severn Basin is given by their large communal monuments. The megalithic tombs of the Severn–Cotswold group in the foothills of the Black Mountains and the adjacent lowlands and on the Cotswold scarp, tell us that farmers were present from the later part of the sixth millenium BP. In the west aerial photographs have shown a henge and a cursus as crop marks on low terraces of the Severn south of Welshpool and a cursus and a massive post circle on the margins of the Radnor Basin. Though undated, it is suggested that these sites might be earlier here than their analogues in England, since dates at the beginning of the fifth millenium BP for henges at Llandegai in North Wales suggest that this type of monument spread from there (Houlder, 1968). Polished stone axes tell us that the Severn Basin was crossed by important communication routes: from Craig Llwyd in North Wales and the Preseli Mountains in the west, axes were distributed throughout the basin and far beyond. In the east of the basin, a crop mark at Wasperton, on a terrace of the Avon, is possibly a henge (Crawford, 1985) and a cursus appears as a crop mark not far away at Wellesbourne (Fennell, 1978).

The farms and farmland of these people are difficult to identify. Land boundaries of some kind are represented by the 'pit alignments', of which a complex in the Four Crosses and Llanymynech area on the gravels of the Severn–Vyrnwy confluence are the best example in the region. Although these are thought to be Neolithic by analogy with dated samples elsewhere, and certainly earlier than the Bronze Age barrows at Four Crosses, their dating within the Neolithic is unknown here. Traces of Neolithic settlements have been identified in excavation of later sites at Wasperton, at Bromfield on the gravels north of Ludlow, where a date in the early part of the fifth millenium BP was obtained (Stanford, 1982), at Sharpstones Hill, near Shrewsbury (Barker, Halden and Jenks in press), at Fridd Faldwyn, near Montgomery, on The Breiddin, overlooking the Severn where it emerges from the Welsh hills (Musson, 1976) and on the nearby Long Mountain (Britnell, 1982). Neolithic occupation at these last two sites is dated to the later part of the fifth millenium BP.

These sites are known either because later burial mounds and hillfort ramparts have protected them from total destruction and have attracted excavation, or because, even if they are almost obliterated, excavation of a later site has resulted in scrutiny of the lower horizons of soil profiles for man-made intrusions. The isolated finds and flint scatters in plough soil which represent the greater part of our knowledge of the distribution of Neolithic presence probably also, in many cases, reflect the former existence of settlements, but the very manner of their discovery demonstrates the degree to which they have been damaged, and they have not been thought to merit excavation. It is therefore impossible to relate these sites to earlier or later phases of farming colonization or to assess the extent to which people from the farming settlements utilized the still forested land for hunting or other wild resource procurement.

The undated flint sites do, however, support the evidence from burial and ritual sites for a lowland preference until towards the end of the period and for the selection of the more easily worked soils, and these factors emphasize that it was a mixed farming economy: pastoralists could have used the clay lands and the mountains. In view of the survey bias mentioned above, the preference for the lighter soils needs substantiation. This is provided by a comparison of the distribution of Neolithic flints with those of the hunter-gatherers of the Mesolithic period. The marl and sandstone lands of Worcestershire and Warwickshire are blessed with the attentions of a number of very assiduous flint collectors, and it is found that whereas both Mesolithic and Neolithic flints occur on the lighter soils, the heavy soils have Mesolithic alone; the farmers shunned them (Barfield, pers, comm.). Moreover, where flints of the two periods are found in the same place, the earlier ones are patinated while the later ones are not, indicating that on these sandstones the forest soils still had a pH high enough to attack silica, whereas by the time the farmers had cleared and used them the soils had become acid. Whether the time gap which this demonstrates means that the Neolithic remains are those of the later part of the period is uncertain. In the south of the basin, where the Megalithic tombs demonstrate the earliest farming presence, the soils are more strongly buffered so no comparison can be made.

Why do we not see more evidence in pollen diagrams of these lowland farmers of the late fourth and the fifth millenium BP? The traditional concept for the period, of shifting agriculture, was derived from too heavy a dependence on the tropical swidden model in the absence of a suitable analogue for primitive agricultural colonization of a temperate forested landscape. It is not necessarily applicable where soils are immature and have a high nutrient reserve. Even the sandstone soils, which did become acid during the Neolithic period, are better in this respect than tropical forest soils. After several thousand years of development under forest most of the soils of the basin would have had a sufficient clay content to inhibit leaching and provide a high cation exchange capacity. These loamy soils have supported agriculture for a further four to five thousand years, so why would they lose fertility so rapidly that Neolithic farmers had to move on within their first few hundred years of exploitation? These were not novices, but experienced farmers with generations of experience acquired as they spread across Europe and the farming system adapted to the edaphic conditions and growing season of the north. That they were not skilled soil managers seems unlikely, particularly since less work is involved in dung-carting than in felling and fencing. In the absence of extensive grazing lands, and with predators about and, in the case of cattle and pigs, the wild males only too eager to undo the effects of selection for desirable domestic characteristics, animals must often have been stalled, tethered or corralled, and dung had to be disposed of somewhere. This again is different from the tropical swidden system, which does not involve the herd animals. Can we therefore envisage an early farming landscape for the lowlands of the Severn Basin in which farms had considerable permanence and replaced each other only on a timescale of centuries rather than decades, rather than a restless shuffling about of fields and pasture? Such farms would be located on river terraces and on the brows and spurs of low hills near springs and small streams, which is where flint finds commonly occur, and would be surrounded by forest largely screening their pollen spectra from valley floor locations. The valley floors, lacking the grasslands on stable floodplains later built from the sediments derived from agricultural erosion, would not yet be attractive. The meadows created naturally by the drainage of former beaver-dammed lakes, postulated as ready-made farmland by Coles and Orme (1984), are simply not there in the pollen evidence from just those valley floor sites which might have been expected to record them.

In these circumstances, not only would much of the agricultural land be invisible palynologically, but such erosion as occurred would contribute little to the fluvial environment. Soil would remain trapped until there was sufficient continuity of arable land and trackways for it to get into the rivers. That soil did move in substantial quantities during the Neolithic period is demonstrated elsewhere. Studies of dry valley fills in the chalklands demonstrate considerable erosion of arable soils, and the inclusion of potsherds in the soil shows that domestic rubbish certainly was disposed of on ploughed fields (Bell, 1983). Such detailed studies of colluvial soils in areas of known settlement have not been done in the Severn Basin, but there is no reason to suppose that they would not show similar evidence.

The pollen diagram from Llyn Mire (Moore and Becket, 1971), does show evidence of agriculture from the elm decline onwards. The soils in this area are predominantly silty clay loams which are now seasonally waterlogged and are used for pasture. These are soils which if, in the early stages of exploitation, they had not developed adverse moisture conditions would have been good arable soils. At Llangorse Lake, in the heartland of the Severn Cotswold chambered tombs, the evidence of forest clearance from the same period coincides with evidence of soil inwash (Jones, Benson-Evans and Chambers, 1985). The soils here are coarse loamy and fine loamy brown earths which even today are only moderately acid and present no drainage problems.

The coincidence of the 'elm decline' with the earliest Neolithic colonization is no longer supported by the archaeological evidence. There are now numerous dates from Neolithic sites which fall before the elm decline, and of cereal pollen below it in pollen diagrams (Edwards and Hirons, 1984). Though palynologists are still tending to use the elm decline as a marker for the first impact of farming on the vegetation, and though it often does coincide with the first appearance of weeds of cultivation and pasture, and of soil erosion, there must be more to it than simply the attack of farmers on the forest. Whereas earlier writers have assumed that if people of that period were using elm fodder for their livestock they could have had the effect observed in the pollen record, Rackham (1980) and Rowley-Conwy (1982) have both made calculations on the scale of the impact, and both, using different routes, have arrived at the conclusion that it would be quite impossible for enough people and livestock to be present—there would have been no room for the forest! The finding of an elm bark beetle in peat deposits on Hampstead Heath, close to the level at which the elm decline occurs (Girling and Greig, 1985), does not prove that disease was the cause of the death of elms, but at least we now know that the disease vector was present, and Rackham suggests that the presence of trees damaged by farmers would facilitate its spread.

Could cause and effect by reversed? Was the spread of agricultural communities facilitated by the changes in forest structure resultant upon the death of elms? As Rackham points out, very large areas must have been involved in these changes, and it could be that the erosion and weeds which we see at this time are the effect of people exploiting the land vacated by the elms. On recent experience, it takes several years for the trees to die and it is a very conspicuous process. As the canopy thins, undergrowth invades and eventually saplings of other trees, as well as elm suckers, form a thicket. The difficulty of felling mature trees with a stone axe is considerable, and ethnographic parallels suggest ringing by axe or fire. Exploitation of areas suffering tree death by disease would be exactly the same as that of areas being killed by ringing, either by putting livestock in to establish pasture from the invading growth or by cultivation. Useful grazing develops with the loss of some 50 per cent of canopy, whereas cereal growing at this latitude demands a considerable area of complete clearance to avoid shading. Perhaps we can postulate progressive establishment of pasture and arable land by the manipulation of the grazing animals and pigs, the former to ensure that clearings remained open and the latter to till the soil ready for arable crops. It is often said that it is impossible to break the sod of a herbaceous ground cover with a simple adze, and spade paring or hoe cultivation are postulated, but pigs will do the job with enthusiasm

## EXPANSION INTO THE UPLANDS

A great deal of discussion of human impact in the uplands begins with the attribution of the elm decline and associated herbaceous pollen to the arrival of farmers at about 4987 BP. There are, however, few pollen diagrams for the Welsh uplands in which dating at this level is secure and the stratigraphy sufficiently expanded for discrimination within the Neolithic timespan, and there is no archaeological evidence for any substantial presence much above 300m at this time, or indeed in the preceding Mesolithic period in Wales. A very few of the Severn Cotswold chambered tombs lie a little above 300m on the southern watershed of the basin, where they seem to be marking significant natural boundaries. Otherwise, their highest locations are very noticeably just within the present boundary of enclosed farmland, and most of them are lower down. Distribution of chance finds of flints confirms this altitudinal limit.

The discovery of a late Neolithic cremation, followed by a settlement dated by radiocarbon to the later part of the fifth millenium BP beneath a Bronze Age burial mound at 370 m on Long Mountain, initiates the archaeological evidence of upland colonization, and it substantiates the palynological record of forest clearance and cereal growing at altitudes above 300 m from this time onwards. The period of rapid change in the cultural attributes of the population in the period between 4087 and 3787 BP marks the transition from late Neolithic to Early Bronze Age. It sees the change from collective burial under long mounds or in chambered tombs to single burial or cremation under round mounds, the use of the distinctive pot type known as beakers, and the introduction of metals. By the time the transitional period is over, we see a proliferation of cairns and stone circles in upland areas which have never again experienced such abundant human presence.

In the lowlands, there are also traces of the 'beaker people', who are not as conspicuous in the Severn Basin as in some other parts of Britain, followed by a great increase in archaeological visibility of the population of the early Bronze Age, the ditches of their round burial mounds surviving the levelling effects of later agriculture in the form of crop marks. There are few upstanding barrows in the lowlands of the Severn Basin, but some do survive, notably at Bromfield, near Ludlow. A large mound at Tenbury Wells, on what must then have been a gravel island in the floor of the Teme valley, could be of this period, and there is an enormous barrow at Drayton in north Worcestershire.

We are still short of evidence of the settlements themselves, but people there were in plenty. It has long been an accepted view in archaeology that the Bronze Age saw a switch to pastoralism, and the expansion into the uplands is taken as one of the indications of this, but it is just at this period that cereal pollen appears very commonly in upland pollen diagrams, and one of the common features of the uplands is the occurrence of 'clearance cairns'. These are heaps of stones which today dot the moorlands in many parts of Britain. They are evidence of the clearance of stone which would interfere with cultivation and reduce the area of soil for the crop. Whether the reduction of pasture area would also be seen as a reason for stone clearance is less clear, but it seems more probable that the cairns are evidence for arable land, because the stones themselves would be an increasing nuisance as cultivation continued and erosion of the plough soil exposed them at the surface. Whether the stones would be a problem at all on first use of the soil may be doubted—it may be that clearance cairns are an indication not of the first use of the land but of the beginnings of its degradation. With the systematization of upland land use, field boundaries were made permanent. Survey in the Severn Basin is only beginning to show the kind of Bronze Age field systems which are known from other upland areas and have been studied in great detail on Dartmoor (Spratt and Burgess, 1985; Fleming, 1983). It would appear that the land organization involved did not develop to the same extent in Wales, but where such field systems are being studied, in the Brecon Beacons just outside the Severn Basin (Briggs, 1985; Leighton, 1986), it is found that the Bronze Age systems disappear as they approach the only other field boundaries of the uplands, those of the medieval period. The later people, who did not expand so far uphill, used the stones for their walls. On the lower ground, it is becoming clear that the agricultural landscape of the Bronze Age has been submerged in that of medieval times, some boundaries being incorporated into the later system, others remaining visible only as faint lynchets, if at all. In the lowlands, where Bronze Age boundaries were banks and ditches rather than walls, and where agriculture has had a more continuous impact, they are only occasionally found as crop marks or as ditches found during excavation of later sites.

The use of bronze and gold had a strong impact on communication routes and, presumably, on political and economic relationships of communities involved in supply, and the rivers and upland routeways of the Severn Basin must have been important. Though tin could only come from south-west England or Brittany, copper sources in Snowdonia were exploited (Northover, 1986), and there are ores in the Shropshire hills which might have been used. A concentration of upland monuments in the Corndon Hill area might be related to the use of these ores as well as to the use of Corndon stone for axes at this period. Welsh gold would have been traded through the basin. Metallurgy must have increased the demand for fuel enormously. Though there is evidence for coppicing before this in the Somerset Levels, it may be doubted whether conservation of woodland was yet regarded as necessary, clearance for farmland providing adequate supplies. Coppicing does, however, provide the right size of the right kind of wood for charcoal production, and we should probably include the coppice cycle as one of the hydrologically significant vegetation patterns from this time on.

## THE RETREAT FROM THE UPLANDS

The uplands were populated for nearly a thousand years and then deserted. The disappearing trick of the late Bronze Age people is archaeologically disconcerting. In part it is a matter of their visibility—stone monuments and burial under mounds went out of fashion, and with the adoption of metal-cutting tools and weapons, which are recycled rather than being discarded and do not survive well in the soil anyway, there are fewer flints to be found. In the uplands the agricultural land does go out of use, the abandoned fields suffering the consequences of neglect and their soils succumbing to waterlogging, podzolization or both. In the lowlands the cremations have no surface markers and are found by chance. At Bromfield, the use of the barrows was followed by establishment of a cremation cemetery which was in use for several hundred years in the later Bronze Age (Stanford, 1982), and in Warwickshire a cemetery has been excavated at Ryton-on-Dunsmore (Bateman, 1978).

Metal objects of this period have frequently been found in rivers and river deposits, and a preoccupation with water is marked by the only common domestic or working site of the period, the burnt mounds. These are heaps of burnt stones, often associated with pits or troughs which could be fed with water from an adjacent small stream, and they can be very common when systematic search of the banks of lowland streams is carried out. It is usually supposed that they are the cooking places of nomadic herdsmen, but it has been suggested that they are sauna baths (Barfield and Hodder, 1981) or washing places for clothes (G. Williams, pers, comm.). We are accustomed to glimpsing prehistoric people in only certain aspects of their existence, but a 'laundry people' may seem the oddest example, unless the development of the fleece of the sheep at this time is a clue (Ryder, 1983). Could these be wool-washing or fulling sites? Burnt mounds are common in the eastern part of the Severn Basin. Less survey has been done in the west, but they do occur in the lowland basins of Central Wales (C. Martin, pers, comm.)

## AGRICULTURAL INNOVATION AND INTENSIFICATION

The first palisades of hillforts were built during the late Bronze Age, and in the following period the Welsh Marches saw the development of one of the densest concentrations in Europe of these heavily defended sites, Stanford (1972) suggested a total occupation of the Marches by hillfort territories, implying a population density whose food requirements would have left little unexploited land in the region. It is for the Iron Age that we begin to have really abundant evidence of where people lived and farmed, with occupation of the lowland river terraces, where arable land on gravel soils is so respon- sive to production of crop marks, showing continuous runs of farmsteads and enclosed fields. Field survey provides abundant evidence that settlement was common on the heavier soils, where aerial photography is unproductive. In the western part of the basin, the predominance of permanent pasture today in lowland areas which were cultivated in medieval times means that these Iron Age farms tend to be more difficult to find, but where they are detected, and on the higher ground where destruction has been less severe, they appear to be more discrete units, separated from their neighbours, rather than the enclosed fields being contiguous. Analysis of the location of these Iron Age settlements in part of the Upper Severn Basin suggests that the upland/ lowland dichotomy in agriculture was well established (Collens, in press).

## CONCLUSIONS

In this survey of prehistoric land use in the Severn Basin, no attempt at complete coverage of the archaeological evidence has been made, but an indication has been given of where farmers were living and what they might have been doing. For large parts of the basin it is now easy to produce print-outs of all known archaeological sites and find spots, variously broken down by period and type; because of the inequalities of preservation and discovery and the uncertainties of dating and time span this is regarded as being an exercise more misleading than useful for our present purpose.

Archaeology cannot describe farmland in any detail and for the earlier part of the prehistoric period can only locate it and measure its extent under the ideal conditions of preservation which sometimes occur in the areas which have been least often exploited. It is therefore impossible to find very much of the earliest farmland, which lies entirely within the zone which has been farmed ever since. Furthermore, there is a strong discrepancy between the archaeological and the palynological evidence which can only be resolved by the collaborative application of archaeological survey and excavation, studies of buried soils, and pollen analysis in selected small catchments. With the spread of agriculture in the uplands, preservation improves and survey is more productive and the new farmland is now in an area which has more obvious opportunities for pollen analysis, and the records of the two fields of study come more into line. Here, however, the relationship between them of mutual predation rather than cooperation has resulted in ideas, such as that of Bronze Age pastoralism, being recycled uncritically, neither field being aware of the extent to which such ideas are derived from its own outdated literature. To find out what kind of land use was being practised in the areas being cleared of forest, which the pollen record shows and which the archaeological survey results demand, detailed cooperative studies of particular localities are again needed.

The vagaries of the archaeological record make the later Bronze Age something of an enigma, and here palynology documents progressive attack on the forest cover by people who persist on being difficult to study archaeologically. Ideas about population collapse at this period cannot be sustained in the face of the vegetational record, and here we should turn for comparison to the post-Roman period, where the archaeological evidence for population reduction, at least in some areas, is matched by evidence for the encroachment of the woodland which emerges into history. Whatever was going on in the later Bronze Age, this kind of reversion of farmland is not apparent, and by the end of that period great agricultural advances had been made. A very high population in the Iron Age was farming a landscape now probably markedly differentiated into a lowland zone having large areas of continuous fields, with ditched boundaries which emphasize the ability to manage drainage, and a highland zone in which the settled valleys gave way to the individual farms of the hills, embedded in a mosaic in which woodland was still well represented, and then to the moorlands, where soils by now acid and waterlogged supported rough grazing.

## REFERENCES

Barfield, L. H. and Hodder, M. (1981). Birmingham's Bronze Age, *Current Archaeology* No. 78, 198–200.
Barker, P. A., Halden, R. and Jenks, W. E. (in press). Excavations on Sharpstones Hill near Shrewsbury, *Transactions of the Shropshire Archaeological Society*.
Bateman, J. (1978). A Late Bronze Age cemetery and Iron Age/Romano-British enclosures, Ryton on Dunsmore, Warwickshire, *Transactions of the Birmingham and Warwickshire Archaeological Society for 1976–77*, **88**, 9–47.
Bell, M. (1983). Valley sediments as evidence of prehistoric land use on the South Downs, *Proceedings of the Prehistoric Society*, **49**, 119–50.
Briggs, C. S. (1985). Problems of the early agricultural landscape in upland Wales, as illustrated by an example from the Brecon Beacons, in Spratt, D. and Burgess, C. (eds), *Upland Settlement in Britain, the Second Millennium and After*, British Archaeological Reports, British Series No. 143, 285–316.
Britnell, W. (1982). The excavation of two round barrows at Trelystan, Powys, *Proceedings of the Prehistoric Society*, **48**, 133–201.
Chambers, F. M. and Jones, M. K. (1984). Antiquity of rye in Britain, *Antiquity*, **58**, 219–24.
Coles, J. M. and Orme, B. J. (1984). Homo sapiens or Castor fiber? *Antiquity*, **57**, 95–102.
Collens, J. (in press). Later prehistoric settlement and soils in the upper Severn valley, *Transactions of the Welsh Soils Discussion Group*.
Crawford, G. (1985). Excavations at Wasperton—5th Interim Report, *West Midlands Archaeology*, **28**, 1–3.
Edwards, K. J. and Hirons, K. R. (1984). Cereal pollen grains in pre-elm-decline deposits: implications for the earliest agriculture in Britain, *Journal of Archaeological Science*, **11**, 71–80.
Fennell, J. F. M. (1978). Flint implements collected at the National Vegetable Research Station, Wellesbourne, Warwickshire, *Transactions of the Birmingham and Warwickshire Archaeological Society for 1976–77*, **88**, 119–23.
Fleming, A. (1983). The prehistoric landscape of Dartmoor, Part 1: North and East Dartmoor, *Proceedings of the Prehistoric Society*, **49**, 195–242.
Girling, M. A. and Greig, J. R. A. (1985). A first fossil record for *Scolytus scolytus* (F.) (Elm Bark Beetle): its occurrence in elm decline deposits from London and the implications for Neolithic elm disease, *Journal of Archaeological Science*, **12**, 347–52.
Houlder, C. (1968). The henge monuments at Llandegai, *Antiquity*, **42**, 216–21.
Jones, M. (1981). The development of crop husbandry, in Jones, M. and Dimbleby, G. W. (eds), *The Environment of Man: the Iron Age to the Anglo-Saxon period*, British Archaeological Reports, British Series No. 87, 95–127.
Jones, R., Benson-Evans, K. and Chambers, F. M. (1985). Human influence upon sedimentation in Llangorse Lake, Wales, *Earth Surface Processes and Landforms*, **10**, 227–35.
Leighton, D. (1986). Upland settlement in Breconshire: new surveys. Paper given at

a conference on 'Wales in the Age of Stonehenge', Gregynog Hall, Powys, March 1986.

Moore, P. D. and Beckett, P. J. (1971). Vegetation and development of Llyn, a Welsh mire, *Nature*, **231**, 363–5.

Musson, C. (1976). Excavations at the Breiddin 1969–1973, in Harding, D. W. (ed.), *Hillforts: Later Prehistoric Earthworks in Britain and Ireland*, Academic Press, London.

Northover, P. (1986). The rise and decline of the Welsh metal industry. Paper given at a conference on 'Wales in the Age of Stonehence', Gregynog Hall, Powys, March 1986.

Rackham, O. (1980). *Ancient Woodland*, Arnold, London.

Rowley-Conwy, P. (1982). Forest grazing and clearance in temperate Europe with special reference to Denmark: an archaeological view, in Bell, M. and Limbrey, S. (eds), *Archaeological Aspects of Woodland Ecology*, British Archaeological Reports, International Series, No. 146, 199–216.

Rudeforth, C. C., Hartnup, R., Lea, J. W., Thompson, T. R. E. and Wright, P. S. (1984). *Soils and their Use in Wales*, Soil Survey of England and Wales, Bulletin No. 11.

Ryder, M. (1983). *Sheep and Man*, Duckworth, London.

Soil Survey of England and Wales (1983). *1:250,000 Soil Map of England and Wales*.

Spratt, D. and Burgess, C. (eds) (1985). *Upland Settlement in Britain, the Second Millennium and after*, British Archaeological Reports, British Series No. 143.

Stanford, S. C. (1972). The function and population of hillforts in the central marches, in Lynch, F. and Burgess, C. (eds), *Prehistoric Man in Wales and the West*, Adams and Dart, Bath, pp. 307–19.

Stanford, S. C. (1982). Bromfield, Shropshire—Neolithic, Beaker and Bronze Age sites, 1966–79, *Proceeding of the Prehistoric Society*, **48**, 279–320.

Palaeohydrology in Practice
Edited by K. J. Gregory, J. Lewin and J. B. Thornes
© 1987 John Wiley & Sons Ltd.

# 13

# River Terraces: The General Model and a Palaeohydrological and Sedimentological Interpretation of the Terraces of the Lower Severn

M. R. DAWSON

*Department of Geography, The University College of Wales, Aberystwyth*

V. GARDINER

*Department of Geography, University of Leicester*

## INTRODUCTION

The term 'river terrace' has been used since the emergence of geomorphology as a field of study. Playfair, for example, in 1802 gave the definition: 'Successive platforms of flat alluvial land, rising above one another, and marking the different levels on which the river has run at different periods of time.' Many different definitions have subsequently been given, embracing notions of morphology, abandonment, incision, sedimentology, chronology, origins as a floodplain, and infrequency of inundation. In attempting to standardize terminology the IGU Commission on Terraces and erosion surfaces stressed that the term applied to the landform and not to the material of which it was made. A recent definition embracing most desirable components is that of Petts and Foster (1985, p. 201): 'Terraces represent former floodplains abandoned when river incision elevates the surface above the channel bed to a height where its frequency of inundation is significantly reduced.' However, particularly for recent terraces, this is perhaps an over-simplified definition, and the frequency of inundation becomes of significance in distinguishing between the channel, the benched floodplain and higher terraces.

River terraces are a fundamental part of fluvial landscapes, and comprise, for example, almost 4 per cent of the total land area of Japan (Yoshikawa, Kaisuka and Ota, 1981). They have, however, been given little more than passing mention in many texts, or even in detailed research, although good introductory reviews of terrace models were given by Howard, Fairbridge and Quinn (1978) and Pounder (1980), and one anthology of research papers (Dury, 1970) concerned terraces explicitly. Terraces are, however, of great practical significance. For early man they provided suitable sites for settlements, being raised above the inundation-prone floodplain on better-drained land, and with a ready water supply being obtainable from the nearby river or from the permeable terrace deposits. Archaeological and historical remains are therefore often associated with terrace deposits. Terraces are also of significance to modern man, in that they often provide a widespread and reasonably abundant supply of sands and/or gravels for building purposes and aggregate, and they may be sources of placer deposits of gemstones (Sivadas, 1986). In some cases equivalent sedimentary bodies in the ancient record may provide suitable reservoir rocks for natural oil and gas deposits.

For research investigations terrace deposits and the mechanisms of terrace formation are of great significance. Terraces provide a series of arbitrary reference levels throughout a basin, giving an approximate relative chronology to which other geological, geomorphological or palaeohydrological events can be related. Terrace deposits may contain fossil organic material from which reconstructions of former climatic and palaeohydrological conditions can be made, and from which absolute dates can be derived. The sedimentological characteristics of terrace deposits can help to reconstruct the morphology of former river channels and, finally, the processes of terrace formation might shed light on some of the ways in which the fluvial system operates.

The first part of this chapter examines some of the general models which have been advanced to explain river terraces as landforms. The terrace suites of the Severn/Avon systems are then briefly described, and a sedimentological interpretation of the lower Severn and Avon terraces is offered, as a contribution towards their palaeohydrological interpretation.

## MODELS OF TERRACE FORMATION

Any attempt to review models of terrace formation is confounded by the multiplicity of features which may be considered according to the definition adopted. In scientific investigations classification is often a useful first step towards understanding. Classification of terraces may rest upon various bases, including the dominant process, symmetry and geological structure. Thus a distinction can be made between *erosional* terraces, these being subdivided according to whether they are cut into bedrock or a former valley fill, and *depositional* features caused by alluviation. They may also be classified as *paired* or *cyclic* terraces, which occur in matched pairs on either side of the valley, and *unpaired* or *non-cyclic* terraces, occurring in discontinuous fragments which cannot be segregated into matching pairs on either side of the valley. In traditional views of terrace formation such classification has often been associated with a process connotation. Paired terraces were explained as the product of lateral planation, sudden rejuvenation and hence rapid incision, whereas unpaired terraces resulted from slower and more continuous rejuvenation, with lateral river erosion remaining of significance. Although most notions of a terrace embrace alluvial features they can, of course, be cut into solid rock, and various kinds of structurally controlled terrace have been described, as for example Ziegler's (1958) rock-defended terrace, where a remnant of valley fill is left on a slope protected by a hard rock floor in a rocky gorge.

As river terraces obviously result from a change in the level of the river channel theories of river terrace formation may be divided according to the stimulus bringing about this change. Thus changes can be instigated by forces which are an integral component of the system itself, or which are external to it. A good but now dated review of the environmental controls of terrace formation is Fairbridge (1968), which contains much bibliographic reference to particular examples.

**External forces**

The main emphasis in Britain in the last century was on external forces, as reviewed by Miller (1883). For example, climatic causes suggested included increases in discharge resulting from inputs of glacial meltwaters (Upham, 1877), increased rainfall (Tylor, 1869), or a decrease in discharge resulting from decreased rainfall (Hayden, 1876). In continental Europe models embracing climatic changes as causes were widely adopted, as exemplified by the classic works of Penck and Bruckner (1909) for the Alps, Bourdier (1958) in France, and also by Fisk (1951) in the USA. Huntington's (1907) Principle, whereby valley degradation takes place in wet climates and aggradation in dry climates, was established as a general model which many followed in constructing terrace chronologies. This generalization is now recognized as being exceedingly simplistic, and not only do erosion and deposition often occur simultaneously in adjacent parts of the valley, but the reverse of Huntington's Principle may apply. For example, Chester and Duncan (1979) showed that terrace deposits in Sicily, which they were able to date from overlying lava flows of Mt Etna, were deposited under a regime of high discharges and episodic rainfall, and suggested that the formation of terraces in Sicily is controlled by variations in the rate of uplift. In addition vegetation acts as a protective cover, inhibiting erosion. Fairbridge (1968) therefore recommended the use of the Langbein–Schumm (1958) rule, assuming that a more complex relationship exists between erosion and precipitation.

There is a temptation to link terrace chronologies with established chronologies of glacial and climatic oscillations, especially when other dating evidence is not available. For example Briggs and Gilbertson (1980) and Cheetham (1976, 1980) ascribed the accumulations within the Thames Basin to a late glacial phase because of their association with periglacial slope deposits, and Green and McGregor (1980) envisaged the Thames as ocillating between a cold-climate braided, terrace-building condition and a meandering, incising condition during interglacials. Fluvial terraces in the Andes have been correlated with the classic Alpine glacial/interglacial model (Sievers, 1888; Tricart and Millies-Lacroix, 1962). However, a perfect association of terrace depositional phases and incision episodes with glacial/stadials and interglacials/interstadials is rarely possible, or even likely. Clayton (1977), for example, quotes the case of a Trent terrace as evidence of interglacial aggradation. Castleden (1980) has refined these notions in developing a model of fluvio-periglacial pedimentation which embraces terrace formation as a response to complex alternations of morphogenetic conditions. Pedimentation or valley widening is ascribed to colder (stadial) periods and incision, clearance or stabilization to warmer ones. Such models are still founded upon an incomplete understanding of how climate is coupled to processes and hence to landform morphogenesis within an evolutionary framework. Difficulties have become exacerbated by the recent progressive disclosure of a much more complex climatic history than was formerly envisaged (Bowen, 1978).

In Britain rather more attention has been paid in this century to relative changes in base-level as an external driving force for river-level changes. This was founded upon early suggestions of a fall in sea level (Hitchcock, 1857) or rise in river level (Darwin, 1846) as possible causative factors. Tectonic movement has also been considered to be the main cause of terracing in areas of more obvious tectonism, such as Japan (Yoshikawa, Kaisuka and Ota, 1981). The most significant influence was the overwhelming impact of Davisian ideas during the early years of the century. For example in 1902 Davis interpreted river terraces in New England as the result of tectonic uplift, within the context of his Cycle of Erosion. These notions developed in the denudation chronology school of geomorphology (Jones, 1924; Baulig, 1935), in which terrace fragments were accepted as being related to river long profiles. They could therefore be used as evidence to reconstruct them, and hence the former relative sea levels to which they were graded. Papers by Jones (1924) and King and Oakley (1936) were the precursors of much subsequent work in which terraces were referred to former sea levels, by extrapolation of the long profile. The work of Kidson (1962) on the Exe exemplifies the approach.

Attempts were made to relate terrace sequences to established sequences of sea-level change, in the same way that they had been linked to patterns of climatic change. Such attempts usually proved less than entirely successful, in view of the uncertainty surrounding sea-level changes; equally river-terrace studies shed little light on the patterns of sea-level change (Clayton, 1977; Rose, 1978). Only in the most favourable circumstances may terraces be traced laterally into marine landforms, thus establishing a direct link between the former floodplain and contemporaneous sea level. For example, Dickson *et al.* (1978) were able to trace a river terrace continuously into a former shoreline. In most situations the complicating factors, including sediment load and discharge variations, are so numerous that no direct link could be made.

Although base level change is perhaps intuitively attractive as a mechanism for propagating incision through a fluvial system, there has rarely been any detailed consideration of how, or indeed whether, such a fundamental adjustment occurred. An assumption was often made that the whole river system incised as a response to base level lowering, without consideration being given to the processes by which this came about, or to the spatial differences in erosion and deposition likely to occur. Rose (1978) showed how the effects of a sea-level change on the development of a river terrace will depend largely upon the sediment discharge of the river, and how no very simple relationship can be derived to link terraces and base level. The complexity of relationships was illustrated by Rose's (1984) study of terraces in Sarawak. Aggradation was attributed to increased sediment discharge due to landslipping, most probably during the last Interglacial, and incision was attributed to a fall in sea level in the last Glacial, as well as to reduced sediment yield caused by reduced precipitation. Again, therefore, the general external model breaks down on a lack of detailed understanding of the processes operating within the system.

**Internal forces**

Explanations of river terraces as resulting from changes in river level brought about by mechanisms intrinsic to the system have lately become popular, although their origins can be traced back to last century. Miller (1883) adopted a multicausational attitude reminiscent of much recent work, and concluded his paper with observations on a miniature stream through a beach runnel, in which terraces were formed under conditions of constant water and sediment discharge. Hitchcock's (1833) paper was one of the first to follow Playfair (1802) in realizing that rivers may make terraces without there necessarily being an external change, and this undoubtedly strongly influenced Miller's review. Geikie (1865) argued that changes in sediment load and size could induce terracing, this conceivably arising as a result of internal changes to the catchment. Challinor (1932) argued in similar vein, and Lewis (1944) showed by flume simulation that terraces could be formed by reduction of sediment load. Field illustrations of how a single external change in a river system can produce a series of terraces have been given by Born and Ritter (1978), Small (1973), Quinn (1957), Ritter (1982) and Haible (1980). However, by far the most complete formulation of this kind or approach has been made by Schumm and co-workers (Schumm, 1977 and references therein; Patton and Schumm, 1981; Womack and Schumm, 1977) as the complex response model. Schumm postulates that terraces are in fact very complex features, resulting partly from external factors such as those reviewed above, and partly from the nature of response in the fluvial system itself. After an initial perturbation to the system, caused by, for example, climatic or base-level change, the reaction of the system is very complex. This complexity arises because the different components making up the system respond at different rates, and because the system is itself a spatial entity in which events in part of the system have inevitable repercussions elsewhere, in an inextricably interlinked fashion. Support for the theory comes from the observations of eroding, stable and depositing reaches within the same system (e.g. Church and Jones, 1982; Schumm, 1961), as well as from the studies of terraces mentioned above.

Laboratory simulations (Schumm, 1976, 1977) have demonstrated how one external stimulus can trigger a complex reaction of cut and fill activity, as the response of the components of the fluvial system to change. Schumm envisaged this ensuing instability as being the channel's response to the existence of intrinsic geomorphological thresholds, with valley slope playing a key role. Thus deposition increases valley slope until an erosional threshold is crossed; then erosion commences, and arroyo formation continues until the gradient is again reduced. Hey (1979a; in press) favours a more complex notion of a 'process-based solution' in which an apparent morphological threshold is in reality a multivariate function of local sediment load, shear stress, bed material size and slope. These are in turn dependent upon events elsewhere in the system, in the way that Schumm envisaged, and at any one location oscillations between erosion and deposition will occur until a gradual reduction in sediment availability damps out the response.

## DIFFICULTIES IN EVALUATING AND APPLYING TERRACE MODELS

The extent to which any of the models outlined above can be applied to a particular set of terrace fragments depends upon a range of complicating considerations. As terraces are acknowledged by most definitions to be floodplain remnants, any model of terrace development must inevitably depend upon a knowledge of floodplain development. The nature of floodplain development is clearly related to the type of the past channel planform, and distinct contrasts in the depositional patterns associated with meandering (Wolman and Leopold, 1957; Allen, 1964, 1965; Bluck, 1971; Lewin and Manton, 1975), anastomosing (Smith and Smith, 1981) and braided planforms may be identified. In the short term channel pattern is controlled by water and sediment discharge and reach slope (Carson, 1984a; Bridge, 1985), which may vary downstream in modern environments over limited distances (Boothroyd and Ashley, 1975; Young and Nanson, 1982; Church and Jones, 1982; Carson, 1984b,c). It is conceivable, therefore, that contrasts in sedimentation may be preserved within the same terrace aggradation. Differences in the sedimentary characteristics of stratigraphic-ally separate terrace aggradations may also be expected as terrace formation can be broadly regarded as a consequence of long-term changes in those factors influencing channel patterns and sedimentation—namely the water:se-diment discharge ratio and base-level controlled reach slope (Richards, 1982). It is thus unlikely that a simple model of terrace aggradation can be generally applicable, and clearly some assessment of the local depositional environment is necessary prior to using the existence of a terrace for a stratigraphic or palaeohydrological interpretation.

Subsequent erosion and modification of the terrace by secondary disturb-ance and solution can also make interpretation difficult. Frye and Leonard (1954) and Ritter and Miles (1973) have pointed out that the surface morphology of a terrace is often a poor indicator of the actual expression of the terrace due to masking by colluvium or as a result of variations in height due to such factors as channel scars or erosion of the terrace scarp. Frye and Leonard (1954) examine the problems of terrace mapping in general, Larue (1982) and McGregor and Green (1983a) have examined post-depositional modification processes in detail, and Colman (1983) points out that channels with relief of up to 3 m are preserved on some terrace surfaces of 100,000 years age, whereas elsewhere much younger terraces have been entirely degraded. One very important feature of Schick's (1974) model is that destruction of terrace fragments occurs by lateral processes during subsequent superfloods; he therefore generates the probabilities of occurrence of a particular number of terraces at any given location. The corollary of his findings is an inherent areal inconsistency in terraces and terrace sequences and, although developed in the context of desert regions, the same general principles undoubtedly apply elsewhere.

One implication of differential preservation of terrace fragments is that correlation of fragments is difficult. Palaeobiological evidence can rarely be used, as the timescale is often very short, the river regimes in which terrace material is deposited are generally unfavourable to life, or its preservation as fossils, and the correlative deposits of phases of incision are usually lost from examination in estuaries or at sea. Although many bases for correlation exist, including morphological evidence, marker horizons, stratigraphic discontinuities, sedimentology, palaeosols, palaeobiology, archaeological and historical artefacts, and association with other deposits, there is no guarantee that use of any of these will be possible, as even morphological correlation rests upon assumptions of an initially continuous surface of a single age. Morphological correlation is also hindered by differential uplift, as suggested for terraces in Virginia by Mills (1986). McGregor and Green (1983b) demonstrate how, in the absence of other evidence, lithological features can be used to confirm lateral continuity, and Latreille and Le Griel (1980) advocate the use of heavy-mineral mineralogy as a guide to correlation. If one assumes that post-depositional modification processes operate at an equal rate throughout the system then such modifications can be turned to advantage as a means of correlation. Thus Alexander and Holovaychuk (1983) showed how the clay mineralogy of terraces along the Cauca River, Colombia, related to the age of the terraces, and Colman (1983) similarly demonstrated progressive changes in indices of terrace preservation with time. Vegetational colonization and succession may also be related to terrace age (Fonda, 1974).

One final factor to consider in evaluating models of terrace evolution is the extent to which the model encompasses other related elements in the landscape. Terraces are, after all, only one component of the entire fluvial system, and their formation is inevitably and inextricably intertwined with other components, as well as with other geomorphological systems. For example, in the Belgium Ardennes, associated features include limestone caves graded to terrace levels, as well as solifluction fans and aeolian sands intimately associated with the terrace deposits (Ek, 1961). Cave systems graded to terraces are also described by Rose (1982) in Sarawak. Dry valley networks (Gardiner, 1983) are often closely related to terraces, as for example in the Kennet Valley of the Thames, where Cheetham (1980) describes them as being graded to a terrace of the Thames, or the Otter Valley in Devon, where Gregory (1971) describes them as dissecting the middle (Middle Pleistocene) terrace, but terminating on the lowest (Late Pleistocene) terrace.

## TERRACES OF THE LOWER SEVERN

Alluvial terrace deposits are present along much of the length of the River Severn. They are most extensively preserved to the south of the Ironbridge Gorge and the Devensian glacial limits where four major terraces and up to three higher level terrace remnants have been identified (Wills, 1983; Hey, 1958; Beckinsale and Richardson, 1964). As shown in Figure 13.1 the lower terraces are traceable as distinct stratigraphic and morphological units along considerable lengths of the Lower Severn, suggesting that they developed as a result of large-scale environmental change rather than as a response to

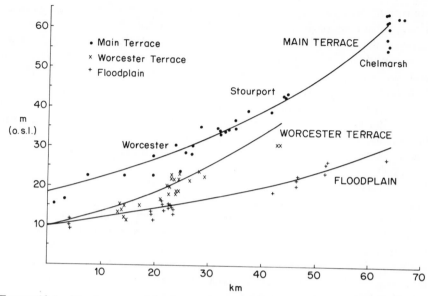

FIGURE 13.1   The terraces of the Lower Severn. Terrace profiles constructed from the surface heights of boreholes logged through known terrace and floodplain deposits and plotted relative to a minimum distance valley centre-line

intrinsic thresholds. Previous work has concentrated on establishing the distribution of the terrace fragments and their stratigraphic relationships.

The oldest deposits, termed the Wooldridge Gravels by Wills (1938), occur in small fragments between Tewkesbury and Gloucester and in the Worcester area. These gravels were attributed to the Anglian glacial stage by Hey (1958), having been previously ascribed a proximal fluvioglacial origin by Wills (1938). Similar deposits are present in the Leadon tributary. Although Wills (1938) associated these gravels with the Wooldridge Gravels, Hey (1958) showed that they lie at a lower elevation and stated that they represented outwash of Wolstonian age in the Severn Valley. Wills (1938) also mapped gravels, distinct from these higher deposits and the lower terraces, at Bushley Green, close to the Severn–Avon confluence. He found that this local terrace lay at a similar altitude to the Avon No. 5 Terrace mapped by Tomlinson (1925), and proposed that it was the product of aggradation at the confluence associated with the development of a late Wolstonian outwash system.

The highest widely distributed terrace deposit, the Kidderminster Terrace, is traceable from the mouth of the Severn as far as Bewdley and thence upstream in the tributary River Stour. There is little evidence of it north of Bewdley along the Severn, although Willis (1924, 1938) suggested that a terrace level above the Main Terrace at Eardington may be an upstream equivalent. The deposits are dominated by Bunter pebbles, although in the Lower Stour there is a notable content of Clent breccia and there are occasional Welsh, Wrekin and, below Worcester, Malvernian erratics (Wills, 1938). Wills (1938) deduced from the distribution of the terrace fragments and the erratic content that the Kidderminster terrace pre-dated the Devensian 'Irish Sea' glaciation, arguing that the absence of terrace fragments from the Severn Valley upstream of the Stour confluence was evidence for a Devensian glacial origin for the diversion of the Severn through the Ironbridge Gorge. The terrace has been correlated directly, at Tewkesbury, with the Avon No. 4 Terrace (Tomlinson, 1925) and with deposits in the Salwarpe Valley (Willis, 1938). Stephens (1970) interpreted the deposits as being of Ipswichian age, but Wills (1938) and Shotton (1973) believed that the deposit had an early Devensian origin.

The most extensive deposit is the Main Terrace which is traceable from Apley Park near Bridgnorth, where it lies approximately 30 m above the present floodplain, to Gloucester, where it descends beneath the modern alluvium. This deposit has a variable width and thickness, but in places has a cross-valley extent in excess of 1 km and a thickness of up to 10 m. Wills (1938) divided the deposit into two units; a lower Main Terrace running into the Ironbridge Gorge (correlated with the Avon No. 2 Terrace); and an upper terrace unit which he associated with the Avon No. 3 Terrace. The local occurrence of two surface levels in the vicinity of Stourport has been supported by Mitchell, Pocock and Taylor (1961). However, there is little evidence for the presence of multiple levels in other areas and Shotton (1953) has shown that the correlation of an upper surface with the Avon No. 3 Terrace is unlikely.

Wills (1938) and Shotton (1977), noting the sizeable proportion of erratic material derived from 'Irish Sea' glacial deposits and the presence of arctic faunal remains, proposed that the terrace was correlated with the 'Irish Sea' glaciation. Wills (1938) argued that the deposition of the gravel commenced during the decay of the ice sheet and continued during the formation of the Ironbridge Gorge. However, sections in gravel workings at Eardington have revealed diamict lithofacies, interpreted as flow till (Shotton, 1977), interbedded with the terrace gravels, which indicates that the aggradation developed, at least partially, as outwash from the ice sheet at its maximal position. This late-Devensian glaciation is known to post-date 30,000 BP (Boulton and Worsley, 1965; Shotton, 1967), and given that Shotton and Coope (1983) determined a date of 12,300 BP for the lowest Severn Terrace, it is believed that the Main Terrace aggradation occurred between 25,000 and 18,000 years BP.

The surface of the Worcester Terrace lies approximately 8 m below that of the Main Terrace, and is traceable from Tewkesbury to Bewdley and also in the Bridgnorth area, although it is not recognizable in the valley between

Bridgnorth and Bewdley. Wills (1938) correlated the terrace with the Avon No. 1 Terrace and with the Uffington Terrace north of the Ironbridge Gorge. However, Coope and Shotton (1981) argued that the Worcester Terrace has no correlative in the Avon Basin, and Shaw (1969) and Jones (1982) have suggested that correlation of terrace fragments between north and south of the Gorge may be unsound, due to the likely presence of stagnant ice in the area downstream of the Uffington Terrace at the time of its formation.

The Worcester Terrace, like the Main Terrace, contains a significant content of Irish Sea erratic material and a number of extremely large clasts have been identified in sites as far south as Stourport (Shotton, 1977) and Grimley. Shotton (1977) and Shotton and Coope (1983) have proposed that the Worcester Terrace is an outwash deposit of the late-Devensian glaciation, possibly related to the Ellesmere (Welsh) re-advance (Beckinsale and Richardson, 1964), although Wills (1938) and Poole and Whiteman (1961) attempted to relate the terrace to a level of Lake Lapworth.

Altitudinally below the Worcester Terrace, there are a number of discontinuous low terraces. Wills (1938) collectively named these the Power House Terrace, and proposed that they represented the upper part of an infilling of a channel deeper than the bed of the present river. Williams (1968) has described a sand and gravel unit, up to 12 m deep, underlying most of the Lower Severn, thickening downstream and thinning towards the valley sides; this Brown (1982) considered to be the first depositional units of the post-glacial valley fill. Below Worcester these deposits underlie the Holocene alluvium, but in the vicinity of Bridgnorth the low terrace deposits lie approximately 8 m above the current floodplain surface and at Stourport the Power House Terrace is evident up to 2.5 m above the floodplain.

The culmination of gravel deposition has been radiocarbon dated at 12570 ± 220 BP (Shotton and Coope, 1983), this date being derived from organic deposits underlying Holocene sands and silts at Stourport. However, as Brown (1982) pointed out, the Power House Terrace cannot be regarded as a unitary sedimentary body and dates relating to the terrace at Stourport are not necessarily applicable to basal gravel units elsewhere. Wills (1938) correlated the Power House Terrace deposits with the Cressage Terrace of the Upper Severn, regarding their deposition as being the result of a pause in the downcutting of the rock barrier at Ironbridge. This association was questioned, however, by Beckinsale and Richardson (1964) who argued that the deposits were related to the Llay re-advance of the Devensian ice sheet.

Although previous work has related the terraces to the glacial chronology, it is clear that, particularly in the Devensian glacial and late-glacial periods, the controls on terrace development were complex. This was, in part, recognized by Wills (1938) who associated the Devensian terraces with the rapid lowering of the thalweg through the Ironbridge Gorge, concurrent with base-level-induced alluviation in the lower reaches. It is significant that Wills

(1938) also recognized the possibility of hydrological change, arguing that a decrease in runoff and flood depths during the Holocene may account for the narrower and lower present floodplain. Subsequent work has questioned the influence of spillway drainage from a pro-glacial lake and has postulated a direct link between terrace aggradation and glacial advances (Beckinsale and Richardson, 1964). In the Lower Severn, however, it is unclear how far such correlations can be extended as they are based on an uncertain, assumed, relationship between glacial advances and periods of high sediment to discharge yields promoting aggradation.

The effect of base-level variation, as an influence upon terrace aggradation and incision in the late-Devensian period, seems to have been slight as the glacial and late-glacial terraces are known to have graded to sea levels up to 30 m lower than present (Wills, 1938; Beckinsale and Richardson, 1964). However, Wills (1938) and Beckinsale and Richardson (1964) postulated fluctuations in sea level to account for the inset nature of Power House Terrace deposits in the present Severn Estuary, relative to the Worcester Terrace, and it is possible that these may have been influential in initiating incision in upstream reaches of the basin. An important de-glacial influence on terrace development, not previously considered (or investigated), may have been isostatic rebound. This is likely to have been differential across the basin away from the ice margin, causing a tectonically controlled north to south increase in river gradients in the late-glacial period. It is feasible that such gradient steepening may have induced a 'threshold' type response inducing terrace incision, while differential uplift may explain the convergent nature of the terraces away from the presumed Devensian ice margin.

## THE FLUVIAL ENVIRONMENT

Previous stratigraphic work has alluded to major changes in discharge magnitudes and channel planform which may have occurred throughout the Pleistocene period and particularly during the late-Devensian and early Holocene. However, the nature of the fluvial environment during the aggradation of the major terraces has largely been a matter for speculation with, for instance, Wills (1938) postulating that the Main Terrace was deposited by a flood-dominated braided channel, while Clayton (1977) favoured the existence of a single meandering channel. Few attempts have been made to reconstruct the depositional environment of the terraces, although Brown (1982, 1983) demonstrated stratigraphic and sedimentological changes in the basal fill. From this evidence he argued that low-sinuosity, bedload-dominated channels, depositing gravel and sand, were progressively replaced during the early and mid-Holocene by more confined and sinuous channels with floodplains constructed from organic silts and clays.

The younger terraces of the Lower Severn have been widely exploited as sources of aggregates, and in places extensive exposures have been produced. By examining the sedimentary structures and their architectural arrangement, as revealed in section, it has been possible to reconstruct aspects of the depositional environment of the Main and Worcester Terraces. These provide evidence for the fluvial conditions present during the Devensian glacial and late-glacial period.

## Main Terrace

The Main Terrace is extensively exposed both in an ice marginal situation and in more distal locations downstream (Figure 13.2). As predicted by previous work (Boothroyd and Ashley, 1975; Miall, 1977; Frazer and Cobb, 1982) the sedimentary attributes of the proximal site contrast with those downstream, although there is little evidence of proximal to distal changes between the three distal sites. The sedimentary sequences are dominated by the presence of coarse-grained lithofacies, with silts and clay deposits being largely absent, forming only local drapes or small-scale channel infills. Internally, the deposits show distinct stratigraphic divisions, marked by major bounding surfaces, and it is possible to identify depositional subenvironments corresponding to channel zone, overbank and alluvial fan deposits.

Channel zone deposits are characterized by coarse-grained, dominantly crudely-stratified gravel lithofacies. At Eardington, a proximal exposure close to the presumed ice margin, such sediments form the lower of two units within the terrace sequence (Dawson, 1985) and include:

1. Coarse gravel units up to 1.2 m thick, comprising crudely planar stratified and tabular cross-bedded lithofacies, interpreted as non-emergent, poorly developed, gravel bars.
2. Planar stratified (upper flow regime) sands and gravels and planar cross-stratified sands, which seem to have developed in areas of shallow flow divergence on the surface of the gravel bar features.
3. Large-scale trough cross-stratified sand and gravel. Individual trough forms may be up to 1 m thick and 5 m wide. Locally, the lamination within the troughs dips at low angles, indicating local increases in flow strength causing a transition to flow over a planar bed.

The gravel bar lithofacies and the large-scale bedforms were seen to be laterally interbedded, and were locally underlain by concave-up scoured surfaces of greater extent than individual lithofacies units, suggesting conterminous formation in a broad shallow channel. In places basal lithofacies in this unit have yielded concentrations of extremely large, subrounded

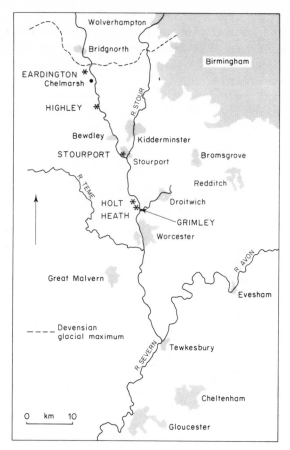

FIGURE 13.2   The Lower Severn. Location of the major exposures in the glacial and lateglacial (Devensian) terrace deposits

clasts ($d > 1$ m). The dimensions of these clasts and the large scale of the trough cross-stratification together indicate the flow depths may have exceeded 5 m.

At distal sites, channel zone lithofacies are notably finer and comprise laterally extensive, tabular sheets up to 2 m thick which overlie erosive planar bounding surfaces (Figure 13.3). These units may be either vertically stacked or offset, rarely extending across the complete width of the terrace. Internally, the units are dominantly composed of poorly defined, tabular or lenticular bodies of framework-supported gravel, up to 0.50 m thick, which possess similar structural attributes to depositional features found in modern gravel-bed rivers (Williams and Rust, 1969; Bluck, 1971, 1974, 1976, 1979;

Rust, 1972; Hein and Walker, 1977). Thicker framework-supported matrix-rich beds, in places showing a lateral transition to tabular cross-bedded units, may be interpreted as riffle or unit bar features forming complex bar platforms. Thinner beds (0.10–0.20 m) of openwork; imbricated matrix-rich disorganized matrix-rich; and tabular cross-bedded gravel, in places dipping at low angles normal to the general transport direction, are likely to have developed in sheet-like form on the surface of complex bars or on lateral within-channel bends. In places these thin-bedded gravel units interdigitate with thin (0.05–0.10 m) sand drapes which may pass laterally into truncated lenticular channel fills, representing former sandy, supra-bar sediments.

At Holt Heath (Figure 13.4), beds of laterally persistent cross-bedded sandy gravels up to 50 m in length and 0.85 m thick, showing convex-up reactivation surfaces, form a distinct sedimentary horizon above an erosive bounding surface. These cross-beds are in turn truncated by channel-like scours, infilled with wedge-like units of tabular cross-bedded gravel. The whole unit is overlain by a laterally persistent massive clay drape and may be interpreted as a flood-depositional sequence.

The dominance of gravel lithofacies comprising the channel-zone units, at both proximal and distal sites, indicates that they were deposited by a coarse-bedload river. The sedimentary characteristics suggest that sedimentation took place in low-sinuosity and possibly braided channels, because (Moody-Stuart, 1966; Jackson, 1978; Bluck, 1979; Bridge, 1985):

1. Fine-grained inner accretionary bank sediments or sand and mud lithofacies forming channel fills, as would be found in sinuous, single-channel, gravel-bed river deposits, are absent.
2. Vertically aggraded, rather than the laterally accreted channel sediments common in sinuous rivers, are dominant.
3. Gravel units possess a tabular and sheet-like geometry suggesting deposition in channels with high width:depth ratios or on bars with low-angle dipping accretionary surfaces, as would be found in braided rivers.

Extensive overbank sedimentation is preserved only at the distal sites in the terrace and include flood or avulsion deposits and a thin-bedded, shallow-flow assemblage. This latter assemblage forms sheet-like units up to 1.5 m thick, composed of thin (0.05–0.25 m) lenticular beds of cross-stratified and ripple-laminated sands; and thin clay stringers. Individual beds are rarely continuous but are occasionally traceable for 15–25 m. Shallow (<0.70 m deep) channel forms may be present, containing vertically and laterally accreted components, the latter being characterized by down-dipping wedges of tabular cross-bedded sands extending from the channel margins. Within some units syndepositional soft-sediment deformation has occurred, as indicated by truncated, disharmonic compressional folds. At Holt Heath (Figure

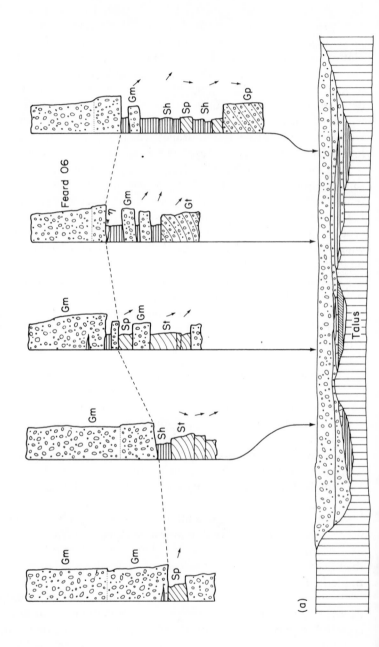

(a)

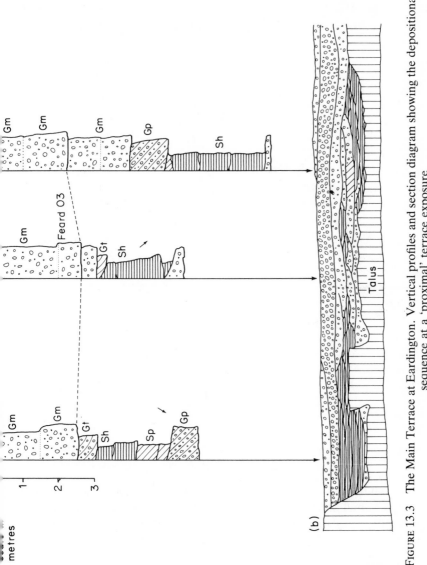

FIGURE 13.3　The Main Terrace at Eardington. Vertical profiles and section diagram showing the depositional sequence at a 'proximal' terrace exposure

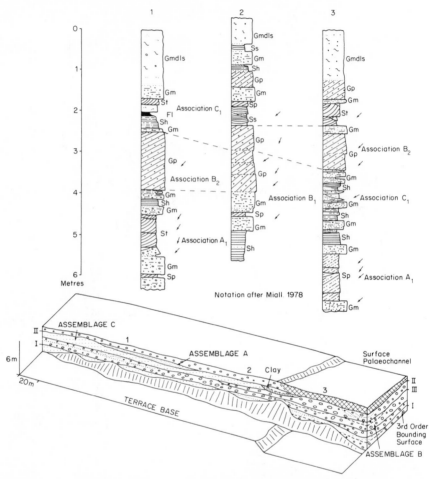

FIGURE 13.4    The Main Terrace at Holt Heath. Block diagram and vertical profiles showing the depositional sequence at a 'distal' terrace exposure

13.4) units of this assemblage occur throughout the sequence, most commonly overlying channel-zone sedimentary units above a conformable horizon, and being truncated by superimposed lithofacies. At other sites this assemblage may succeed flood deposits, being most commonly located towards the terrace surface.

The flood and avulsion deposits are dominantly composed of vertically aggraded cosets (0.05–1.25 m thick) of cross-stratified pebbly sand. Individual sets show both trough and planar cross-stratified forms (0.10–0.45 m high). Gravel lags are common at set and coset boundaries, and there may be a transition to planar or ripple-laminated sets towards the top of the cosets.

Typically, the units overlie shallowly concave-up erosive horizons and were observed in two stratigraphic positions:

1. Underlying (sometimes conformably) a channel-zone body, but resting above a more extensive erosive surface which truncates a thin-bedded overbank unit. Here the units form flat-topped, lenticular bodies up to 2 m thick and 90 m wide. These seem to have been produced by local, channelized, flow avulsion into an overbank area, prior to a large-scale migration of the channel zone.
2. Overlying channel-zone units in extensive sheet-like bodies. In places the units are laterally continuous with a channel-zone body above the same major bounding surface. These units may be interpreted as having resulted from overbank flood flows away from the channel zone, transporting and reworking mainly sand-grade sediments.

At proximal locations, particularly where steep-gradient tributaries join the main Severn Valley, alluvial fan deposits have prograded over valley axis sediments. At Eardington a lower unit comprising channel-zone sediments is overlain by extensive coarse-grained assemblage, dominantly composed of beds of framework-supported gravel showing an imbrication direction approximately normal to the valley axis. These have a poorly developed structure and an indistinct tabular geometry defined by welded vertical contacts. In places these units show a lateral, downflow, transition to erosively bounded wedge-like units of planar cross-stratified sand and fine gravel up to 15 m wide and 0.75 m thick. Towards the terrace surface, channel forms have been eroded into the underlying lithofacies. These contain conformable fills of massive sand with gravel lags. Although the structural characteristics of the unit are poorly defined, there is a clear similarity between the lithofacies assemblage described here and facies models erected for proximal alluvial fan deposits (Miall, 1978; Rust, 1978a,b). The gravel units may be considered to be superimposed complex bars, while the sand units represent the truncated remains of supra-platform or bar-lee sedimentation and slough channel-fill deposits. The upper unit at this site may therefore be considered to be the remnants of an alluvial fan which aggraded eastwards from the Mor Brook tributary into the main Severn Valley.

The architectural structure within the terrace at distal sites may be considered to have resulted from the relative migration of the channel and overbank zones across the valley floor during the terrace aggradation. The limited lateral extent of the coarse gravel units within the terraces indicates that the channel zone occupied only part of the valley floor and that there were extensive overbank areas, particularly towards the valley sides. Higher rates of aggradation within the channel zone relative to the overbank areas as postulated by Bridge and Leeder (1979) and Bryant (1983), would lead to an elevation difference and eventual avulsion under flood conditions to a new position on the valley floor. Such a process would account for the offset stacking of the channel-zone units within the terrace. The transition from sandy, cross-stratified, avulsion deposits to channel-zone deposits above the same major erosive horizon indicates that the initial avulsion may have been in the form of a crevasse splay, subsequently enlarged and occupied by the migrating channel zone (Figure 13.5).

While a braided channel planform seems to have persisted throughout the terrace aggradation, the channel form seems to have changed prior to the incision of the terrace aggradation. Large, non-braided, palaeochannels are preserved towards the surface of the terrace at both Stourport and Holt Heath, indicating, as postulated by Maizels (1983) elsewhere, the concentration of flow into large single-channel forms. As shown by the palaeohydrological calculations (p. 296) and the occurrence of large-scale cross-stratification in the channels at Stourport, terrace incision may have taken place without a large-scale change in discharge conditions.

**Worcester Terrace**

The Worcester Terrace is less well exposed than the Main Terrace, the only major section, at present, occurring at Grimley, although Shotton (1977) briefly described a section at Stourport. Like the distal exposures in the Main Terrace, the sedimentary sequence contains a number of erosive bounding surfaces which separate channel-zone units. However, the lithofacies characteristics of some of these channel-zone bodies contrast markedly with those of the Main Terrace, while depositional conditions seem to have varied during the terrace aggradation, such that a threefold stratigraphic division is apparent within the exposure (Figure 13.6).

The basal assemblage (A) occurs as a distinct unit towards the valley sides, resting directly on the bedrock and pinching out towards the valley centre. The unit is dominantly composed of laterally traceable gravel beds which show a well-imbricated, framework-supported structure, and thin and fine upwards. Bed thicknesses decrease from in excess of 1 m to 0.15–0.30 m near to the contact with the overlying lithofacies assemblage. Where gravel beds rest directly upon each other the contact is usually welded; frequently, however, the bounding surfaces are marked by a thin stringer of clay or planar-bedded sands and locally there are thin beds of planar cross-bedded sand and fine gravel. Palaeocurrent directions, as indicated by pebble imbrication, closely parallel the orientation of the valley axis.

Assemblage B truncates, and is inset into, the basal assemblage. It forms the major sedimentary body within the terrace exposure and is predominantly composed of cosets of cross-stratified sand and fine gravel lithofacies up to 2 m thick. The boundaries between the cosets are usually planar or slightly

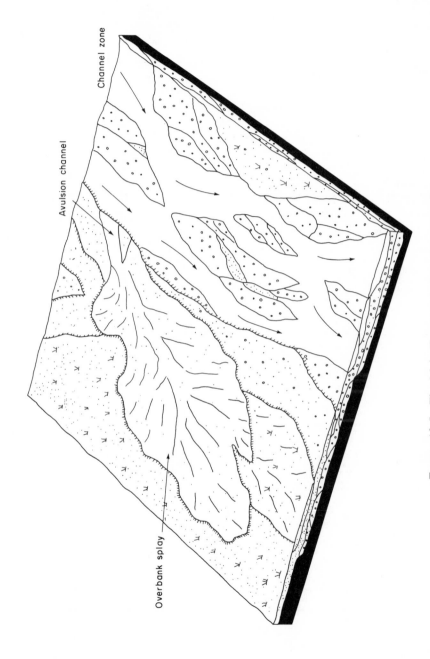

FIGURE 13.5    The Main Terrace: A depositional model

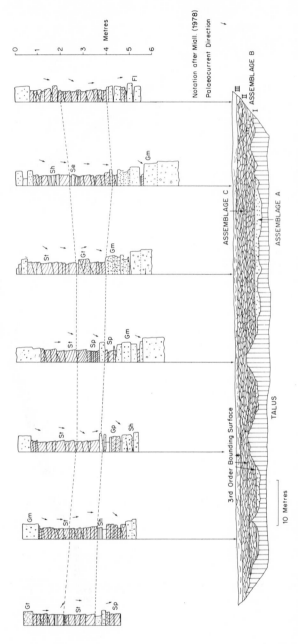

FIGURE 13.6 The Worcester Terrace at Grimley. Vertical profiles and section diagram showing the depositional sequence within one site in the Worcester Terrace

concave-up and can often be traced along the section far in excess of 50 m. Typically, coset boundaries are overlain by a gravel lag and may be erosive into a unit of fine planar-laminated or massive sandy clay sediments up to 0.30 m thick. Within the cosets two vertical associations may be identified:

1. Vertically aggraded channel fills. These are dominantly composed of trough cross-bedded sand, produced by three-dimensional dune migration, but which locally show transitions to upper flow regime planar-bedded and cross-stratified fine gravel, and ripple-laminated sands.
2. Isolated bar sequences comprising tabular cross-bedded units of sand or fine gravel up to 1 m thick, overlain by inset, thin planar-laminated or planar-cross-bedded sand lithofacies. This association seems to have resulted from the migration of a large-scale bedform with a tabular avalanche face, such as a linguoid bar, modified by small-scale bedforms on its upper surface.

Palaeocurrent directions indicated by the cross-stratification are usually consistent within cosets, but show slight variations between cosets, although the mean direction closely parallels the valley axis.

The dominantly sandy lithofacies assemblage B is vertically truncated along a major bounding surface, traceable along the length of the exposure. This is distinguished for much of its extent by a massive sandy-clay unit up to 0.50 m thick, resting on a gravel lag horizon. A similar bounding surface and fine-grained unit separates two units of the overlying lithofacies assemblage C. Towards the valley centre units of assemblage C are mainly composed of beds of crudely planar-stratified gravel and tabular cross-stratified sand and gravel up to 0.60 m thick. The latter lithofacies shows two forms; large sets aligned parallel to the general transport direction, grading from crudely planar-stratified units; and small wedge-like units occupying erosive channel forms, orientated normal to the main down-valley palaeocurrent direction. These gravel lithofacies interdigitate towards the valley centre with cosets of cross-stratified sand, which share many similarities to the underlying litho-facies assemblage B. However, in contrast to that assemblage, the cosets are composed almost entirely of trough cross-bedded and ripple-laminated sands, set heights are small (0.10–0.30 m) and gravel lags are infrequent.

The three assemblages may each be regarded as channel-zone units, although it is evident that the channel planform and possibly the reach slope varied during the terrace aggradation. The basal assemblage resembles channel-zone units described from the Main Terrace, although the laterally continuous nature of individual beds indicates that it developed without extensive reworking by channel migration. The thinning and fining upwards seem to have resulted from declining flow depths and velocities during deposition, although flow stage must have fluctuated as waning flow drapes separate individual beds. The stratification evident is analogous to that which Rust (1972, 1984) postulated would develop during the emergence of a complex bar form. However it is possible that the unit may have been rapidly deposited as part of a flood sequence.

The contrasting sedimentary characteristics of the overlying assemblage B were clearly produced by aggradation in a sand-bed channel. The dominance of vertically accreted channel bedforms, an absence of lateral accretion structures and a low palaeocurrent variance (<60°) is evidence of a low-sinuosity environment, while the presence of cross-cutting, subplanar coset boundaries indicates that flow occurred in channels with low width to depth ratios. These characteristics suggest the occurrence of sandy braided conditions, and the stratification within the sequence closely resembles sandy braided sequences described from the ancient record (Moody-Stuart, 1966; Campbell, 1976; Cant and Walker, 1976; Allen, 1983). The laterally persistent cosets of trough cross-bedded sand seem to have developed as a result of channel-floor bedform migration, the coset boundaries either being the channel floor itself or localized reactivation horizons within a much larger channel. Campbell (1976) favoured the former interpretation. Larger scale bar forms are represented by the tabular cross-bedded units with superimposed small bedforms, these stratification types resembling linguoid bar sequences described by Smith (1970), Collinson (1970) and Blodgett and Stanley (1980). However, extensive complex bars, as described by Cant and Walker (1976, 1978) and Allen (1983) seem to have been absent.

The interdigitation of cross-stratifieds and beds and gravel units in assemblage C clearly indicates the co-existence of sandy bedforms and gravel bars within the same channel zone. The sedimentary attributes of the gravel units are similar to those produced by complex bar development in a braided reach (Rust, 1972, 1984; Bluck, 1976, 1979; Ramos and Sopena, 1983), although, unlike units in the Main Terrace, component sedimentary units (unit bars) seem to have had well-developed foreset margins. This indicates relatively low aggradation rates and high water depths to grain size ratios (Hein and Walker, 1977), suggesting that the stratification sequence developed through the sudden introduction of gravel sediments into a previously sandy braided channel, although a similar interdigitation between channel-floor dune forms and complex gravel bars in coarse braided outwash has been described by Eynon and Walker (1974).

Unlike the Main Terrace, extensive overbank deposits are absent. However, as previously noted, laterally extensive sandy clay units, often resting on a gravel lag horizon, are associated with the major bounding surfaces within the terrace. Similar, thinner and less extensive, fine-grained beds occur at coset boundaries within assemblage B. The form, internal structure, and position within the lithofacies sequence of these beds closely resemble those of truncated intra-formational mudstones described by Cant and Walker (1976) and Allen (1983), which were interpreted as forming on the surface of inactive bar forms and in an overbank position within ancient sandy braided river systems. The fine-grained units observed within the Worcester Terrace may, therefore, have had a similar origin, with the thinner beds in assemblage B representing supra-bar sediments, while the more extensive units are remnants of overbank sedimentation.

The causes of the changes in channel planform during the terrace aggradation at Grimley are unclear. The contrast in the sedimentary attributes between the basal assemblage A and the overlying sandy assemblage B indicates that there may have been an abrupt change in channel planform, although it is more likely that there is a significant gap in the sedimentary record. However, the changes in the channel planform later in the aggradation seem to have been more progressive. While it is possible that these variations were caused by a local depositional control, such as a base-level produced backwater effect (which would influence grain size diminution in the upstream reach), the three-fold stratigraphic division closely resembles that described by Shotton at the upstream Stourport site. This suggests that there may have been a larger scale variation in sediment supply and discharge, affecting considerable lengths of the Lower Severn during deglaciation.

## PALAEOHYDROLOGICAL CONDITIONS

Palaeodischarges occurring during the deposition of the terrace sediments may be estimated using empirical and theoretical relationships derived for open channel flow. Various methodologies have been proposed (Church, 1978; Maizels, 1983b; Costa, 1983) but they cannot be regarded as being particularly precise due to the difficulties in determining former channel conditions and because of the procedural assumptions (see Maizels, 1983) which have to be introduced during the calculations.

Table 13.1 outlines the relationships used to estimate palaeodischarges in the Lower Severn Basin from gravel sedimentary units. The techniques closely follow those used by Cheetham (1976, 1980), Church (1978) and Maizels (1983a,b), being based on field measurements of former channel-bed grain size distributions, channel slopes and where possible channel geometry. Critical flow depths are derived from grain size using Shields (1936) and Du Boys (1879) relationships, although in contrast to previous studies a value for Shields dimensionless shear stress ($\theta$) is estimated from the empirical relationship of Andrews (1983). Calculated values for flow depth and measured values for grain size and slope are then substituted into the Darcy–Weisbach equation to derive a mean flow velocity. The Hey (1979a) method for determining the friction factor ($f$) is used, although this is not

TABLE 13.1  Open channel flow relationships used in the calculation of palaeodischarges

---

Critical depth

$d_c = t_c/yS$

where:  $t_c$    $= \theta(y_s - y_f)D$
         $\theta$    $= 0.0834 \ (D_i/D_{50})^{-0.872}$
         $d_c$    $=$ critical depth
         $y, y_f$   $=$ specific weight of fluid
         $y_s$    $=$ specific weight of sediment
         $D$    $=$ grain size
         $S$    $=$ slope

Critical flow velocity

   1: Manning–Limerinos approach

$$V_c = \frac{d^{2/3} \ S^{1/2}}{n}$$

   where:  $n$   $= \dfrac{0.113^{1/6}}{1.16 + 2.0 \ \log(d/D^{84})}$

   2: Darcy–Weisbach–Hey approach

$$V_c = \frac{(8g \ d \ S)}{f}^{1/2}$$

   where:  $\dfrac{1}{\sqrt{f}}$   $= 2.03 \ \log \ \dfrac{a \ d}{3.5 \ D_{84}}$
           $a$    $= 11.26$ (assumed)

Channel discharge

$cQ = Vc \ A$

where:  $A$   $= W \cdot D$ (flow area)
         $W$   $= D \cdot W/D$ (ratio derived from analogue studies)
             $=$ (channel width)

Reach-wide discharge

$tQ = cQ/pQ$

where:  $pQ$   $=$ percentage discharge in main channel
              $= 71\%$ (derived from Mosley, 1983)

---

strictly applicable to sediments exposed in section as a value for the channel form parameter (*a*) has to be assumed. However, it should be noted that a similar assumption about the channel form has previously been introduced in deriving flow depths using the Du Boys (1879) relationship.

Examination of the sedimentary characteristics of gravel channel-zone bodies in both the Main and Worcester terraces indicates deposition by rivers of braided planform. Calculation of discharge for an assumed single channel is likely, therefore, to be inappropriate. Furthermore, distinct channel forms are rarely apparent, and thus assumptions have to be introduced concerning the form of the channels, and the distribution of discharge between them.

Given that flow depths have been previously determined, flow widths can be approximated from width:depth ratios in modern environments. Although channel geometry may be estimated using empirical relations (e.g. Schumm, 1960; Briggs, 1983), these may poorly represent palaeochannel forms (Riley, 1975) and hence the analogue approach is felt to be more valid. Pickup and Higgins (1979) showed that width:depth ratios in braided environments are log normally distributed and consideration of data presented by Church (1972), for large-scale braided rivers, confirms this distribution, a median value being 70. Calculation of discharges using the continuity equation ($Q = VA$) with such estimated flow widths produces an approximation to a single-channel discharge. It is assumed here that the gravel unit sampled was deposited within a distinct channel at a riffle site where relatively high width:depth ratios are apparent.

Derivation of a reach-wide discharge requires further simplifying assumptions. As the gravel samples were taken from the coarsest gravel unit on a given sedimentary horizon it is assumed that the unit was deposited in the largest channel. Mosley (1983) showed that in the braided Rakaia River (New Zealand), the main channel carried between 47 and 93 per cent of the flow, independent of total discharge, with a mean value of 71 per cent. Correction of the channel discharges using these values produces an approximation to total reach-wide discharges.

In the calculation of palaeodischarges an estimate of the error range associated with the techniques may be made. Five sources of error may be identified, including variations in the:

(a) value of Shields coefficient derived using the Andrews (1983) relationship;
(b) suspended sediment concentration—a possible range of variation may be determined from modern analogue studies;
(c) calculated friction factor value, as derived from Hey's (1979) equation, including differences caused by changing channel form;
(d) palaeochannel form, particularly the width:depth ratio, as is apparent in modern environments (Church, 1972; Rice, 1979);
(e) percentage discharge in the main channel.

In addition error is introduced in assuming steady uniform flow conditions and by substituting flow depth for hydraulic radius in the Du Boys' (1879) relationship. The cumulative effect of these five sources of error may be considerable—up to 3 orders of magnitude, with variation in the width: depth ratios providing the major contribution.

Values characterizing discharges in the glacial and de-glacial periods were calculated for two sites: Holt Heath (Main Terrace) and Grimley (Worcester Terrace). Observations in modern environments (Church, 1972; Nordseth, 1973; Dawson, 1982), indicate that the highest bedload transport rates in braided rivers occur when complex bars are submerged and that the most extensive and coarsest sedimentary units are produced at values approaching the mean annual flood. The calculated palaeodischarges are therefore thought to approximate former mean anual flood values, similar assumptions having been made by Cheetham (1975), Church (1978), and Briggs (1983).

Examination of the sedimentary characteristics of the terraces shows that it is possible to identify variations in the sedimentary conditions that occurred during their deposition. Discharges are therefore presented for individual sedimentary units, although it should be noted that the difference in the values calculated for the separate units is considerably smaller than the error range associated with each estimate. Grain size characteristics were determined from percentage by number of samples taken from the coarsest gravel beds within a given channel-zone unit, while palaeoslopes were surveyed from the terrace surface, corrected for channel sinuosity.

Table 13.2 shows calculated flow characteristics and a possible error range for the selected sedimentary units. Although there is considerable uncertainty in the techniques, the derived values indicate that mean annual flood discharges during an initial stage in the deposition of the Worcester Terrace exceeded similar discharges during the formation of the Main Terrace by 4 to 7 times. This evidence would seem to suggest that palaeodischarges were greatest during the deglacial period, increasing, after the formation of the Main Terrace, to a maximum prior to the aggradation of the Worcester

TABLE 13.2    Palaeodischarge estimates for the Lower Severn in the late-Devensian period

| | Channel depth (m) | Channel width (m) | Channel velocity $(ms^{-1})$ | Channel discharge $(m^3s^{-1})$ | Total discharge Mean | Max $(m^3s^{-1})$ | Min |
|---|---|---|---|---|---|---|---|
| **Main Terrace** | | | | | | | |
| Surface palaeochannel | 1.15 | 62 | 2.0 | 144 | 203 | — | — |
| Coarsest unit | 1.3 | 92 | 2.1 | 264 | 365 | 3655 | 21 |
| Basal unit | 1.05 | 74 | 1.8 | 143 | 201 | 2015 | 11 |
| | | | | | | | |
| **Worcester Terrace** | | | | | | | |
| Surface unit | 1.6 | 116 | 2.1 | 401 | 556 | 5642 | 32 |
| Basal unit | 2.5 | 173 | 2.5 | 1066 | 1502 | 14992 | 86 |

Terrace. Mean annual flood discharges seem to have remained relatively constant during the aggradation of the Main Terrace, but declined during the deposition of the Worcester Terrace. It is therefore possible that this latter aggradation is a partial response to this decline in discharges during a period of high sediment availability.

It is possible to compare the calculated mean discharge values with present-day mean annual flood values recorded at Bewdley (397 $m^3$ $s^{-1}$). This comparison indicates that mean annual flood discharges did not exceed present-day flood values, and Worcester Terrace discharges were only 3.8 times greater. These relative values are considerably less than those predicted by other palaeohydrological studies of late Devensian terraces or valley forms in southern Britain (Dury, 1964, 1965; Cheetham, 1975, 1976, 1980; Clarke and Dixon, 1981; Briggs, 1983). However, Jones (1982) found that the relative decrease in discharges in the Upper Severn since the final stages of deglaciation was similar to the decrease noted here and it seems likely that the decline in discharges in the Lower Severn was not as marked as in other basins. This may be related to the considerable expansion in basin area since the glacial period.

The development of the terraces studied was therefore controlled by a complex interaction between channel sediment inputs, discharge and changes in relative base level. Channel planform during the aggradation of both terraces was characterized by low-sinuosity, often braided, channels although notable differences are apparent in the sedimentary sequences of the two terraces and between proximal and distal sites in the same terrace. The Main Terrace is dominantly composed of gravelly channel-zone sediments, with interbedded coarse gravel and sand deposits locally overlain by alluvial fans, at proximal locations showing a downstream transition to sequences where channel-zone and sandy overbank units are distinctly segregated. Channel planform seems to have remained constant during the formation of the Main Terrace, although a transition to a single-channel planform is likely to have occurred during terrace incision. In contrast channel planform changed from a gravelly braided channel during the initial aggradation of the Worcester Terrace to a largely sand-bed low-sinuosity channel. Discharges appear to have been greatest during the deglacial period, increasing after the formation of the Main Terrace to a maximum prior to aggradation of the Worcester Terrace.

## CONCLUSIONS

River terraces have been studied scientifically for over a century. The terraces of some rivers have been examined in immense detail with, for example, the lithostratigraphy, palaeobiology and palaeogeography of the Middle Thames being documented by a recent survey (Gibbard, 1985). Yet despite such advances we can only begin to make crude estimates of the magnitudes of hydrological change which have occurred during terrace formation, and the precise mechanisms by which terraces have been produced remain equally obscure. Progress is, however, being made, and it is clear that not only can a knowledge of the sedimentary characteristics of terrace deposits aid in hydrological retrodiction, but palaeohydrological studies can help in formulating a general model of terrace formation.

## ACKNOWLEDGEMENTS

We should like to acknowledge the award of a Natural Environment Research Council research studentship (to MRD), and the help given by participants in the IGCP Project 158 during the tenure of this studentship.

## REFERENCES

Alexander, E. B. and Holovaychuk, N. (1983). Soils on terraces along the Cauca River, Colombia: II. The sand and clay fractions, *Soil Science Society of America Journal*, **47**. 721–7.

Allen, J. R. L. (1964). Studies in fluvial sedimentation: six cyclothems from the Lower Old Red Sandstone, Anglo-Welsh Basin, *Sedimentology*, **3**, 163–98.

Allen, J. R. L. (1965). Fining upwards cycles in alluvial successions, *Geological Journal*, **4**, 229–46.

Allen, J. R. L. (1983). Studies in fluviatile sedimentation; bars, bar complexes and sandstone sheets (low-sinuosity braided streams) in the Brownstones (L. Devonian), Welsh Borders, *Sedimentary Geology*, **33**, 237–93.

Andrews, E. D. (1983). Entrainment of gravel from naturally sorted river-bed material, *Bulletin, Geological Society of America*, **94**, 1225–31.

Baulig, H. (1935). The changing sea level, *Transactions, Institute of British Geographers*, **3**, 1–46.

Beckinsale, R. P. and Richardson, L. (1964). Recent findings on the physical development of the lower Severn Valley, *Geographical Journal*, **130**, 87–105.

Blodgett, R. H. and Stanley, K. O. (1980). Stratification, bedforms and discharge relations of the Platte braided river system, Nebraska, *Journal of Sedimentary Petrology*, **50**, 139–48.

Bluck, B. J. (1971). Sedimentation in the meandering River Endrick, *Scottish Journal of Geology*, **7**, 93–138.

Bluck, B. J. (1974). Structure and directional properties of some valley sandur deposits in southern Iceland, *Sedimentology*, **21**, 533–54.

Bluck, B. J. (1976). Sedimentation in some Scottish rivers of low sinuosity, *Transactions, Royal Society of Edinburgh*, **69**, 425–56.

Bluck, B. J. (1979). The structure of coarse grained braided stream alluvium, *Transactions, Royal Society of Edinburgh*, **70**, 181–221.

Boothroyd, J. C. and Ashley, G. M. (1975). Process, bar morphology and sedimentary structures on braided outwash fans: North-Eastern Gulf of Alaska, in Jopling, A. V. and McDonald, B. C. (eds), *Glaciofluvial and Glaciolacustrine Sedimentation.*, Soc. Econ. Min. Paleont. Spec. Publ. No. 23, 193–222.

Born, S. M. and Ritter, D. F. (1978). Modern terrace development near Pyramid

Lake, Nevada, and its geologic implications, *Bulletin, Geological Society of America*, **81**, 1233–42.

Boulton, G. S. and Worsley, P. (1965). Late Weichselian glaciation of the Cheshire-Shropshire basin, *Nature*, **207**, 704–6.

Bourdier, F. (1958). Origine et succès d'une théorie géologique illusoire: l'eustatisme applique aux terrasses alluviales, *Revue de géomorphologie dynamique*, **10**, 16–19.

Bowen, D. Q. (1978). *Quaternary Geology*, Pergamon, Oxford.

Bridge, J. S. (1985). Perspectives: Paleochannel patterns inferred from alluvial deposits, *Journal of Sedimentary Petrology*, **55**, 579–706.

Bridge, J. S. and Leeder, M. R. (1979). A simulation model of alluvial stratigraphy, *Sedimentology*, **26**, 617–44.

Briggs, D. J. (1983). Palaeohydrological analysis of braided stream deposits, in Briggs, D. J. and Waters, R. S. (eds), *Studies in Quaternary Geomorphology*, Procs. VIth British-Polish Seminar, Geobooks, Norwich, 49–62.

Briggs, D. J. and Gilbertson, D. D. (1980). Quaternary processes and environments in the upper Thames valley, *Transactions, Institute of British Geographers*, **5**, 53–66.

Brown, A. G. (1982). Late Quaternary palaeohydrology, paleoecology, and floodplain development of the Lower River Severn, unpublished Ph.D. thesis, University of Southampton.

Brown, A. G. (1983). Floodplain deposits and accelerated sedimentation in the lower Severn Basin, in Gregory, K. J. (ed.), *Background to Palaeohydrology*, Wiley, Chichester, pp. 375–98.

Bryant, I. D. (1983). Facies sequences associated with some braided river deposits of late Pleistocene age from southern Britain, in Collinson, J. D. and Lewin, J. (eds), *Modern and Ancient Fluvial Systems: Sedimentology and Processes*, Int. Assoc. Sedimentologists, Spec. Publ. No. 6, 267–75.

Campbell, C. V. (1976). Reservoir geology of a fluvial sheet sandstone, *Bulletin, American Association of Petroleum Geologists*, **60**, 1009–20.

Cant, D. J. and Walker, R. G. (1976). Development of a braided fluvial model for the Devonian Battery Point Sandstone, Quebec, *Canadian Journal of Earth Sciences*, **13**, 102–19.

Cant, D. J. and Walker, R. G. (1978). Fluvial processes and facies sequences in the sandy braided South Saskatchewan River, Canada, *Sedimentology*, **25**, 625–48.

Carson, M. A. (1984a). The meandering-braided river threshold: a reappraisal, *Journal of Hydrology*, **73**, 315–34.

Carson, M. A. (1984b). Observations on the meandering-braided river transition, the Canterbury Plains, New Zealand. Part One, *New Zealand Geographer*, **40**, 12–19.

Carson, M. A. (1984c). Observation on the meandering-braided river transition, the Canterbury Plains, New Zealand. Part Two, *New Zealand Geographer*, **40**, 89–99.

Castleden, R. (1980). Fluvioglacial pedimentation: a general theory of fluvial valley development in cool temperate lands illustrated from western and central Europe, *Catena*, **7**, 135–52.

Challinor, J. (1932). River terraces as normal features of valley development, *Geography*, **17**, 141–7.

Cheetham, G. H. (1975). Late Quaternary paleohydrology, with reference to the Kennet Valley, unpublished Ph.D. thesis, University of Reading.

Cheetham, G. H. (1976). Palaeohydrological investigations of river terrace gravels, in Davidson, D. A. and Shackley, M. (eds), *Geoarchaeology: Earth Science and the Past*, Duckworth, London, pp. 335–43.

Cheetham, G. H. (1980). Late Quaternary palaeohydrology: the Kennet valley case-study, in Jones, D. K. C. (ed.), *The Shaping of Southern England*, I.B.G. Spec. Pub. 11, Academic Press, 203–23.

Chester, D. K. and Duncan, A. M. (1979). Interrelationships between volcanic and alluvial sequences in the evolution of the Simeto river valley, Mount Etna, Sicily, *Catena*, **6**, 293–315.

Church, M. (1972). *Baffin Island Sandurs: A Study of Arctic Fluvial Processes*, Geol. Surv. Can. Bull. No. 216, 208 pp.

Church, M. (1978). Paleohydrological reconstructions from a Holocene valley fill, in Miall, A. D. (ed.), *Fluvial Sedimentology*, Can. Soc. Petrol. Geol. Mem. No. 5, pp. 743–72.

Church, M. and Jones, D. (1982). Channel bars in gravel-bed rivers, in Hey, R. D., Bathurst, J. C. and Thorne, C. R. (eds), *Gravel-Bed Rivers*, Wiley, Chichester, pp. 291–324.

Clarke, M. and Dixon, A. J. (1981). The Pleistocene braided river deposits in the Blackwater Valley area of Berkshire and Hampshire, England, *Proceedings, Geologists Association*, **92**, 139–57.

Clayton, K. M. (1977). River terraces, in Shotton, F. W. (ed.), *British Quaternary Studies, Recent Advances*, Clarendon Press, Oxford, pp. 153–68.

Collinson, J. D. (1970). Bedforms of the Tana River, Norway, *Geografisker Annaler*, **52A**, 31–55.

Colman, S. M. (1983). Progressive changes in the morphology of fluvial terraces and scarps along the Rappahannock River, Virginia, *Earth Surface Processes and Landforms*, **8**, 201–12.

Coope, G. R. and Shotton, F. W. (1981). The Devensian terraces of the Severn below Ironbridge and of the Warwickshire Avon, in *Palaeohydrology of the Temperate Zone*. I. G. C. P. Project 158a Fluvial Environments, Severn Basin Summary Reports and Maps.

Costa, J. E. (1983). Paleohydraulic reconstructions of flash flood peaks from boulder deposits in the Colorado Front Range, *Bulletin, Geological Society of America*, **94**, 986–1004.

Darwin, C. (1846). *Observations on South America*, Smith Elder, New York.

Davis, W. M. (1902). River terraces in New England, *Bulletin, Museum of Comparative Zoology*, **38**, Geological Series V, 281–346.

Dawson, M. R. (1982). Sediment variation in a braided reach of the Sunwapta River, Alberta, unpublished M.Sc. thesis, University of Alberta, 186 pp.

Dawson, M. R. (1985). Environmental reconstructions of a late Devensian terrace sequence. Some preliminary findings, *Earth Surface Processes and Landforms*, **10**, 237–46.

Dickson, J. A., Stewart, D. A., Thompson, R., Turner, G., Baxter, M. S., Drndarsky, N. D. and Rose, J. (1978). Palynology, palaeomagnetism and radiometric dating of Flandrian marine and freshwater sediments of Loch Lomond, *Nature*, **274**, 548–53.

Du Boys, P. (1879). Etudes du régime du Rhone et l'action exercée par les eaux un lit a fond de graviers indéfiniment affouillable, *Annales des Ponts et Chausses*, Ser. 5, **18**, 141–95.

Dury, G. H. (1964). *Principles of Underfit Streams*, U.S.G.S. Professional Paper 452-A, 67 pp.

Dury, G. H. (1965). *Theoretical Implications of Underfit Streams*, U.S.G.S. Professional Paper 452-C, 43 pp.

Dury, G. H. (ed.) (1970). *Rivers and River Terraces*, Macmillan, London, 283 pp.

Ek, C. (1961). Conduits souterrains en relation avec les terraces fluviales, *Annales Société Géologique Belgique*, **81**, 313–40.

Eynon, G. and Walker, R. G. (1974). Facies relationships in Pleistocene outwash gravels, Southern Ontario: A model for bar growth, *Sedimentology*, **21**, 43–70.

Fairbridge, R. W. (1968). Terraces, fluvial-environmental controls, in Fairbridge, R. W. (ed.), *Encyclopaedia of Geomorphology*, Reinhold, New York, pp. 1124–38.

Fisk, H. N. (1951). Loess and Quaternary Geology of the lower Mississippi Valley, *Journal of Geology*, **59**, 333–56.

Fonda, R. W. (1971). Forest succession in relation to river terrace development in Olympic National Park, Washington, *Ecology*, **55**, 927–42.

Frazer, G. S. and Cobb, J. C. (1982). Late Wisconsinian proglacial sedimentation along the West Chicago moraine in north-eastern Illinois, *Journal, Sedimentary Petrology*, **52**, 473–92.

Frye, J. C. and Leonard, A. R. (1954). Some problems of alluvial terrace mapping, *American Journal of Science*, **252**, 242–57.

Gardiner, V. (1983). Drainage networks and palaeohydrology, in Gregory, K. J. (ed.), *Background to Palaeohydrology*, Wiley, Chichester, pp. 257–77.

Geikie, A. (1965). *The Scenery of Scotland*, Macmillan, London, 360 pp.

Gibbard, P. L. (1985). *The Pleistocene History of the Middle Thames Valley*, Cambridge University Press, Cambridge.

Green, C. P. and McGregor, D. F. M. (1980). Quaternary Evolution of the River Thames, in Jones, D. K. C. (ed.), *The Shaping of Southern England*, Academic Press, London, pp. 177–202.

Gregory, K. J. (1971). Drainage density changes in south-west England, in Ravenhill, W. L. D., and Gregory, K. J. (eds), *Exeter Essays in Geography*, University of Exeter, Exeter, pp. 33–54.

Gregory, K. J. and Walling, D. E. (1973). *Drainage Basin Form and Process*, Edward Arnold, London, 456 pp.

Haible, W. W. (1980). Holocene profile changes along a California coastal stream, *Earth Surface Processes and Landforms*, **5**, 249–64.

Hayden, F. V. (1876). Notes descriptive of some geological sections of the country about the headwaters of the Missouri and Yellowstone rivers, *Bull., U.S. Geol. and Geogr. Surv. of the Terr.*, **2**, 197–209.

Hein, F. J. and Walker, R. G. (1977). Bar formation and the development of stratification in the gravelly braided Kicking Horse River, British Columbia, *Canadian Journal of Earth Sciences*, **14**, 562–70.

Hey, R. D. (1979a). Dynamic process response model of river channel development, *Earth Surface Processes*, **4**, 59–72.

Hey, R. D. (1979b). Flow resistance in gravel bed rivers, *American Society of Civil Engineers, Journal of the Hydraulics Division*, **105**, HY4, 365–79.

Hey, R. D. (in press). River dynamics, flow regime and sediment transport.

Hey, R. W. (1958). High level gravels in and near the lower Severn Valley, *Geological Magazine*, **95**, 161–8.

Hitchcock, E. (1833). *Report on the Geology, Mineralogy, Botany and Zoology of Massachusetts*, J. S. and C. Adams, Amherst, 700 pp.

Hitchcock, E. (1857). *Illustrations of Surface Geology*, Smithsonian Institute, Washington.

Howard, A. D. (1968), Fairbridge, R. W. and Quinn, J. H. Terraces, Fluvial—Introduction, Fairbridge, R. W. (ed.), *Encyclopaedia of Geomorphology*, Reinhold, New York, pp. 1117–24.

Huntington, E. (1907). Some characteristics of the glacial period in non-glaciated regions, *Bulletin, Geological Society of American*, **18**, 351–88.

Jackson III, R. G. (1978). Preliminary evaluation of lithofacies models for meandering alluvial streams, in Miall, A. D. (ed.), *Fluvial Sedimentology*, Canadian Society of Petroleum Geologists, pp. 543–76.

Jones, M. D. (1982). The palaeogeography and palaeohydrology of the River Severn, Shropshire during the Late Devensian glacial stage and the Early Holocene, unpublished M. Phil. thesis, University of Reading.

Jones, O. T. (1924). The Upper Towy drainage system, *Quarterly Journal of the Geological Society of London*, **80**, 568–609.

Kidson, C. (1962). Denudation Chronology of the River Exe, *Transactions, Institute of British Geographers*, **31**, 43–66.

King, W. B. R. and Oakley, K. P. (1936). The Pleistocene succession in the lower part of the Thames valley, *Proceedings, Prehistorical Society*, **2**, 52–76.

Langbein, W. B. and Schumm, S. A. (1958). Yield of sediment in relation to mean annual precipitation, *Transactions, American Geophysical Union*, **39**, 1076–84.

Larue, J.-P. (1982). L'évolution morphologique des terrasses alluviales, *Norois*, **29**, 365–84.

Latreille, G. and Le Griel, A. (1980). Reconstruction du système des terrasses de la Loire dans les bassins de Roanne et de Digoin a l'aide des mineraux lourds, *Revue de Géologie Dynamique et de Géographie Physique*, **22**, 223–8.

Leopold, L. B., Wolman, M. G. and Miller, J. P. (1964). *Fluvial Processes in Geomorphology*, W. H. Freeman, San Francisco.

Lewin, J. and Manton, M. M. M. (1975). Welsh floodplain studies: The nature of floodplain geometry, *Journal of Hydrology*, **25**, 37–50.

Lewis, W. V. (1944). Stream trough experiments and terrace formation, *Geological Magazine*, **81**, 241–53.

Maizels, J. K. (1983a). Proglacial channel systems: Change and thresholds for change over long intermediate and short timescales, in Collinson, J. D. and Lewin, J. (eds), *Modern and Ancient Fluvial Systems*, Int. Assoc. Sed. Spec. Pub. No. 6, 251–66.

Maizels, J. K. (1983b). Palaeovelocity and palaeodischarge determination for coarse gravel deposits, in Gregory, K. J. (ed.), *Background to Palaeohydrology*, Wiley, Chichester, pp. 101–40.

McGregor, D. F. M. and Green, C. P. (1983a). Post-depositional modification of Pleistocene terraces of the River Thames, *Boreas*, **12**, 23–33.

McGregor, D. F. M. and Green, C. P. (1983b) Lithostratigraphic subdivision in the gravels of the proto-Thames between Hemel Hempstead and Watford, *Proceedings of the Geologists' Association*, **94**, 83–5.

Miall, A. D. (1977). The braided river depositional environment, *Earth Science Reviews*, **3**, 1–62.

Miall, A. D. (1978). Lithofacies types and vertical profile models in braided river deposits: a summary in Miall, A. D. (ed.), *Fluvial Sedimentology*, Can. Soc. Petrol. Geol. Mem. No. 5, 597–604.

Miller, H. (1883). River terracing: its methods and their results, *Procs. Royal Phys. Society of Edinburgh*, **7**, 263–306.

Mills, H. H. (1986). Possible differential uplift of new river terraces in Southwest Virginia. *Neotectonics*, **1**,

Mitchell, G. F., Pocock, R. W. and Taylor, J. H. (1961). *Geology of the Country around Droitwich, Abberley and Kidderminster*, Geological Survey of Great Britain, Memoir No. 182, 137 pp.

Moody-Stuart, M. (1966). High and low sinuosity stream deposits with examples from the Devonian of Spitsbergen, *Journal of Sedimentary Petrology*, **36**, 1102–17.

Mosley, M. P. (1982). Analysis of the effect of changing discharge on channel morphology and instream uses in a braided river. Ohau River, New Zealand, *Water Resources Research*, **18**, 800–12.

Mosley, M. P. (1983). The response of braided rivers to changing discharge, New Zealand, *Journal of Hydrology*, **22**, 18–67.

Nordseth, K. (1973). Fluvial processes and adjustments on a braided river. The Islands of Koppangsoyne on the River Glomma, *Norsk Geografisk Tidskrift*, **27**, 77–108.

Patton, P. C. and Schumm, S. A. (1981). Ephemeral-stream processes: implications for studies of Quaternary Valley Fills, *Quaternary Research*, **15**, 24–43.

Penck, A. and Bruckner, E. (1909). *Die Alpen um Eiszeitalter*, Tauchnitz, Leipzig, 3 vols.

Petts, G. E. and Foster, I. D. L. (1985). *Rivers and Landscape*, Edward Arnold, London, 274 pp.

Pickup, G. and Higgins, R. J. (1979). Estimating sediment transport in a braided gravel channel: Kawerong River, Bougainville, Papua, New Guinea, *Journal of Hydrology*, **40**, 287–97.

Playfair, J. (1802). *Illustrations of the Huttonian Theory of the Earth*, Wm. Creech, Edinburgh, 528 pp.

Poole, E. G. and Whiteman, A. J. (1961). The glacial drifts of the southern part of the Shropshire-Cheshire basin, *Quarterly Journal of the Geological Society of London*, **117**, 91–130.

Pounder, E. J. (1980). Theories of river terrace development: a review, *Brighton Polytechnic Geographical Society Magazine*, **8**, 24–34.

Quinn, J. H. (1957). Paired river terraces and Pleistocene glaciation, *Journal of Geology*, **65**, 149–66.

Ramos, A. and Sopena, A. (1983). Gravel bars in low sinuosity streams (Permian and Triassic, central Spain), in Collinson, J. D. and Lewin, J. (eds), *Modern and Ancient Fluvial Systems*, Int. Assoc. Sed. Spec. Publ. No. 6, 301–12.

Rice, R. J. (1979). The hydraulic geometry of the Sunwapta River Valley Train, Jasper National Park, Alberta, unpublished M.Sc. thesis, University of Alberta.

Richards, K. S. (1982). *Rivers: Form and Process in Alluvial Channels*, Methuen, London, 358 pp.

Riley, S. J. (1975). The channel shape–grain size relation in eastern Australia and some palaeohydrology implications, *Sedimentary Geology*, **14**, 253–8.

Ritter, D. F. (1982). Complex river terrace development in the Neviana Valley, near Healy, Alaska, *Bulletin, Geological Society of America*, **93**, 346–56.

Ritter, D. F. and Miles, C. (1973). Problems of stream terrace correlation and reconstruction of geomorphic history caused by colluvium, *Geografisker Annaller*, **55a**, 85–91.

Ritter, D. F., Kinsey, W. F., III, and Kauffman, M. E. (1973). Overbank sedimentation in the Delaware River Valley during the last 6000 years, *Science*, **179**, 374–5.

Rose, J. (1978). River terraces and sea level change, *Brighton Polytechnic Geographical Society Magazine*, **3**, 13–29.

Rose, J. (1982). The Melinau River and its terraces, *Cave Sciences—Transactions, British Cave Research Association*, **9**, 113–27.

Rose, J. (1984). Alluvial terraces of an equatorial river, Melinau drainage basin, Sarawak, *Zeitschrift für Geomorphologie*, **28**, 155–77.

Rust, B. R. (1972). Structure and process in a braided river, *Sedimentology*, **18**, 221–46.

Rust, B. R. (1978a). The interpretation of ancient alluvial successions in the light of modern investigations, in Davidson-Arnott, R. and Nicklin, W. (eds), *Research in Fluvial Sedimentology*, Procs. 5th Guelph Symposium on Geomorphology, 1977, Geobooks, Norwich, pp. 67–105.

Rust, B. R. (1978b). Depositional models for braided alluvium, in Miall, A. D. (ed.), *Fluvial Sedimentology*, Can. Soc. Petrol. Geol. Mem. No. 5, 605–96.

Rust, B. R. (1984). Proximal braidplain deposits in the middle Devonian Malbaie Formation of Eastern Gaspé, Quebec, Canada, *Sedimentology*, **31**, 675–96.

Schick, A. P. (1974). Formation and obliteration of desert stream terraces—a conceptual analysis, *Zeitschrift für Geomorphologie, Supplementband*, **21**, 88–105.

Schumm, S. A. (1960). *The Shape of Alluvial Channels in Relation to Sediment Type*, U.S.G.S. Professional Paper No. 352-B, 17–30.

Schumm, S. A. (1961). Effect of sediment characteristics on erosion and deposition in ephemeral stream channels, *U.S. Geological Survey, Professional Paper 352-C*, 31–70.

Schumm, S. A. (1976). Episodic erosion: a modification of the geomorphic cycle, in Melhorn, W. N. and Flemal, R. C. (eds), *Theories of Landform Development*, pp. 69–85.

Schumm, S. A. (1977). *The Fluvial System*, Wiley, New York.

Shaw, J. (1979). Aspects of glaciagenic sedimentation, with special reference to the Area around Shrewsbury, unpublished Ph.D. thesis, University of Reading.

Shields, A. (1936). Anwendung der Ahnlickzeitsmechanik und Turbulenz Forschung aut die Geshiebwgung, *Berlin Preuss Versuchsnstalt für Vasser Erd und Schiffbau*, No. 26, 26 pp.

Shotton, F. W. (1953). Pleistocene deposits of the area between Coventry, Rugby and Leamington and their bearing on the topographic development of the Midlands, *Philosophical Transactions of the Royal Society*, Ser. B, **237**, 209–60.

Shotton, F. W. (1967). Age of the Irish Sea Glaciation of the Midlands, *Nature*, **215**, 1366.

Shotton, F. W. (1973). The English Midlands, in Mitchell, G. F., Penny, L. F., Shotton, F. W. and West, R. G. (eds), *A Correlation of Quaternary Deposits in the British Isles*, Geol. Soc. Lond. Spec. Rep. No. 4, 18–22.

Shotton, F. W. (1977). *The English Midlands*, INQUA Excursion Guide A2, Xth INQUA Congress, Birmingham, 51 pp.

Shotton, F. W. and Coope, G. R. (1983). Exposures in the Power House Terrace of the River Stour, Wilden, Worcestershire, England, *Proceedings of the Geologists' Association*, **94**, 33–44.

Sievers, W. (1888). Die Kordilliere von Merida Nebst Bemerkungen uber das Karibische Gebirge, *Geog. Abhandl*, **3**, 238.

Sivadas, K. M. (1986). Geomorphology and ore characteristics—gems of Kerala, in Gardiner, V. (ed.), *Proceedings of the First International Conference on Geomorphology*, Wiley, Chichester, pp. 263–272.

Small, R. J. (1973). Braiding terraces in the Val D'Herens, Switzerland, *Geography*, **158**, 129–35.

Smith, D. G. and Smith, N. D. (1981). Sedimentation in anastomosed river systems: examples from alluvial rivers near Banff, Alberta, *Journal of Sedimentary Petrology*, **50**, 157–64.

Smith, N. D. (1970). The braided stream depositional environment: Comparison of the Platte River with some Silurian clastic rocks, north-central Appalachians, *Bulletin, Geological Society of America*, **81**, 2993–3014.

Steel, R. J. and Thompson, D. B. (1983). Structures and textures in Triassic braided stream conglomerates in the Sherwood Sandstone Group, North Staffordshire, England, *Sedimentology*, **30**, 341–68.

Stephens, N. (1970). The lower Severn Valley, in Lewis, C. A. (ed.), *The Glaciation of Wales and Adjoining Regions*, pp. 107–17.

Tomlinson, M. E. (1925). River terraces of the lower valley of the Warwickshire Avon, *Quarterly Journal of the Geological Society of London*, **81**, 137–63.

Tricart, J. and Millies-Lacroix, A. (1962). Les terrasses Quaternaires des Andes Vénézueliennes, *Société Géologique de France Bulletin*, **7**, 201–8.

Tylor, A. (1869). On Quaternary gravels, *Quarterly Journal of the Geological Society of London*, **5**, 57–100.

Upham, W. (1877). The northern part of the Connecticut Valley in the Champlain and terrace periods, *American Journal of Science*, **14**, 459–70.

Williams, G. E. (1968). Formation of large-scale trough cross-stratification in a fluvial environment, *Journal of Sedimentary Petrology*, **38**, 136–40.

Williams, G. J. (1968). The buried channel and superficial deposits of the lower Usk and their correlation with similar features in the lower Severn, *Proceedings of the Geologists' Association*, **79**, 325–48.

Williams, P. F. and Rust, B. R. (1969). The sedimentology of a braided river, *Journal of Sedimentary Petrology*, **39**, 649–79.

Wills, L. J. (1924). The development of the Severn Valley in the neighbourhood of Ironbridge and Bridgnorth, *Quarterly Journal of Geological Society of London*, **80**, 274–314.

Wills, L. J. (1938). The Pleistocene development of the Severn from Bridgnorth to the sea, *Quarterly Journal of the Geological Society of London*, **94**, 161–242.

Wolman, M. G. and Leopold, L. B. (1957). *River floodplains: some observations on their formation*. U.S.G.S. Professional Paper 282–C, 87–107.

Womack, W. R. and Schumm, S. A. (1977). Terraces of Douglas Creek, North Western Colorado, *Geology*, **5**, 72–6.

Yoshikawa, T., Kaisuka, S. and Ota, Y. (1981). *The Landforms of Japan*, University of Tokyo Press, Tokyo, 222 pp.

Young, R. W. and Nanson, G. C. (1982). Terrace formation in the Illawarra region of New South Wales, *Australian Geographer*, **15**, 212–19.

Ziegler, J. M. (1958). Geological study of Shamsir Ghar Cave, southern Afghanistan, and report of terraces along Panjshir valley near Kabul, *Journal of Geology*, **66**, 16–27.

Palaeohydrology in Practice
Edited by K. J. Gregory, J. Lewin and J. B. Thornes
© 1987 John Wiley & Sons Ltd.

# 14

# Long-term Sediment Storage in the Severn and Wye Catchments

A. G. BROWN

*Department of Geography, University of Leicester*

## THE STORAGE COMPONENT IN LONG-TERM SEDIMENT BUDGETS

The drainage basin, as an integral part of global sedimentary and geochemical cycles, involves erosional, transportational and sedimentary or storage processes. In tectonic depressions and internal drainage basins terrestrial sediments may be stored for millions of years before removal, while some colluvial or fluvial deposits may remain in storage for only a few years before remobilization. Thus the drainage basin not only facilitates the transport and removal of sediment but stores it in a variety of locations for varying periods of time. Such stored sediment may subsequently be eroded, adding to the sediment yield of the basin. It is therefore clear that the sediment yield of any basin can be higher or lower than hillslope and bedrock erosion in the basin, as illustrated by Trimble (1975). Hillslope and bedrock erosion includes the erosion of *in situ* soils, weathered rock and bedrock, but not of stored sediment or cumulative soils (i.e. it is primary erosion). If the total sediment yield exceeds primary erosion this can only continue until sediment storage is reduced to zero. Therefore total sediment yields are to some extent affected by the volume of the storage component.

The storage of sediment in any basin is described by the sediment budget equation which can be written:

$$\frac{\delta Si}{\delta t} = I_{it} - Q_{it}$$

in which $Si$ the storage volume, $I$ = the sediment input or erosion, $Q$ = the sediment output or yield, $i$ = the reach or catchment segment and $t$ = the time period. It is the $\delta Si/\delta t$ term that palaeohydrological studies may attempt to quantify by the dating of sedimentary units. In contemporary studies such as that by Dietrich and Dunne (1978) $I_{it}$ and $Q_{it}$ may be measured. The basin may be divided into a series of reaches and/or geomorphic compartments such as hillslope soils, colluvium, alluvium and lake sediments. These storage compartments may be sites of both erosion and sedimentation—eroding fields are almost always sites of deposition. Both the efficiency (measured as fraction retained) and the residence time of sediment in these stores will vary according to factors such as surface topography, climate and vegetation. If the primary erosion and total sediment yield of a basin are known or can be estimated, the sediment delivery ratio ($SD_{it}$) can be calculated:

$$SD_{it} = 100 \ Q_{it}/I_{it}$$

Despite the difficulties involved in the collection of this data and the wide scatter of values obtained, many studies have shown a decrease in the ratio with increasing catchment area of the general form:

$$SD_{it} = 0.36A^{-0.2}$$

where $A$ = the catchment area in km$^2$ (US Soil Conservation Service, 1971 and ASCE, 1975). However, most of the data are derived from the Mid-West, USA, and may not be representative of other climatic and physiographic regions and the variation displayed by different curves is as much as 70 per cent for any given basin size. Ratios calculated from Slaymaker's (1972) data for ten small catchments in the Upper Wye Valley vary from 38 to 213 per cent with the average for all sites being 108 per cent, suggesting slight overall degradation. However, sediment delivery ratio–area relationships that have been estimated do show the overwhelming importance of sediment storage in relation to measurable sediment yield for catchments of $10^3$–$10^4$ km$^2$ in area. Indeed the most detailed study to date by Trimble (1983) showed that for a relatively small catchment (Coon Greek, Wisconsin 360 km$^2$) sediment yields after European settlement fell to as low as 6 per cent of the primary erosion.

The ratio can also be regarded as the inverse of the catchment's storage efficiency as in the percentage trap efficiency($Cs$) of a lake. This has been shown to vary with the capacity/watershed ratio (Brune, 1953) where:

$$Cs = 100 \left[ 1 - \frac{1}{1 + (0.1C/A)} \right]$$

where $C$ = the capacity for storage and $A$ = the catchment area. Since the volume of colluvial and alluvial storage increases with catchment area, and from geometrical considerations, we can regard, $C$ as positively related to $A$ and since

$$SD_{it} = 100\text{-}Cs$$

it becomes apparent that the sediment delivery ratio will decrease with increasing catchment area due to the downstream increase in storage capacity. However, the relationship will vary as storage capacity is a function of topographic, ecological and other factors. Because capacity changes over time (generally decreasing) sediment delivery ratios will change over time. Indeed Trimble's (1983) data show a change in the sediment delivery ratio from 5.3 per cent in 1853–1938 to 6.6 per cent in 1938–75. Boyce (1975) has pointed out that, as generally stated, the ratio–catchment area curves conflict with Playfair's law of accordant tributary junctions. Without altering the relationship over a long time period Boyce (1975) has suggested that the answer lies in the decrease in the average land slope with increasing catchment area which causes a decrease in the efficiency of overland flow in these low-slope, low-sediment-producing downstream areas. Roehl (1962) had previously shown that the ratio is also related to the basin relief–length ratio. However, the ratio–basin area relationship could also be due to variation in the length of time taken for eroded sediment to reach a stream channel due to the redistribution of sediment and the construction of non-alluvial stores and lowland subcatchment storage. This is partially the result of the generally negative relationship between catchment area and natural drainage density or texture ratio (Gardiner, Gregory and Walling, 1977).

Investigations of the long-term sediment budget of any catchment must be based upon some appreciation of the spatial variation of erosion rates and the spatial distribution of different types of sediment stores. This involves the collation of many individual pieces of geomorphic research for basins the size of the Severn and the Wye.

External factors (*sensu* Schumm, 1977) will also cause changes in sediment delivery ratios, including climatic change via increasing or decreasing erosion and transport efficiency in relation to storage capacity. The spatial arrangement of storage sites is important as basins with similar total storage volumes can have very different sediment output due to storage linkages and distance from active channels. There are some sites that can be regarded as sinks for at least one interglacial cycle such as infilling kettle holes (see Figure 14.1) whereas others are relatively transient, such as within-channel storage of fine sediment. The density of the channel network in relation to the spatial focii of erosion and sediment storage will affect sediment output. Schumm (1979) has suggested that it is when high-magnitude events coincide with slopes that are near the threshold of instability that very high sediment yields are achieved or when temporary extension of the network taps a new sediment source.

This chapter looks at the largest UK catchment from a sediment storage perspective and tries to assess how the catchment has functioned through

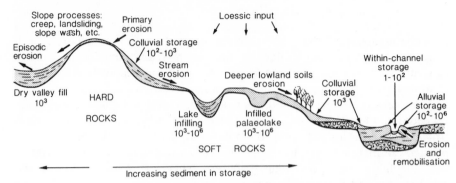

FIGURE 14.1    Storage components of the sediment budget with broadly applicable sediment residence times in years

the Holocene, thus illustrating long-term changes in its sediment budget. To do this it is necessary to investigate both the controls on the spatial variation of erosion during the last 10,000 years as well as the spatial distribution and history of sediment storage sites. To this end the chapter uses data derived from two principal sources; soil maps of the basin produced by the Soil Survey of England and Wales and detailed geomorphic and palaeoecological investigations within the basin by a number of authors.

## SPATIAL VARIATIONS IN SEDIMENT SUPPLY

During the Holocene the erosion of soil from hillslopes varied spatially within the Severn and Wye catchments due to the interaction of constant factors such as topography and certain soil characteristics with more variable factors such as rainfall erosivity and vegetation cover. Some of these factors can be estimated for the past while others cannot. It can be assumed that contributing slopes have remained relatively constant over the last few thousand years as suggested by simulation studies and the degree of preservation of Pre-Holocene landforms. We can also assume that geologically inherited soil characteristics have remained constant although there are two problems here. Firstly, for the geomorphologist, the pattern of geological outcrops changes over time, and as sediments are stripped off so different sediments are revealed; as soil erosion proceeds soil characteristics change (Kirkby, 1980). Secondly, changes in land use and pedogenic processes differentially alter the erodibility of different soil types. Erosivity in the past is difficult to estimate, because it is normally measured by the kinetic energy (KE) index and requires values for both the intensity and the duration of rainfall. Palaeo-climatic data tend to be aggregate or average in nature as exemplified by Lamb, Lewis and Woodroffe's (1966) estimates of Holocene rainfall. The traditional view of Holocene climatic change suggests a shift from a more continental climate to a maritime circulation during the Atlantic period. Although this led to wetter conditions erosivity may well have decreased and Lockwood (1979) has suggested that this period may have witnessed a decrease in stream discharges due to greater evapotranspiration under closed canopy woodland. Indeed it seems that Holocene climatic changes were subtle in nature and may have had relatively little effect on erosion except where vegetation cover was removed. The most convincing evidence of Holocene climatic change comes from very sensitive ecosystems such as marginally located raised bogs (Barber, 1982). Although absolute values of KE probably varied during the Holocene the spatial pattern of KE is unlikely to have altered much from the present pattern as mapped by Morgan (1980) as it is partly topographically determined. The majority of the Severn and Wye catchments have a mean KE > 10 (total kinetic energy for rainfall intensities ≥10 mm hr$^{-1}$) under 1100 J m$^{-2}$ with over 1300 J m$^{-2}$ only in the Welsh mountains.

The erodibility of soils is positively related to the silt and fine sand content and inversely related to organic matter content except for organic soils (Evans, 1980). Of the 296 Soil Associations mapped by the Soil Survey of England and Wales, 12 are regarded as having a risk of water erosion. Table 14.1 lists those mineral soil associations with a risk of water erosion that lie within the Severn and Wye catchments. In this context water erosion refers to above-average soil loss or accelerated erosion well above the replacement--weathering rate and the observations listed in Table 14.1 with each association give evidence of this. It should be noted that this evidence is often of a depositional nature, indicating soil erosion in the past. These soil associations have been combined and remapped in Figure 14.2a and they cover 16 per cent of the Severn Basin and 22 per cent of the Wye Basin. The map shows that these soils are largely restricted to the lowland and central areas of the catchments, and they are generally developed on sedimentary rocks such as Devonian, Carboniferous and Trias sandstones and marls. The percentage area covered by erodible soils varies greatly from one catchment to another; some have high coverage such as the Lugg (38 per cent) and others little or no coverage. This map and Morgan's (1985) soil-erosion-risk map suggest potentially high rates of erosion in lowland Herefordshire and Worcestershire to the west of the River Severn and east Worcestershire and Shropshire to the east of the Severn. Despite the combination of data sources used by Morgan (1985), including rainfall erosivity, land capability and soil erodibility, the pattern shown in Figure 14.2A is predominantly affected by lithology and the erodibility of soils under arable cultivation. Spatial variations in erosion are generally revealed by the analysis of gauged suspended load as used at a global scale by Walling and Webb (1983). However, few good-quality, long-duration records exist for stations in the Severn and Wye catchments. Major problems are associated with the measurements that do

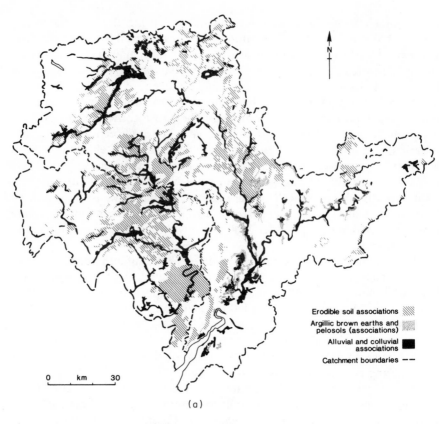

FIGURE 14.2(a)    The Distribution of erodible and cumulative soil associations in the Severn and Wye basins

exist, summarized in Table 14.2, including short duration, variable methodology, and variable sampling frequency. However, the values of sediment yield do suggest very variable outputs from small upland catchments including the highest figures from the Upper Severn, perhaps due partly to the erosion of sheep-hollows (Slaymaker, 1972), and high output from the Wye and Avon catchments. Mitchell (1979) regards the relatively high figure for Farlow Brook, a tributary of the Wye, as being due to the availability of sediment, especially easily transported fine sand from Old Red Sandstone soils. Both the Wye and the Avon are regarded as dirty rivers and this has been demonstrated, using remote sensing (Brown, Gregory and Milton unpub. data). Further discussion of the suspended sediment output of the Severn Basin can be found in Chapter 9.

From studies of contemporary erosion in the region it seems likely that it

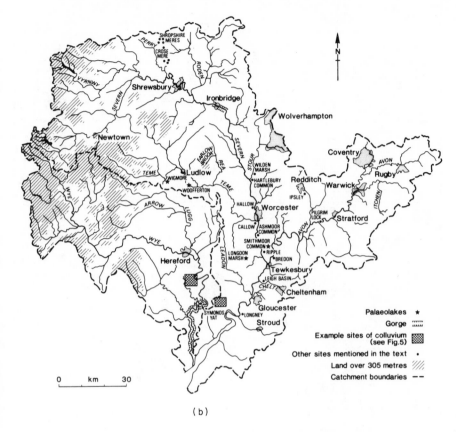

(b)

FIGURE 14.2(b)  Location map including sites mentioned in the text

is the combination of arable cultivation, soils susceptible to erosion and poor cultivation practices that leads to the highest erosion rates. Reed (1979) has shown that over 95 per cent of cases of soil erosion in the West Midlands and East Shropshire between 1967 and 1976 were associated with soil compaction, downslope cultivation and/or the length of the fallow period. During the Holocene the most variable of the factors implicated in accelerated soil erosion would have been vegetation cover and land use practices. It is generally agreed that lack of vegetation cover becomes of prime importance below about 70 per cent cover (Hudson, 1971) as could have been the case when man started to burn, clear and cultivate significant tracts of land. Prior to this it can be assumed that cover was generally over 100 per cent (i.e. multilayered vegetation cover and a litter layer) limiting slopewash erosion.

TABLE 14.1 Soil associations with a risk of water erosion and alluvial/colluvial soil associations which lie within the Severn/Wye Basins and are mapped in Figure 14.2A

| | | Erodible soil associations | | Alluvial/colluvial associations | | |
|---|---|---|---|---|---|---|
| Association | Risk* | Subgroup | Field evidence | Association | Subgroup | Substrate |
| Barton | Wa | Typical brown earths | Capping | Adventurers 3 | Earthy eutr-amorphous peat | Fens, tufa, river alluvium |
| Bridgenorth | Wa, Wi | Typical brown sands | Gullying (esp. Kidderminster and Bridgenorth area) | Compton | Pelo-alluvial gley | Reddish river alluvium |
| Bromsgrove | Wa | Typical brown earths | Gullying, especially on steep slopes | Conway | Typical alluvial gley | River alluvium |
| Eardiston 1 | We | Typical brown earths | Gullying, shallow on brows, slaking | Enborne | Typical alluvial gley | River alluvium |
| Esrick | We | Typical argillic brown earths | Gullying, rilling, slaking | Fladbury 1 | Pelo-alluvial gley | River alluvium |
| Munslow | Wa | Typical brown earths | Hedge accumulations, gullying, capping | Fladbury 2 | Pelo-alluvial gley | River alluvium |
| Newport 1 | We, Wi | Typical brown sands | Rilling and slaking | Fladbury 3 | Pelo-alluvial gley | River alluvium |
| Whimple 3 | Wa | Stagnogleyic argillic brown earths | Smearing, compaction, poaching | Frome | Calcareous alluvial gley | Chalky and gravelly river alluvium |
| Winter Hill | Ws, Wa | Raw oligo-fibrous peat soils | Peat erosion, deep gullies and haggs | Hollington | Typical alluvial gley | Reddish river alluvium |
| Worcester | Wa | Typical argillic pelosols | Poaching, structural damage | Lugwardine | Typical brown alluvial soils | Reddish river alluvium |
| | | | | Middelney | Pelo-alluvial gley | River alluvium over peat |
| | | | | Teme | Typical brown alluvial soils | River alluvium |
| | | | | Wharfe | Typical brown alluvial soils | River alluvium |
| | | | | Willingham | Gleyic rendzina like alluvial soils | Lake marl, tufa and peat |

TABLE 14.2   Measured and estimated sediment yields for the Severn and Wye catchments and subcatchments

Measured rates of sediment yield

| Basin | Area km² | Sediment yield t km⁻²yr⁻¹ | Method * | No. of years of record | Author |
|---|---|---|---|---|---|
| Upper Wye[a] | 0.08 | 11 | S | 1 | Oxley (1974) |
| Upper Severn[b] | 6.6 | 431 | S,B | 4 | Slaymaker (1972) |
| Farlow Brook (Salop.) | 20 | 112 | S | 3 | Mitchell (1979) |
| Chelt (Glos.) | 34 | 17 | S | 1 | SES |
| Frome (Glos.) | 198 | 2 | S | 1 | SES |
| Avon (Worcs.) | 258 | 160 | S | — | Fleming (1969) |
| Leadon | 293 | 3 | S | 1 | SES |
| Wye | 4040 | 74 | S | 1 | SES |
| Severn | 9983 | 11 | S | 1 | SES |

Estimated minimum mean rates of sediment yield from storage

| | Area km² | From min. depth of 1.25 m t km⁻²yr⁻¹ | From accumulation of 0.1 mm yr⁻¹ t km⁻²yr⁻¹ |
|---|---|---|---|
| Llandewi | 111 | 2.0 | 2.9 |
| Perry | 181 | 18.4 | 26.5 |
| Wye | 4040 | 8.1 | 11.7 |
| Severn | 9983 | 5.3 | 7.7 |
| Ripple | 19 | 23.7 | 20–140[c] |

* S = suspended sediment sampling, B = bedload sampling
[a] Average of 2 small catchments
[b] Average of 12 small catchments
[c] Derived using local accumulation rates from Brown and Barber (1985) for three different time periods in the Holocene (see their Figure 8)
SES Severn Estuary Survey and Systems Panel

The principal source of evidence for changes in vegetation cover during the Holocene is pollen analysis. Unfortunately the great majority of pollen evidence comes from the upland fringes of the basin and relates to comparatively little of the basin area and certainly not the most favourable areas for prehistoric agriculture. However, there are now a number of lowland diagrams from small lakes (Beales, 1980; Barber and Twigger, Chapter 11 this volume), terrace depressions and floodplains (Brown, 1983a) which give some indication of the timing and magnitude of lowland changes in vegetation. A pollen diagram cannot directly indicate vegetation cover but from the ecological assemblage and indicator species inferences about cover can be made. For example, an assemblage zone dominated by *Quercus/Corylus* with low NAP, or pasture, can be assumed to have a cover of over 100 per cent but not a cleared landscape with evidence of arable cultivation. In addition, several pollen types are associated with winter and summer

TABLE 14.3   Generalized vegetation changes in the Severn and Wye catchments taken from pollen analytical studies

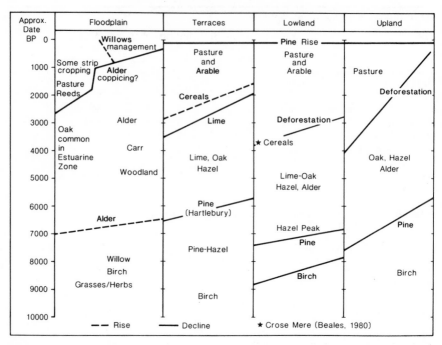

| Approx. Date BP | Floodplain | Terraces | Lowland | Upland |
|---|---|---|---|---|
| 0 | Willows management | | Pine Rise | |
| 1000 | Some strip cropping    Alder coppicing? | Pasture and Arable | Pasture and Arable | Pasture |
| 2000 | Pasture Reeds | Cereals | | Deforestation |
| 3000 | Alder | Lime | Deforestation | |
| 4000 | Oak common in Estuarine    Carr | | ★ Cereals | Oak, Hazel |
| 5000 | Zone    Woodland | Lime, Oak Hazel | Lime-Oak Hazel, Alder | Alder |
| 6000 | Alder | Pine (Hartlebury) | | |
| 7000 | | | Hazel Peak | Pine |
| 8000 | Willow | Pine-Hazel | Pine | |
| 9000 | Birch    Grasses/Herbs | | Birch | Birch |
| 10000 | | Birch | | |

–– Rise   —— Decline   ★ Crose Mere (Beales, 1980)

cereals (see Behre, 1981, for a list) and presumably some bare ground, so that used in conjunction with assemblages, changes in ground cover can be inferred.

The traditional emphasis on deforestation as a cause of accelerated erosion may be unwarranted since the response will vary depending upon methods used, subsequent land use and timing. Ring-barking, burning or felling and interplanting without stump removal may well not induce accelerated soil erosion especially if done gradually. There will also be lag-effect caused by the processes of soil degradation, opportunities for soil storage and distance from channels. It is outside the scope of this chapter to discuss in detail the Holocene vegetation history of the basin but based on over 20 sites Table 14.3 presents a generalized picture of vegetation changes relevant to soil exposure on terraces and slopes close to the floodplain, the lowlands and the uplands. Rational limits have been used and the event-line slope represents time transgression *sensu* Smith and Pilcher (1973).

Deforestation in the uplands spanned a considerable period of time from the Neolithic to as late as the medieval period and was often gradual in nature (Wiltshire and Moore, 1983). In the lowland and terrace environment deforestation seems to have occurred over a shorter period essentially between 4000 BP (Beales, 1980) and around 3000 BP and to have been more sudden and complete. Floodplain deforestation occurred later, often in two stages, around 2500–2000 BP and 1500–1000 BP. In the lowlands indicators of arable cultivation and bare ground appear post-5000 BP; although earlier Mesolithic cultures probably caused changes in vegetation they were of limited extent and of short duration. One lowland terrace site at Hartlebury Common (Brown, 1984) gives strong evidence of intensive arable cultivation and open ground after 2600 BP. There is an increase in Gramineae, *Plantago coronopus*, *Plantago media/major* and *Rosaceae* and the appearance of *Potentilla* type (probably *Potentilla erecta*) and cereal-type pollen which reaches unusually high percentages (5 per cent TLP) suggesting arable cultivation close to the site on what are regarded by the Soil Survey (1984) as soils susceptible to both water and wind erosion. From many pollen sites it appears that by 3000 BP much of the lowlands within the catchment had been deforested, probably with the exception of outliers like the Dean and Wyre forests, and a substantial amount of land was under arable cultivation. In the uplands, although there was deforestation and some arable cultivation in the valleys, most of the cleared land was under pasture or reverted to heathland with the growth of blanket peat (Wiltshire and Moore, 1983) which suggests relatively high shrub and grass vegetation cover and the accumulation of organic matter at a faster rate than soil erosion and decomposition.

## SPATIAL VARIATIONS IN THE STORAGE COMPONENT

The size of the Severn and Wye catchments combined (about 15,000 km²) precludes the use of direct field survey for the identification and quantification of the storage component. However, the Soil Survey 1:250,000 soil map of England and Wales based on 1:25,000 scale field survey with one profile description approximately every 5 km², gives considerable information on soil and sediment storage. Table 14.2 lists those associations mapped in Figure 14.2(a) as sediment sinks. They are generally cumulative alluvial and colluvial soils of Holocene age. These soils have a minimum depth of 1.25 m (Avery, 1980) and occur on floodplains, low terraces, marshes, old lake beds, valley bottoms and slope bases; they include tufa, algal marl and recent colluvial deposits with gleyed subsurface horizons. The map shows how the great bulk of these soils are in the central and lowland portion of the basins associated with the floodplains of the Middle and Lower Severn, Lower Avon (Worcestershire), Middle and Lower Teme and the Middle Wye. Areas of notable absence include gorge sections such as that at Ironbridge on the Severn and the Symonds Yat section of the Wye. However, the resolution of the field survey and final map means that these soil types are generally omitted from catchments smaller than about 50 km². Overall these soils cover

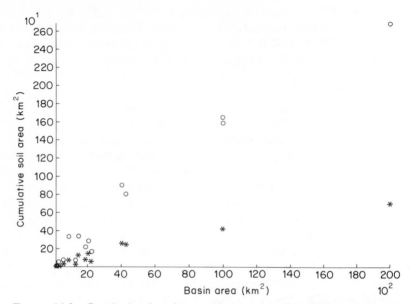

FIGURE 14.3   Graph showing the cumulative area of erodible and alluvial soil associations against basin area. Data taken from the 1:250,000 soil map of England and Wales. Stars are alluvial soils, circles are erodible soils

6.5 per cent of the Wye Basin and 4.2 per cent of the Severn. Figure 14.3 shows the cumulative area of these soils plotted against drainage basin area. The increase in storage area with increasing drainage basin area is relatively slow in comparison to the increase in area covered by erodible soils, but this is partially offset by the increase in depth of alluvial storage which is the order of 0.025 m km$^{-1}$ of main channel (Brown, 1983a). Warren and Cowie (1976) used this technique to investigate the relationship between soil types and landscape location in the Elwy valley, north-east Wales, and showed the Aled series on alluvium is not present until basins are at least of 5$^{Th}$ order, using the Shreve system.

Colluvial storage is undoubtedly under-represented in Figure 14.2(a) as it is the most difficult component to measure. It can be defined as poorly stratified or unstratified deposits accumulated from slopewash or downslope creep during the Holocene (Avery, 1980). This material fills in hollows, mantles breaks of slope at slope-bases, fills dry valley bottoms and builds up against obstructions such as walls or hedges. Indeed much of the evidence of erosion cited in Table 14.1 was colluvial sedimentation. This is because the highest rates of colluvial storage are associated with the most erodible soils and high erosion rates. Little information is available for the Severn and Wye catchments concerning colluvial storage but some idea of extent can be gained by looking at the incidence of soil associations and series known to be predominantly colluvial in origin. These are often a subtype of the Brown earth group called creep or colluvial brown earths and also colluvial rendzinas. This approach is discussed more fully later (p. 323).

Also included in Figure 14.2a are several large areas of sediment storage associated with palaeolakes. Lakes are rare in the catchments and are limited to a few in North Wales and very small lakes (meres) in the Shropshire lowlands. However, in the past larger areas of the basin were covered by freshwater, including small, possibly interconnected, proglacial lakes in the Shropshire lowlands, in the Wem district (Crompton and Osmond, 1954), the Perry–Roden catchments and the Vyrnwy junction. To the south-west were lakes Wigmore and Woofferton in the Upper/Middle Teme valley (Eyles, 1973; Cross, 1971), lake Longdon near Tewkesbury (Brown, 1983a) and possibly others in the Hereford–Lugg basin. Although many, but not all, of these lakes were dry by the Holocene they have continued to function as sites of alluvial and colluvial deposition. The remainder of the area of stored sediment in Figure 14.2(a) is fine-grained alluvial deposition associated with the major river valleys. Examples of these sediment stores will now be looked at in detail.

**Palaeolakes and Holocene sediment storage**

These sites vary in size from a few hectares (e.g. Smithmoore Common in the Ripple Basin) to tens of square kilometres. The most important area for palaeolake deposits is the Shropshire–Cheshire plain. As well as the development of deep kettle holes in a stagnating ice environment there was the extensive deposition of lake clays and *Chara* marl as seen below the Perry and Roden floodplains. A soil survey of Powys has provided evidence of a large proglacial lake at the Severn–Vyrnwy confluence (Thompson, 1979) and Humphreys (1979) discovered lacustrine deposits in the nearby Newtown reach of the Severn Valley associated with deglaciation.

A second area is in the Welsh Borderland, the Upper/Middle Teme and Wye. Proglacial lake Woofferton covered 38 km² (Cross, 1971) most of which is now a site of alluvial deposition and palaeolake Wigmore (Eyles, 1973) is now overlain by peat and alluvial deposits.

Many of these sites silted up during the early post-glacial, becoming marshes, but continued to receive large amounts of sediment, judging by the thickness of post-lacustrine sediments, until they were drained. A good example of this is the small (6 km²) palaeolake Longdon 8 km north of Tewkesbury (see Figure 14.2B) midway between the Malvern Hills and the Severn. During the 1970s the Severn-Trent Water Authority considered this site for a pumped storage reservoir due to its closed depressional shape. The topography of the site was mapped at 0.5 m intervals and subsurface

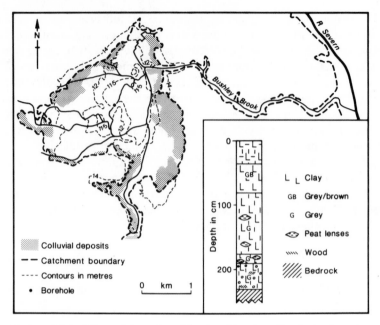

FIGURE 14.4   Map of Longdon Marsh with generalized stratigraphy
from the main core

explorations were conducted (STWA, 1976). From this data and work by
the author it is known that lacustrine and alluvial deposits fill the marsh area
and the Bushley Brook valley (Figure 14.4). They consist of soft to firm
fissured grey/brown clays and silty clays overlying soft light-grey and grey-
blue clays with some silt. At the base brown/black peats and blue/green
organic silts and clays were frequently encountered, resting on either sand
and gravel or soliflucted materials. Analysis of the grey silty clay from the
main core revealed remains of 15 species of diatoms. The only central and
most common species was *Cyclotella meneghiniana* with the next most
common species being *Synedra ulna*, *Cymbella ventricosa* and *Pinnularia
viridis*; the complete species list can be found in Brown (1983a). The spectrum indicates fairly shallow calcareous water (pH 8+) with considerable
stream inputs as most of the species except for *Cyclotella meneghiniana* had
many broken valves. The abundance of *Synedra ulna* and *Pinnularia viridis*
along with the presence of *Navicula lanceolata* suggests water of fairly high
mineral content. The diatoms confirm the presence of a shallow alkaline
lake. Grain size analysis of the sediments showed very fine, well-sorted
unimodal clayey silts with a $D^{50}$ of 8 phi (and 20 per cent clay) coarsening
upwards to a unimodal medium silt, probably of alluvial origin. The sedi-

ments are also calcareous (approx. 5 per cent) and organic matter content varies from 4 to 17 per cent by weight excluding the present soil horizons. It is concluded that the site was occupied by a lake covering about 6 km² during the Early and possibly Mid-Holocene. Using data provided by the STWA and topographic evidence of a shoreline at 14 m OD it can be estimated that the lake would have had a live volume of around $40 \times 10^5$ m³. This volume must have decreased due to sedimentation until the site became a marsh receiving only alluvial sedimentation. It was drained during the 1860s but still floods regularly today.

A very similar lake at Smithmoor Common seems to have existed in the Ripple catchment. From correlation of its pollen spectrum with a nearby (1 km) $^{14}$C dated diagram it seems that the lake clays were deposited during the Early to Mid-Holocene but continuing in part until historical times. Indeed a lake 4 hectares in size, named Shuttcock Lake, existed at the time of the Domesday Survey (Gray, 1947).

Undoubtedly many other similar small palaeolakes existed in the lowland portions of the basin, providing significant sediment sinks during the Early to Mid-Holocene, and they receive some sediment as alluvium and colluvium today.

**Colluvial storage**

Colluvial storage is the most difficult of all the sediment stores to estimate. This is, firstly, because colluvial storage is an inherent component of the distribution and pattern of soil cover, as most soils towards the base of slopes have or in the past have had significant colluvial inputs; secondly, because it occurs in relatively minor volumes in many locations; and, thirdly, because the boundaries between the processes of colluviation and alluviation are often not clear. However, both archaeologists and soil surveyors have produced evidence of the chronology and spatial extent of colluvial deposition in the UK. The Soil Survey regard colluvium as resulting from slopewash or downslope creep during the Holocene particularly as a result of accelerated erosion following clearance of natural vegetation by man (Avery, 1980). Five characteristics are used in the identification of colluvium and these are: a gradual or irregular decrease in organic carbon content with depth; inclusion of charcoal, pottery or other artifacts; weakly differentiated horizons; stratigraphic relationships with older deposits; and slight or no evidence of gleying. Colluvium was originally included in the Soil Classification for England and Wales as a subtype of the Brown earth group but on the recent 1:250,000 maps it is included as colluvial rendzinas and colluvial brown calcareous earths. However, colluvial deposits tend to occur in narrow disjointed strips

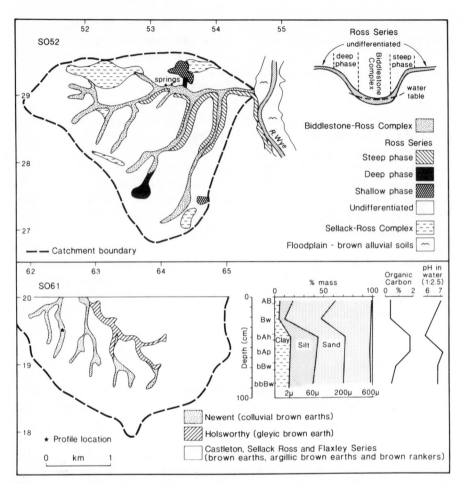

FIGURE 14.5   Examples of colluvial sedimentation. Spatial pattern and cross-section of the Biddlestone–Ross complex (top) and spatial pattern and profile characteristics of the Newent series (bottom), from Whitfield (1971) and Cope (1973). For general location see Figure 14.2(b)

and are therefore often below the resolution of maps at this scale. Maps at 1:25,000 can often demarcate colluvial soil series. On the Cinderford sheet (S061) 80 ha or 0.8 per cent of the map is covered by the Newent series which is a coarse loamy reddish colluvium and on the Ross-on-Wye sheet (S052) 470 ha or 4.7 per cent is covered by the Biddlestone series mapped as the Biddlestone–Ross complex (see Figure 14.5). This is a fine sandy colluvium derived from Devonian sandstone and is developed in narrow strips along valley sides and bottoms and it is over 1 m deep. As Figure 14.5 shows, it grades into the Ross series (deep phase) illustrating the under-representation of the colluvial component because the deep and steep phases of many soils are probably colluvial in origin. Indeed these samples and the frequent identification of a deep phase of a soil series, usually on moderate slopes in the West Midlands, suggests much of the lowland catchment has a coverage of at least 5 per cent colluvial storage.

Additional evidence of colluvial deposition is available from archaeological investigations which frequently expose colluvial sediments. Archaeologists have focused on colluvial deposits directly related to land use especially in association with lynchets, headlands, ridge and furrow, plough marks, and more recently slope bases and dry valley fills (Bell, 1982). Bell (1982) has illustrated the information that can be derived concerning land use practices and soil erosion from trenches in colluvial deposits and has shown that colluvial deposition significantly precludes records of accelerated alluviation with a clustering of dates between 4600–2500 BP. Due to the uneven distribution of archaeological sites and dry conditions, colluvial deposits on carbonate rocks have received most attention, including colluvial rendzinas in the Cotswolds. However, similar processes undoubtedly occurred on non-carbonate rocks as do dry valleys (Gregory, 1966). Colluvial processes have also received considerable attention in the Ardennes both from a process and a chronological point of view (Kwaad, 1977; and Kwaad and Mücher, 1977). Work by these authors has shown how the landscape is composed of a mosaic of soils of different depositional history and age, often caused by man's effects on erosion and colluviation. In the USA the importance of the accumulation of wedges of colluvium in bedrock hollows and their evacuation by landslides in mountainous areas has been recently documented by Dietrich and Dorne (1984) and Marrion (1985).

**Alluvial storage**

In recent years there has been increased interest in the role of alluvial storage in sediment budgets (Trimble, 1975) and particularly its role in the alteration of the sediment delivery ratio. Approximately 690 km$^2$ or 5 per cent of the Severn and Wye catchment is covered by Holocene alluvial deposits. The alluvial soils mapped in Figure 14.2A and used to calculate Figure 14.3 were generally typically brown alluvial soils and calcareous or pelo-alluvial gley soils. Common Associations include the Wharfe and Teme which like the Trent include several Soil Series restricted to specific floodplain environments (Figure 14.6). From Figure 14.3 we can see that the area covered increases at a relatively slow rate with increasing drainage basin size. This is partly due to the increase in thickness and partly due to the limitations imposed upon alluvial deposition by pre-existing erosional topography, in the case of the Severn confinement by Pre-Holocene terraces. The limitations of scale mean that catchments under 20 km$^2$ are rarely recorded as having alluvial

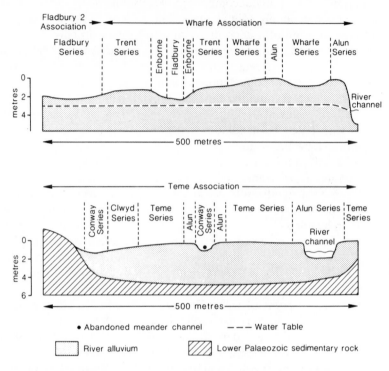

FIGURE 14.6   Examples of alluvial soil associations from the River Trent (top) and the River Teme (bottom) showing the floodplain environments of different soil series. The Teme association is mapped in Figure 14.2(a) and included in Figure 14.3. Redrawn from Soil Survey (1984)

soil types in them; however, on larger scale maps alluvial soils can be found in catchments of this size indicating that the areas in Figure 14.3 are probably underestimates. The distribution of Holocene alluvial deposits is concentrated around the middle and lower reaches of the major river valleys, especially the Wye, Teme, Severn and Avon. However, the floodplains are not regular curvilinear features but, due to variations in floodplain width and slope, there are some areas of extensive flat and backswamp development such as the Hereford Basin and the Leigh (Chelt) Basin as well as narrow floodplain reaches with little oportunity for long-term sediment storage such as from the Ironbridge gorge to Worcester on the Severn, some 60 km. Since the residence time of alluvial storage is dependent on the migration rate of the river relative to the floodplain width, broad expanses of floodplain such as that at tributary junctions may provide the oldest and most continuous records of floodplain alluviation. Radiocarbon dates from the base of several floodplain cross-sections and cores indicate that all the fine sediment is of Holocene age.

An example of a site of extensive alluvial deposition is the large backswamp area developed at the junction of the River Chelt with the Severn above Gloucester, which is called the Leigh Basin. It is frequently inundated and acts as a natural flood-storage site, reducing flooding in the Gloucester reach. Cores from the basin revealed approximately 2 m of fine sediments over an area of 8 km². The silts, clay and fine sand are of overbank origin as the geometry of the basin precludes migration across it by either the Chelt or the Severn. The sediments are also CaCO₃-rich due to inputs from the Chelt which drains a catchment predominantly underlain by oolitic limestone. They are also organic-rich (6–35 per cent) with a major organic layer at 50 cm with freshwater mollusca indicating marshy conditions. This site has not been dated but other similar sites have and the accumulation rates are shown in Figure 14.7 along with dates from the Severn at Hallow (near Worcester), the Avon at Pilgrim Lock and the Arrow at Iplsey published by Shotton (1978). The data show how accumulation rates vary from as low as 0.1 mm yr⁻¹ to as high as 5 mm yr⁻¹ depending upon the location of the core, with higher rates being more common closer to the present channel, suggesting relative channel stability during the Mid and Late-Holocene. Investigations of the main valley fill from Worcester to Gloucester show an increase in thickness of silts, silty clay and organic sediments from 3–4 m to 6 m+ and it is the topmost silty clay unit which thickens the most downstream (Beckinsale and Richardson, 1964; Brown, 1983b). This unit has been dated at Hallow as post-2600 BP by Shotton (1978) but the degree to which it is time-transgressive is not known. The mineralogy of this unit indicates it was derived from the local Keuper and Bunter lowlands (Brown, 1983a). Investigations and cross-sections have also revealed 2–7 m of sand silt and clay at Bredon in the Avon valley and in the Warwickshire Itchen (Dury, 1952). Accelerated Late-Holocene alluviation has been reported from the upper Thames by Robinson and Lambrick (1984) and two smaller catchments in Sussex (Burrin, 1985). However, from this type of work it is extremely difficult to calculate the volumes of storage involved; crude estimates based upon the lowest accumulation rate recorded and using a measured mean bulk density of 1.8 g cm⁻³ give 8 t km² yr⁻¹ which is comparable to the estimated annual yield of suspended sediment of around 10 t km⁻² yr⁻¹ (see Table 14.2 for further details) suggesting that during the Holocene as much sediment has gone into storage as has left the basin. Suspended sediment yield is limited by hydrological regime and discharge which may not have changed dramatically during the Holocene whereas using the accelerated rate of floodplain accumulation observed at several sites between 2000–3000 BP of around 1 mm yr⁻¹ estimated storage would have been 10 times greater and an order of magnitude larger than present suspended sediment yield.

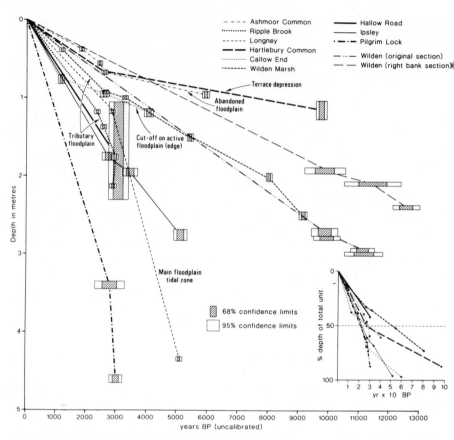

Figure 14.7   Radiocarbon dates of floodplain sediments and accumulation rates. For location of the sites see Figure 14.2(b). Data from published sources (Shotton, 1978 and Shotton and Coope, 1983) and the author

## CASE-STUDY OF A LOWLAND SUBCATCHMENT

Variation in Holocene sediment yield has also been estimated from a small lowland subcatchment of the Severn drained by Ripple Brook (for location see Figure 14.2. In this study Brown and Barber (1985) used palaeoecological evidence to infer the causes of changes in sedimentation within the catchment. Pollen analysis was used to give a record of vegetation at the extralocal scale of the catchment by using small sites, and diatoms were used to give evidence of changing hydrological conditions on the floodplain and upstream. Radiocarbon dates (Figure 14.6) from vertically accreted floodplain deposits allowed the calculation of inorganic accumulation rates. These show a dramatic increased in sediment deposition during the late Bronze Age and early Iron Age (2900–2300 BP) due to deforestation and cultivation of the catchment slopes and resultant soil erosion. The lag of full sedimentary response by some 200–300 years after deforestation suggests the cause is cultivation along with progressive soil deterioration. The soils within the catchment were suceptible to structural damage, waterlogging and slope wash erosion and they are mapped in Figure 14.2A. From the calculated increases in sediment storage, estimates of catchment erosion were made (Figure 14.8) which vary from around 20 to 140 tons $km^{-2}$ $yr^{-1}$. As can be seen from Table 14.2 this is high by comparison with modern rates but typical of accelerated rates. It also must be pointed out that these crude figures are likely to be underestimates rather than overestimates because of the omission of the majority of colluvial storage, eroded alluvial storage and the solution load.

Shotton (1978) has shown similar accelerated erosion at three sites in the lowland part of the catchment caused by arable cultivation around 2600 BP and it seems that arable cultivation rather than deforestation was the trigger, although the relative role of changes in cultivation practices and subtle climatic change is not known. The erosional and sedimentary responses to changes in land use were dependent upon geologically inherited soil characteristics with the greatest erosion and sedimentation in the lowland portions of the catchments.

## CONCLUSIONS AND IMPLICATIONS

Traditional views of drainage-basin dynamics tend to emphasize the erosional component of the upland headwaters and the transportational role of the middle zones (Schumm, 1977, p. 3). However, estimated sediment delivery ratios and the importance of the storage element in the middle and lower zones of basins indicate that, in time, these zones can become focii of remobilized sediment from stores of different residence times. The length of time involved is dependent upon hydrological and geomorphological conditions and their modification by man. Following the case-study of alluvial storage in the Mid-West USA by Trimble (1983) one might expect much of the Severn's floodplain sediment rapidly deposited around 2000–3000 years ago to be remobilized by increased fluvial action due to increases in flood magnitude and frequency but reduced slope erosion. However, the canalization, channelization and regulation of British lowland channels has prevented this from occurring, artificially lengthening the residence time of sediment in floodplain storage. Embankment has also led to a reduction in flooding and presumably to a decrease in floodplain accumulation rates but an increase in within-channel sedimentation, especially in lower reaches. The effect is part of the complex response of alluvial systems as formulated by Schumm (1977) and it will produce an alternation of erosional and depositional phases without any changes in base level, climate or slope erosion,

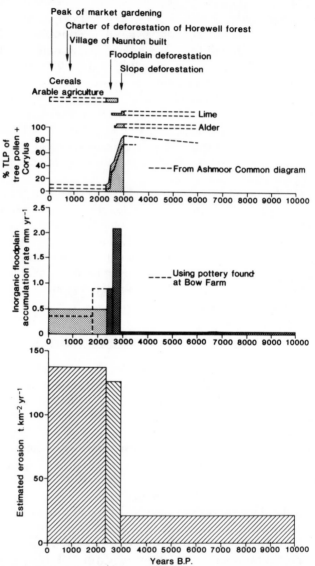

FIGURE 14.8 Floodplain accretion rates and estimated erosion rates for the Ripple Brook catchment with additional information on vegetation changes within the basin. Ashmoore Common is a similar floodplain pollen site 8 km north of Ripple (see Figure 14.2b). Reproduced from Brown and Barber (1985) by permission of Academic Press Inc.

leading to changes in the slope of the sediment delivery ratio curve. Indeed Trimble and Lund's (1982) diagrams of hysteresis between upstream sediment supply and downstream sediment yield imply this. Each phase could be regarded as being a short-term equilibrium or dynamic metastable equilibrium and major threshold or catastrophic events such as channel avulsion may be needed to change the erosional/depositional balance in one location relative to another but only when the store is sufficiently close to instability (Schumm, 1979). What this produces is what can be called punctuated equilibrium, where a time series of different, relatively short but varying climatic equilibrium states exists, separated by major threshold changes. The quantities of stored sediment in lowland basins such as that of the Severn, in relation to present sediment yields, suggests that man altered the relationships between the erosion, transport and storage processes during the Mid and Late-Holocene. In the process deep and fertile or potentially fertile, alluvial and colluvial soils have been created. Quantification of sediment budgets requires the identification of sources and stores, and the determination of transport rates and residence times. Radiometric dating can provide the residence times and rates but the extent of the sediment stores in the Severn Basin is poorly known. Calculations based on these rates suggest that as much sediment has gone into storage as has left the Severn and Wye basins during the Holocene. The storage component is not spatially uniform and in particular it is likely that the majority of hillslope erosion went into proximal colluvial storage.

## ACKNOWLEDGEMENTS

The author thanks NERC for providing radiocarbon dates and K. Moore for drawing the illustrations.

## REFERENCES

American Society of Civil Engineers (1975). *Sediment Engineering*, SCE, New York.

Avery, B. W. (1980). *Soil Classification for England and Wales (Higher Categories)*, Technical Monograph 14, Soil Survey, Harpenden.

Barber, K. E. (1982). Peat-bog stratigraphy as a proxy climate record, in Harding, A. F. (ed.), *Climatic Change in Later Prehistory*, Edinburgh University Press, pp. 103–13.

Beales, P. W. (1980). The Late Devensian and Flandrian vegetational history of Crose Mere, Shropshire, *New Phytol.*, **85**, 133–61.

Beckinsale, R. P. and Richardson, L. (1964). Recent findings on the physical development of the lower Severn valley, *Geog. J.*, **130**, 87–105.

Behre, K-E. (1981). The intepretation of anthropogenic indicators in pollen diagrams, *Pollen et Spores*, **23**, 225–45.

Bell, M. (1982). The effects of land use and climate on valley sedimentation, in Harding, A. F. (ed.), *Climatic Change in Later Prehistory*, Edinburgh University Press, pp. 127–42.

Boyce, R. C. (1975). Sediment routing with sediment-delivery ratios, *US Agric. Serv. Dept.*, ARS-S-40, 61–5.

Brown, A. G. (1983a). Late-Quaternary, palaeohydrology, palaeoecology and floodplain development of the Lower River Severn, unpublished Ph.D. thesis. University of Southampton.

Brown, A. G. (1983b). Floodplain deposits and accelerated sedimentation in the lower Severn basin, in Gregory, K. J. (ed.) *Background to Palaeohydrology*, Wiley, Chichester, pp. 375–98.

Brown, A. G. (1984). The Flandrian vegetational history of Hartlebury Common, Worcestershire, *Proc. Birm. Nat. Hist. Soc.*, **25**, 89–98.

Brown, A. G. and Barber, K. E. (1985). Late Holocene palaeoecology and sedimentary history of a small lowland catchment in central England, *Quat. Res.*, **24**, 87–102.

Brune, G. M. (1953). Trap efficiency of reservoirs, *Trans. Am. Geophys. Union*, **34**, 407–18.

Burrin, P. J. (1985). Holocene alluviation in southeast England and some implications for palaeohydrological studies. *Earth Surface Processes and Landforms*, **10**, 257–74.

Burrin, P. J. and Scaife, R. G. (1984). Aspects of Holocene valley sedimentation and floodplain development in southern England, *Proc. Geol. Ass.*, **95**, 81–96.

Cope, D. W. (1973). *Soils in Gloucestershire. Sheet S082 (Norton)*, Soil Survey Record 13, Harpenden.

Crompton, E. and Osmond, D. A. (1954). *Soils of The Wem District*, Soil Survey Memoir 1, Harpenden.

Cross, P. (1971). Glacial and preglacial deposits, landforms and river diversions in the Teme Valley near Ludlow, unpublished Ph.D. Thesis. University of London.

Dietrich, W. E. and Dorne, R. (1984). Significance of thick deposits of colluvium on hillslopes: A case-study involving the use of pollen analysis in the coastal mountains of northern California, *J. of Geol.*, **92**, 133–46.

Dietrich, W. E. and Dunne, T. (1978). Sediment budget for a small catchment in mountainous terrain, *Zeitschrift für Geomorphologie N.F. Suppl. Bd.*, **29**, 191–206.

Dury, G. H. (1952). The alluvial fill of the Warwickshire Itchen near Bishops Itchington, *Proc. Coventry Nat. Hist. and Science. Soc.*, **2**, 180–5.

Evans, R. (1980). Characteristics of water-eroded fields in lowland England, in *Assessment of Erosion*, M. de Boodt and D. Gabriels (eds), Wiley, Chichester, pp. 77–87.

Eyles, N. (1973). The Vale of Wigmore, Shropshire—its Quaternary geomorphology: an analysis of pro-glacial Lake Wigmore, unpublished B.Sc. diss., University of Leicester.

Fleming, G. (1969). The Clyde Basin hydrology and sediment transport. unpublished Ph.D. thesis, Strathclyde University, Glasgow.

Gardiner, V., Gregory, K. J. and Walling, D. E. (1977). Further notes on the drinage density–basin area relationship, *Area*, **9**, 117–21.

Gray, E. F. (1947). *Ripple Waterway*, unpublished manuscript, Ripple Church Records.

Gregory, K. J. (1966). Dry valleys and the composition of the drainage net, *J. Hydrol.*, **4**, 327–40.

Humphreys, A. M. (1979). The Quaternary deposits of the Upper Severn Basin and adjoining areas, unpublished Ph.D. thesis, University of Wales.

Kirkby, M. J. (1980). The problem, in M. J. Kirkby and R. P. C. Morgan (eds), *Soil Erosion*, Wiley, Chichester, pp. 1–16.

Kwaad, F. J. P. M. (1977). Measurements of rainsplash erosion and the formation of colluvium beneath forest in the Luxembourg Ardennes, *Earth Surface Processes and Landforms*, **2**, 161–73.

Kwaad, F. J. P. M. and Mücher, H. J. (1977). The evolution of soils and slope deposits in the Luxembourg Ardennes near Wiltz, *Geoderma*, **17**, 1–37.

Lamb, H. H., Lewis, R. P. W. and Woodroffe, A. (1966). Atmospheric circulation and the main climatic variables between 8000 and 0 BC; meteorological evidence, in *Proc. Int. Symp. on World Climate 8000 BC to 0 BC*, Royal Met. Soc., London, pp. 174–217.

Lockwood, J. G. (1979). Water balance of Britain, 50,000 BP to the present day, *Quat. Res*, **12**, 297–310.

Marrion, D. C. (1985). Colluvium in bedrock hollows on steep slopes, Redwood Creek drainage basin, northwestern California, in Jungerius, P. D. (ed.), *Soils and Geomorphology, Catena*, Suppl. 6, 59–68.

Mitchell, D. J. (1979). Aspects of the hydrology and geomorphology of the Farlow Basin, Shropshire, unpublished M.Sc. thesis, University of Birmingham, 2 vols.

Morgan, R. P. C. (1980). Soil erosion and conservation in Britain, *Prog. in Phys. Geog.*, **4**, 24–47.

Morgan, R. P. C. (1985). Assessment of soil erosion risk in England and Wales. *Soil Use and Management*, **1**, 127–31.

Oxley, N. C. (1974). Suspended sediment delivery rates and the solute concentration of stream discharge in two Welsh catchments, in *Fluvial Processes in Instrumental Watersheds*, Inst. Brit. Geogs. Sp. Publ. 6, 141–54.

Reed, A. H. (1979). Accelerated erosion of arable soils in the United Kingdom by rainfall and run-off, *Outlook in Agriculture*, **10**, 41–8.

Robinson, M. A. and Lambrick, G. H. (1984). Holocene alluviation and hydrology in the upper Thames basin, *Nature*, **308**, 809–14.

Roehl, J. W. (1962). Sediment source areas, delivery ratios, and influencing morphology factors, *Int. Ass. Scientific Hydrol. Committee of Land Erosion*, 202–13.

Schumm, S. A. (1977). *The Fluvial System*, Wiley, New York.

Schumm, S. A. (1979). Geomorphic thresholds: the concept and its applications, *Trans. Inst. Brit. Geogs.*, New Series 4, 485–515.

Shotton, E. W. (1978). Archaeological inferences from the study of alluvium in the lower Severn–Avon valleys, in *Main Effect on the Landscape: The Lowland Zone*, S. Limbrey and I. G. Evans (eds), C. B. A. Res. Dept. 21, pp. 27–32.

Shotton, N. and Coope, G. R. (1983). Exposures in the Power House terrace of the river Stour, Wilden, Worcestershire, England, *Procs. Geol. Assoc.*, **94**, 33–44.

Slaymaker, H. O. (1972). Patterns of present sub-aerial erosion and landforms in Mid-Wales, *Trans. Inst. Brit. Geogs*, **55**, 47–68.

Soil Survey (1984). *Soils and Their Use in Midland and Western England*, Bulletin 12, Harpenden.

Severn-Trent Water Authority (1976). *Report on Longdon Marsh Reservoir*, 2 vols. Binnie and Partners, unpublished Report.

Thompson, T. R. E. (1979). *Soils in Powys*, Soil Survey Memoir, Harpenden.

Trimble, S. W. (1975). Denudation rates: can we assume a steady state? *Science*, **188**, 1207–8.

Trimble, S. W. (1983). A sediment budget for Coon Creek basin in the driftless area, Wisconsin, 1853–1977, *Am. J. Science*, **283**, 454–74.

Trimble, S. W. and Lund (1982). Soil conservation and the reduction of erosion and sedimentation in the Coon Creek basin, Wisconsin, *US Geol. Surv. Prof. Pap.*, 1234.

United States Soil Conservation Service (1971). Sediment sources, yields and delivery ratios, in *National Engineering Handbook*, Section 3, Chapter 6, US Dept. of Agriculture, Washington DC.

Walling, D. E. and Webb, B. W. (1983). Patterns of sediment yield, in Gregory, K. J. (ed.), *Background to Palaeohydrology*, Wiley, Chichester, pp. 69–100.

Warren, A. and Cowie, J. (1976). The use of soil maps in education, research and planning, in *Soil Survey Interpretation and Use*, Welsh Soils Discussion Group Report 17, D. A. Davidson (ed.), 1–4.

Whitfield, W. A. D. (1971). *Soils in Herefordshire II. Sheet SO52 (Ros-on-Wye West)*, Soil Survey, Record 3, Harpenden.

Wiltshire, P. E. J. and Moore, P. D. (1983). Palaeovegetation and palaeohydrology in upland Britain, in Gregory, K. J. (ed.), *Background to Palaeohydrology*, Wiley, Chichester, pp. 433–52.

Palaeohydrology in Practice
Edited by K. J. Gregory, J. Lewin and J. B. Thornes
© 1987 John Wiley & Sons Ltd.

# 15

# The Evolution of European Rivers—A Complex Response

## L. STARKEL

*Institute of Geography, Department of Geomorphology and Hydrology,
Polish Academy of Sciences, Krakow*

## INTRODUCTION

The European subcontinent is located in the temperate forest zone, excluding the northernmost fragments of the treeless tundra and the south-eastern corner belonging to the steppe zone. But due to latitudinal as well as longitudinal extension, Europe shows distinct differences in temperature and humidity. The temperate forest zone is succeeded by the boreal forest zone to the north, by deciduous forest on the western middle part, and by mediterranean forest in the south. This chapter provides a context for studies of the Severn Basin in preceding chapters.

Palaeogeographic changes occurring since the time of the maximum extent of the last ice sheet are evidenced in sediments, morphology and various biotic remains. Between 20,000 and 18,000 BP much of Europe was covered with large ice sheets and by a wide periglacial zone giving way to the south into the steppe zone. The refuges of forest vegetation were restricted to Eastern Europe (on the permafrost) and to the lower mountain belt in the south (Beug, 1982; Grichuk, 1982; Starkel, 1979). The most dramatic changes occurred between 13,000 and 8000 BP, when the morphogenetic system of the cold stage was replaced by a warm one with the dominance of westerly winds and an oceanic climate. All these variations as well as the second-order climatic fluctuations have been reflected in the history of European fluvial systems (Starkel, 1983).

## CLIMATIC MODELS OF EUROPEAN FLUVIAL SYSTEMS

Changes in fluvial systems in Europe during the last 15 ka were not uniform over space or through time (Schumm, 1965; Starkel, 1979). Such changes were related not only to the regime of river discharge ($Qw$) and sediment load ($Qs$), but also were expressed in the simultaneous changes of channel gradients and lengths of rivers, caused by such supplementary factors as deglaciation, eustatic transgression, and glacioisostatic uplift. In a north to south transect across Europe four main types of river system evolution can be distinguished:

1. River systems recently developed on areas formerly covered by ice sheets and affected by various tectonic movements.
2. River systems of the former periglacial zone, partly influenced by ice sheets.
3. River systems of the former periglacial zone with the lower sections influenced by eustatic sea level changes.
4. River systems of the former cold steppe and forest-steppe zone, usually deeply influenced by differentiated tectonic movements as well as by early agricultural activities.

Each of the four types is outlined in the following sections.

### River systems in areas previously covered by ice sheets

In the areas previously covered by the Scandinavian ice sheet most river valleys were created after the period of deglaciation. Many such river basins are still now in the young stage of evolution according to the terminology proposed by Falkowski (1975). This terminology assumes that the channel form, gradient and floodplain are far from an equilibrium profile, and the river valley is composed of reaches without gradients where swamps, mires or transfluent lakes are separated by reaches with steep gradients, and sometimes by incised epigenetic ones with rapids (e.g. Koutaniemi, 1979; Sundborg 1956). In the extensive plains of Fennoscandia there are still many basins which have not been integrated into the present drainage systems. An additional factor is glacial rebound, which increases upstream and causes the steepening of the river gradient and so induces downcutting (Koutaniemi, 1983). A reverse trend is visible at the coast of the Bothnian Gulf, where river lengths have extended (Mansikkaniemi, 1985).

### River systems of the former periglacial zone partly influenced by the ice sheet

During the cold stage these rivers were fed by the catchments located in the periglacial zone as well as by meltwaters from the ice sheet. Therefore after the ice margin had retreated and the vegetation spread out there was a rapid change which caused the formation of underfit streams or underdeveloped channels contrasting with wide alluvial plains created by glacifluvial waters (cf. Dury, 1965; Kozarski, 1983). The best example of this type is the basin of the Upper Volga and its discharge decrease is reflected in the regression of the Caspian Sea (Kvasov, 1976). In the case of the Oder, Vistula, Niemen and western Dvina rivers, their lower reaches were blocked by ice. Subsequently when they flowed to the low Baltic depression this change of course caused increased downcutting supported also by glacial rebound (Mescheriakov and Fedorova, 1961; Voznyachuk and Valczyk, 1978; Eberhards and Miidel 1984). Therefore only in the upper reaches of these basins is the reflection of second-order hydrological fluctuations particularly remarkable (Starkel, 1982, 1983).

The other subtype is represented by the upper valley reaches of the Danube, Rhein and other Alpine rivers, the headwaters of which were covered by an extensive mountain glaciation. Rapid melting of valley glaciers during the Lateglacial caused the accretion of the river length upstream and at mountain forelands the formation of incised fills, which reflect Holocene variations in the hydrological regime and in the advances of glaciers (cf. Patzelt, 1977; Becker, 1982; Schreiber, 1985).

### River systems of the former periglacial zone

The clearest sequence of changes reflecting the decrease of $Qw$ and $Qs$, associated with the retreat of permafrost and with the invasion of forest communities as well as with second-order climatic fluctuations which were later superimposed, are visible in the river basins of the former periglacial zone. The model of change from a braided to a meandering river system (Schumm, 1965) was recognized in Europe by Dury (1965) and by Falkowski (1975). Falkowski also added the transitional phase of large paleomeanders (Falkowski, 1975). The exact dating of the changes from braided to meandering river systems at 13–12 ka BP, and from large to small paleomeanders at the beginning of the Holocene was possible mainly in the Vistula and Oder river basins (Kozarski and Rotnicki, 1977; Starkel *et al.*, 1981; Szumanski, 1983). Second-order fluctuations, reflected in the sequence of cut and fill, were early recognized in the river valleys of the Weser (Luttig, 1960) the Wisloka and the San (Starkel, 1960) and the Kama (Gorecky, 1964). However, the relation of those phases with climatic rhythmic fluctuations and with oscillations of mountain vertical belts was established and dated for the river valleys of Southern Poland (cf. Starkel, 1983), of Southern Germany (Becker and Schirmer, 1977; Becker, 1982) as well as in the Rhein–Maas system (Paulissen, 1973; Brunnacker, 1978). Short phases with a high flood frequency are mainly reflected in the straightening and widening of river channels and/or in avulsions. These are especially well visible in valleys which have their headwaters in highlands with much higher annual precipitation. Alternatively, in the smaller lowland valleys periglacial aggrading braided rivers were replaced by slight downcutting of channels and later by organic aggradation (Munaut and Paulissen, 1973; Vanderberghe and Bohncke 1985).

River systems of the former periglacial zone affected by differentiated tectonic movements can be exemplified by aggradation in the subsidence basins along the Danube valley (Vaskovsky, Vaskovsky and Schmidt, 1979; Somogyi, 1975; Borsy and Felegyhazi, 1983) and by the downcutting in the Iron Gate (Brunnacker, 1971). Many rivers, including the Rhein, Elbe, Schelde and Seine, were much longer in the past and their lower courses were submerged during the Flandrian transgression. This reduction in length influenced the later stages of development when deltas were formed and second-order fluctuations were reflected in the interfingering of marine and fluvial deposits (Hageman, 1969; Mojski, 1984). Some deltas are growing very fast and give rise to prolongation of river channels as in the case of the river Danube.

**River systems of Southern Europe**

The river systems of the former cold steppe and forest-steppe zone are restricted to the mountains of the Alpine orogeny in Southern Europe with an intensive uplift, and to the subsidence basins surrounding them. The chronostratigraphy of alluvial fills is so far recorded in less detail in this part of Europe. Deposits of the last cold stage are dissected. The younger fill represents the last 2–3 millenia and reflects extensive deforestation, cultivation and overgrazing (Vita-Finzi, 1975; Davidson, 1980; Cremaschi, 1981). In many cases the alluvial fans of the mountain forelands (Budel, 1977; Cremaschi, 1981) have their continuation in the deltas expanding towards the Mediterrenean Sea (Veggiani, 1974).

## HUMAN IMPACT ON THE TRANSFORMATION OF EUROPEAN FLUVIAL SYSTEMS

In the Holocene history of European river valleys the significance of human activity decreases towards the north and east. The first reason for this decrease is the spatial pattern of relief and rainfall distribution. The heaviest precipitations causing intensive runoff and floods are observed in the southern half of Europe where there are many mountain ranges. Also the agricultural revolution started in the south. Therefore the Neolithic and later phases of deforestation and extensive agriculture caused intensive soil erosion in this zone and the river channels and alluvial plains have been totally transformed. The western and central European uplands were later affected by intensive cultivation, which usually is connected with the Roman period or with medieval times (Schimer, 1973; Havlicek, 1983). In both parts of Europe these accelerated processes were superimposed on the rhythmic fluctuations of fluvial activity (Starkel, 1983). During the Little Ice Age this caused the transformation of many river reaches from meandering to braided channel patterns (Falkowski, 1975).

The East-European plains (cf. Khotinsky, 1984) as well as areas of northern Europe were affected by human activity much later and to a lesser extent. Low river gradients and a lower precipitation caused only a slight rise in the deposition of suspended material on floodplains.

The regulation of river channels and the construction of dams and canals led to a partial or total change of the natural hydrologic regime. A natural consequence of the reduction of water level amplitudes and a decrease of the sediment load is the downcutting going on in natural as well as in artificial river channels (Gregory, 1977).

## REFERENCES

Becker, B. (1982). Dendrochronologie und Palaoekologie subfossiler Baumstamme aus Flussablagerungen, ein Beitrag zur nacheiszeitlichen Auenentwiklung im südlichen Mitteleuropa, *Mitteil. der Kommission fur Quartarforschung*, Osterreich, Akad. der Wiss., **5**, Wien, 120 pp.
Becker, B. and Schirmer, W. (1977). Palaeoecologic study of the Holocene valley development of the River Main, Southern Germany, *Boreas*, **6**, 4, 303–21.
Beug, H. J. (1982). Vegetation history and climatic changes in central and southern Europe, in Harding, A. F. (ed.), *Climatic Changes in Later Prehistory*, Edinburgh University Press, pp. 85–102.
Borsy, Z. and Felegyhazi, E. (1983). Evolution of the network of water courses in the North-Eastern part of the Great Hungarian Plain from the end of the Pleistocene to our days, *Quaternary Studies in Poland*, **4**, Poznan, 115–24.
Brunnacker, K. (1971). Geologisch-pedologische Untersuchungen in Lepenski Vir am Eisernen Tor, *Fundamenta*, Reihe A, **3**, 20–32.
Brunnacker, K. (1978). Der Niederrhein im Holozan, *Fortschritte der Geologie Rheinland und Westfalen*, **28**, 399–440.
Budel, J. (1977). *Klima-Geomorphologie*, Gebruder Borntrager, Berlin–Stuttgart, pp. 1–304.
Cremaschi, M. (1981). Il quadro geostratigrafico dei depositi archeologici del fiume Parana. Il Neolitico e l'Eta del Rame, *Ricerca a Spilamberto e S. Cesario 1977–1980, Cassa die Risparmio di Vignola*, 29–41.
Davidson, D. A. (1980). Erosion in Greece during the first and second millennia BC, in R. A. Cullingford, D. A. Davidson and J. Lewin (eds), *Timescales in Geomorphology*, Wiley, Chichester, pp. 143–58.
Dury, G. H. (1965). Theoretical implications of underfit streams, *US Geol. Surv. Prof. Paper* 152-C.
Eberhards, G. and Miidel, A. (1984). Main features of the development of river valleys in the East Baltic, *Proceedings of the Acad. of Sci. of the Estonian SSR, Geology*, **33**, 3/4, 136–45.

Falkowski, E. (1975). Variability of channel processes of lowland rivers in Poland and changes of the valley floors during the Holocene, *Biuletyn Geologiczny UW*, **19**, Warszawa, 45–78.

Gorecky, G. I. (1964). *Alluvial Deposits of the Great Anthropogenic Rivers of Russian Plain* (in Russian), Moscow, 249 pp.

Gregory, K. J. (ed.) (1977). *River Channel Changes*, Wiley, Chichester.

Grichuk, V. P. (1982). Vegetation of Europe during Late Pleistocene (in Russian), in *Paleogeography of Europe during the Last One Hundred Thousand Years* (Atlas-monograph), Moscow, pp. 92–109.

Hageman, B. P. (1969). Development of the western part of the Netherlands during the Holocene, *Geologie en Mijnbouw*, **48**, 42, 373–88.

Havlicek, P. (1983). Late Pleistocene and Holocene fluvial deposits of the Morava River (Czechoslovakia), *Geolog. Jahrbuch*, **A71**, 209–17.

Khotinsky, N. A. (1984). Holocene vegetation history, in Velichko, A. A. (ed.), *Late Quaternary Environments of the Soviet Union*, University of Minnesota Press, pp. 179–200.

Klimek, K. and Starkel, L. (1974). History and actual tendency of flood plain development at the border of the Polish Carpathians, *Nachrichten Akad. Gottingen*, in Report of Commission on present-day Processes IGU, pp. 185–96.

Koutaniemi, L. (1979). Outline of the development of relief in the Oulanka river valley, North-eastern Finland, *Acta Universitatis Ouluensis*, A, **82**, 3, 29–38.

Koutaniemi, L. (1983). Complexity of solar variability, hydrology and climatic conditions as evidenced in the case of the Oulujoki and Kemijoki River Basins, Northern Finland, *Fennia* **161**, 2, 289–301.

Kozarski, S. (1983). River channel changes in the middle reach of the Warta Valley, Great Poland Lowland, *Quaternary Studies in Poland*, **4**, 159–69.

Kozarski, S. and Rotnicki, K. (1977). Valley floors and changes of river channel pattern in the North Polish Plain during the Late-Wurm and Holocene, *Quaestiones Geographicae*, **4**, Poznan, 51–93.

Kvasov, D. D. (1976). Paleohydrology of Eastern Europe during Valdai Period (in Russian), in *Problemy paleogidrologii*, Moskva, pp. 260–6.

Luttig, G. (1960). Zur Gliederung des Auelehmes in Flussgebiet der Weser, *Eiszeitalter u. Gegenwart*, **11**, Ohringen, 39–50.

Mansikkaniemi, H. (in press). Development of river valleys originated by emergence, S-Finland. Case-study of the river Kyronjoki, *Proceedings of the IGCP Project 158—A meeting in Belgium*, Sept. 1986.

Mescheriakov, J. A. and Fedorova, R. W. (1961). Age and genesis of the terraces of Western Dvina river (in Russian), *Materialy Wsiesoj. Soviesch. po izutch. Chetvertich. perioda 2*, Moskva, pp. 32–46.

Mojski, J. E. (1984). Lithostratigraphical units of the Holocene in the NW part of Vistula Delta Plain, *Geolog. Jahrbuch*, A, **71**, 171–86.

Munaut, A. V. and Paulissen, E. (1973). Evolution et paleoécologie de la vallée de la petite Nethe au cours du post-wurm (Belgique). *Annales de la Société Géologique de Belgique*, **96**, 2, 301–48.

Patzelt, G. (1977). Der zeitliche Ablauf und das Ausmass postglazialer Klimaschwankungen in den Alpen, in *Dendrochronologie und postglaziale Klimaschwankungen in Europa*, Erdwiss, Forschung 13, Wiesbaden, pp. 249–59.

Paulissen, E. (1973). De morfologie en de Kwartairstretigrafie van de Maasvallei in Belgisch Limburg, *Verhandel. Kon. Vlaamse Acad. Kl. Wetenschappen*, 35, 127, Brussel, 266 pp.

Schirmer, W. (1973). The Holocene of the Former Periglacial areas, *Eiszeitalter u. Gegenwart*, **23/24**, 306–20.

Schirmer, W. (1983). Criteria for the differentiation of late Quaternary river terrace, *Quaternary Studies in Poland*, **4**, 199–205.

Schreiber 1985, Das Lechtal zwischen Schongau und Rain im Hoch-, Spat- und Postglazial, *Geolog. Inst. Univ. Koln, Sonderveroeffent-lichungen*, **58**, 1–192.

Schumm, S. A. (1965). Quaternary palaeohydrology, In Wright, H. E. and Frey, D. G. (eds), *The Quaternary of the United States*, Princeton University Press, pp. 783–94.

Somogyi, S. (1975). Contribution to the Holocene history of Hungarian river valleys, *Biul. Geologiczny*, UW, Warszawa, **19**, 185–93.

Starkel, L. (1960). Rozwoj rzezby Karpat fliszowych w holocenie (sum. Development of the relief of the Polish Carpathians in the Holocene). *Prace Geogr. IGPAN*, **22**, 1–239.

Starkel, L. (1979). Typology of river valleys in the temperate zone during the last 15,000 years, *Acta Univ. Ouluensis*, Series A, **82**, Geologica, **3**, 9–18.

Starkel, L. (1983). The reflection of hydrologic changes in the fluvial environment of the tempeate zone during the last 15000 years, in Gregory, K. J. (ed.), *Background to Palaeohydrology: A Perspective*, Wiley, Chichester.

Starkel, L. (ed.), Alexandrowicz, S. W., Klimek, K., Kowalkowski, A., Mamakowa, K., Niedziakkowska, E. and Pazdur, M. (1981). The evolution of the Wisloka valley near Debica during the Lateglacial and Holocene, *Folia Quaternaria*, **53**, 1–91.

Sundborg, A. (1956). The River Klaralven. A study of fluvial processes, *Geogr. Ann.*, **38**, 2–3.

Szumanski, A. (1983). Paleochannels of large meanders in the river valleys of the Polish Lowland, *Quaternary Studies in Poland*, **4**, 207–16.

Vanderberghe, J. and Bohncke, S. (1985). The Weichselian Late glacial in a small lowland valley (Mark river, Belgium and the Netherlands), *Bulletin de l'Assoc. Franc. pour l'étude du Quater.*, **2–3**, 167–75.

Vaskovsky, E., Vaskovsky, I. and Schmidt, Z. (1979). Formation, structure and composition of Holocene sediments of the Zitny ostrov island, Danube lowland, Czechoslovakia, *Acta Univ. Ouluensis*, Series A, **82**, Geologica, **3**, 155–63.

Veggiani, A. (1974). Le variazioni idrografiche del basso corso del fiume Po negli ultimi 3000 anni, *Estratto da Padusa-Rivista del Centro Polesano di Studi Stovici Archeologici et Etnografici–Rovigo*, 1–2, pp. 1–22.

Vita-Finzi, G. (1975). Chronology and implication of Holocene alluvial history of the Mediterranean Basin, *Biul. Geol. U.W.*, **19**, 137–47.

Voznyachuk, L. N. and Valczyk, M. A. (1978). Morphology, geology and development of the Niemen valley in Neopleistocene and Holocene (in Russian), *Inst. of Geochemistry and Geophysics*, Minsk, 210 pp.

Wasylikowa, K., Starkel, L., Niedzialkowska, E., Skiba, S. and Stworzewicz, E. (1985). Environmental changes in the Vistula valley at Pleszow caused by Neolithic man, *Przeglad Archeologiczny*, **33**, 19–55.

Palaeohydrology in Practice
Edited by K. J. Gregory, J. Lewin and J. B. Thornes
© 1987 John Wiley & Sons Ltd.

# 16

# *Conclusions: Palaeohydrological Synthesis and Application*

K. J. GREGORY

*Department of Geography, University of Southampton*

J. LEWIN

*Department of Geography, University of Aberystwyth*

## GENERAL PROBLEMS

Palaeohydrological synthesis is confronted by at least four major problems. *First* of all, there are the data available. As indicated by Cullingford, Davidson and Lewin (1980), when considering environmental evolution we have to depend upon the information available through the windows in time which dating techniques and the field evidence available provide. For example, our knowledge of Lateglacial and early Holocene alluvial sediments is so far rather thin, partly at least because later erosion has removed the evidence and possibly because suitable sedimentary evidence has yet to be discovered. A further difficulty arises in the identification of such sediments particularly in the sandy gravel terrace sediments of the uplands where there may be little with which to date or reconstruct palaeoenvironments. Thus the information available and the analysis undertaken using that information are major constraints upon any synthesis of palaeohydrology. Such constraints are compounded by the technology and conceptual ideas available at a particular time.

A *second* problem arises because for the purposes of analysis it has been conventional to subdivide the scientific problem of the hydrology of the past amongst several component disciplines. This tendency to fragment studies of environmental change has been recognized as a reductionist approach and one which has predominated during the last two decades. There are often communication problems between disciplines, and misunderstandings are likely to arise where researchers are working essentially on the same topic but in isolation from one another. The existence of such compartmentalization is the reason why it has been necessary to employ a multidisciplinary approach, and palaeohydrological investigations should be most effective when contributions are made by scientists from a broad spectrum of disciplines. We have been particularly conscious that each discipline is most aware of particular aspects of palaeohydrology. For instance, sedimentologists dealing with former bedload sediments are especially concerned with flood flows whose varying magnitudes through time may have proved competent to transport sediments of contrasting sizes. Geomorphologists have been especially concerned with the magnitude of bankfull discharges in palaeochannels. In complete contrast, palaeobotanists may have been concerned with hydrological factors leading to groundwater fluctuations in mires and with humidity changes affecting humification in peats. Each of these approaches gives a different perspective on hydrologies of the past, and there is a continuing need for a more holistic approach.

*Thirdly* there is the problem that it is also artificial to separate the past from the present. Although the last two decades have seen a refreshing concern for the investigation and analysis of contemporary processes, one of the consequences of this interest has been the realization that there is no hard and fast boundary line that can be drawn between what is contemporary and what is past. Although it is therefore possible to distinguish several timescales (Cullingford, Davidson and Lewin, 1980) that are appropriate for different investigations, one of the outstanding challenges for the investigation of physical environment is to undertake analysis in such a way that information from one timescale can be translated to another. One feature of recent developments in the study of environmental change has been to employ a greater knowledge of contemporary processes to assist interpretation and understanding of the past (Gregory, 1985). This is encouraged because, for example, rare hydrologic events such as hurricanes on the east coast of the United States or the Lynmouth flood in 1952 in Britain are not likely to be regarded as contemporary events unless the contemporary timescale is extended to several centuries. It is because of the problem of the lack of a clear distinction between present and past that a retrospective approach has become more popular, as indicated in the Introduction (p. 9).

The importance of linking present and past is one of the reasons why this volume has adopted a three-fold approach, with a linking 'historical' section between present and prehistoric palaeohydrology. Earlier definitions of palaeohydrology suggested that it was concerned with hydrology before records became available (e.g. Schumm, 1968), but we now feel that historical data provide a most useful opportunity to link manifest hydrological variations with the environmental fluctuations which have produced them. Although the last century or so has not seen hydrological fluctuations on the scale of our 15,000-year time period as a whole, it is beginning to be possible to see just what environmental changes are needed or are likely in future to change or affect hydrological conditions. It has to be recognized that all three of the problems discussed so far relate to the fact that time and space present a continuum and that it is we who subdivide them for purposes of study. We can never be in possession of all necessary data, but we have also to be aware that our selection and focus, in particular on either 'the present' or 'the past', may impose limitations on the value of our interpretations or expectations.

In addition to these three problems there is a *fourth* which arises because any investigation takes place over a period of years during which other research developments must be acknowledged. IGCP project 158 has covered the ten years from 1977 to 1986 and in that decade there have been significant advances in science which have a bearing upon any attempt to achieve palaeohydrological synthesis. Such advances may be thought of as of three major kinds. There has been progress within the broad area of palaeohydrology and some has been generated by the project itself as referred to in the Introduction and the preceding chapters. However, other advances in palaeohydrology have been achieved independently and particular contributions made by V. R. Baker (1973, 1979, 1983) have been extremely important in indicating how large events such as the Missoula flood, the palaeohydrology of large tropical rivers, and speculations about the palaeohydrology of other planets can enhance our understanding. In addition there has been specific progress exemplified by the way in which G. P. Williams (1983) collated all the equations available for estimating former discharges. This provided an extremely valuable summary of the numerical methods available and underlined the need to cross-check results by using several methods. There have been, secondly, developments in the study of environment which although not exclusively palaeohydrological have a significance for palaeohydrological synthesis. In advancing our understanding of how environment works, important progress has been made from studies of river channel behaviour in relation to sediments (e.g. Collinson and Lewin, 1983), and by focusing on the interaction between force and resistance in fluvial systems (e.g. Graf, 1979) and upon the opportunities for analysing changes of power distribution in fluvial systems (Graf, 1983). Important studies have also focused upon patterns of sediment production at a world scale (Walling and Webb, 1983) or at the regional scale. The latter is exemplified by studies undertaken by S. W. Trimble in the USA which have illuminated the significance of sediment storage in drainage systems in recent time (Trimble, 1983).

On the whole, it must be said that much of the seminal work relevant to palaeohydrology and published in the last decade has pointed to the complexities of palaeohydrological environments and to the need to attempt reconstructions with considerable caution. For example, early reconstructions based on flume-derived relationships between sediment of uniform sizes and the flow parameters of transporting discharges must now be regarded with some scepticism. Contemporary hydraulic studies are revealing the consider-able reach-scale variations in sediment transport and the important effects of natural sediments of mixed sizes (e.g. Bridge and Jarvis, 1976; Dietrich, Smith and Dunne, 1979; Andrews, 1983; Reid, Frostick and Layman, 1985). Very precise palaeohydrological reconstructions are probably illusory, though what can be reliably achieved has been shown by Church in particular (Church, 1978; Ryder and Church, 1986).

The way in which work and power can be utilized throughout studies of the physical environment has been explored in a number of publications during the last 10 years. In summarizing these and in outlining the way in which such approaches apply to the physical environment (Gregory, 1987), it was suggested that it is possible to adopt a more energetic approach to the physical environment. In terms of contemporary environment it is therefore possible to seek to express the power of nature in consistent units such as the watt and the joule and also to establish budgets of energy expenditure and to assess the relative importance of different controls upon the contemporary budget. Although such approaches have been developed particularly in relation to contemporary environments, it is desirable that they should be extended to assist in the interpretation of past environments as well. It should therefore be possible to envisage palaeohydrology on an energetic basis and to see how it is that power has altered during the last 15,000 years in the context of this IGCP Project.

A focus upon energetics in relation to environmental change is in keeping with more energetic approaches which have become more extensive throughout other scientific disciplines. In the biological and physical sciences there has been a longer established focus upon energy, energy transfer and transfer functions and indeed there has now been a return to the more holistic approach in science and acknowledgement that it is necessary to look for interrelationships between different scientific approaches. This is not in any way minimizing the difficulties of adopting such an holistic approach.

## THE DATA BASE

In this volume we have predominantly been concerned with the 15,000 km$^2$ catchments of the Severn and Wye. Though this is not a large area by world standards, it has to be remembered that the largest UK river by area (the Thames) is just under 10,000 km$^2$, and that the most copious river (the Tay) achieves a mean annual discharge of only just over that of the Wye and Severn combined.

The present-day nature of this drainage area and its climatic and hydrolog-ical characteristics have been analysed in Chapters 3–5. Viewed in world terms, the area is one of unusual geological complexity which possesses an intricate evolutionary history bound up with the fluctuations of the Quat-ernary ice sheets. As discussed in Chapter 3, this area may be seen as part of a circumglobal belt of river systems that were catastrophically flooded and transformed by ice invasion. The Severn suffered disruptions comparable, in nature if not in extent, to those of the Great Lakes drainage systems and the Lake Missoula floods in North America and those involved in the complex histories of the European rivers draining to the North Sea. The result, as far as Severn and Wye are concerned, has been a coming-together of subcatchments with contrasted relief, lithology and morphological inheritance. There are equally strong contrasts between the present climate and hydrology of the Palaeozoic uplands and the lowlands of Shropshire and of the lower Severn and Avon basins (Chapters 4 and 5).

Data available for palaeohydrological reconstruction, and reported or used in this volume, are shown in Figure 16.1 and Table 16.1. Sites have been broadly classified into six groups according to the main purpose for which they were examined. Appendix I gives the references for the work undertaken at each of the sites. Rainfall and runoff gauging sites are not shown in Figure 16.1, because they were considered systematically in Chapters 4–7 earlier on.

These contemporary data provide the basis of present-day river management and allow secular climatic and runoff trends to be identified. It is most important to appreciate, both in management and in palaeo-reconstructions, that even a time period of several decades can incorporate fluctuations which individually do not betoken a 'change' from some static and stable average. Fluctuations are to be expected in the present environment. Although data are limited, Chapter 7 suggests broadly that climatic factors, rather than land use change or river management practice, are probably the more significant in determining flood hydrology fluctuations at present. However, changes in land (Figure 7.4) and river (Table 7.2) management can also be responsible for hydrological change. Because the much longer timespan of our study period incorporates deforestation and the onset of agriculture (as discussed in Chapter 12), they must certainly have done so.

Concerning the data, it has to be appreciated that despite the best efforts of a number of researchers, spatial and temporal coverage is uneven and limited. For example, the Plynlimon catchments of the Institute of Hydrology have provided invaluable information on upland hydrology, but only since 1968, with no other data of comparable quality elsewhere in our study area. The longest flow record used by Higgs in Chapter 7 was for Bewdley and covered 60 years; other data he was able to use were for 40 years duration or less. Although such data are as good as can be expected in the UK, the scope for spatial or temporal analysis is nonetheless limited.

Information on sediment transport and channel change, derived from field survey and from historical maps, is also not as comprehensive as one would like. Thus the data for catchments in the Wye system discussed in Chapter 9 (See Figure 13.1 and Table 9.3) are broadly in line with available data from the United Kingdom (Newson, 1986, Figure 1). But there is no routine

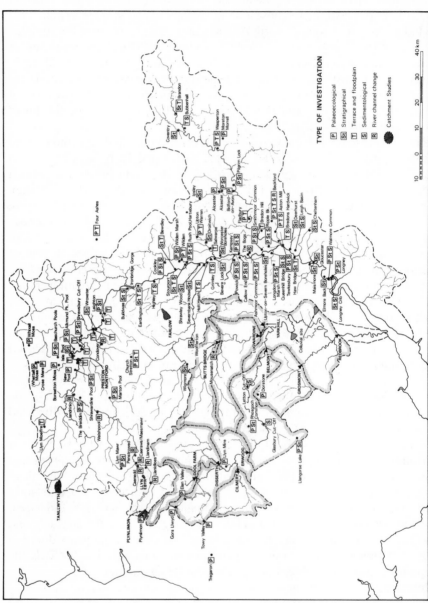

FIGURE 16.1  Data base for Severn and Wye Basins. The classification is explained in the text and the sources

TABLE 16.1   References for sites in Figure 16.1

| | |
|---|---|
| Alcester | Osborne (1971) (ref S. Limbrey, 1980) |
| Alkmund Park Pool | work in progress (Barber and Twigger, 1987) |
| Ashmoor Common | Brown (1981) 158; Brown (1983a) |
| Aston Mill | work in progress (Limbrey, 1980; Dawson 1986a,b) |
| Baschurch Pools | Twigger, Ph.D. in prep. |
| Beckford | work in progress (Limbrey, 1980) and Brown (1983b) Dawson (1986a) |
| Belmont | Mitchell (1983) |
| Bevere Lock | Brown (1983a) |
| Bewdley | Burrin (1980) cited: 158, Dawson and Gardiner (1987) |
| Bidford-on-Avon | Shotton (1978) (Limbrey, 1980) also work in progress |
| Bogs | Brown (1983a) |
| Boreatton Moss | work in prep., Twigger |
| Brandon | Shotton (1968) |
| Bredon Hill | Gerrard (1980) (hillslope failure) and Burrin (1980) (borehole) |
| Bredons Hardwick | Dawson (1986a,b) |
| Breidden | work in progress Limbrey (1980) |
| Bubbenhall | Dawson (1986a) |
| Buildwas | Gerrad (1980) Shaw (1986) Gorley (1970) |
| Butts Bridge | Mitchell (1983) |
| Caenedd | Lewin in Brown (1983b) |
| Caersws | Lewin (1980) Thorne and Lewin (1979) also this volume |
| Callow End | Brown (1982) Brown (1983a) |
| Cheltenham | Brown (1983a) |
| Church Stretton | Osborne (1972), Rowlands and Shotton (1971) (cited: Brown, 1983a) |
| Cilmery | Mitchell (1983) |
| Condover | Hughes (1977) |
| Cookley | Work in progress (Limbrey, 1980) |
| Coventry | Shotton (1965) |
| Crose Mere | Beales (1980) (ref Barber and Twigger, 1987) |
| Ddol Farm | Mitchell (1983) |
| Deerhurst | Beckinsale and Richardson (1964) (cited Brown, 1983a |
| Disserth | Mitchell (1983) |
| Dorstone | work in progress (cited: Limbrey, 1980) |
| Droitwich | Brown (1983b) |
| Eardington | Coope and Shotton (1981) Dawson and Gardiner (1987) |
| Elan Valley | Moore and Chater (1969) (cited: Limbrey, 1980) (cited Brown, 1983c) |
| Elmore Back | Brown (1983a) |
| Erwood | Mitchell (1983) in Barber and Gregory (1983) |
| Farlow Basin | Mitchell (1980) |
| Fladbury | Coope and Shotton (1981) |

TABLE 16.1—*continued*

| | |
|---|---|
| Glasbury | Mitchell and Gerrard this volume |
| Gloucester | Burrin (1980), Brown (1983a) |
| Gors Llywd | Brown (1983b) |
| Grimley | Coope and Shotton (1981) Dawson (1986a) Dawson and Gardiner (1987) this volume |
| Grosmont | Mitchell (1983) |
| Hartlebury Common | Brown (1983a) |
| Haw Bridge | Burrin (1980) also cited Brown (1983a) |
| Highley | Dawson and Gardiner (1987) Dawson (1986) |
| Holt Heath | Dawson and Gardiner (1987) this volume, Dawson (1986) Dawson and Bryant (1987) |
| Ipsley | Burrin (1980) |
| Ironbridge | Coope and Shotton (1981) |
| Isle Pool | work in progress (Barber and Twigger) |
| Kempsey | Brown (1983a) |
| Leigh Basin | Brown (1983a) |
| Leighton | Brown (1983b) |
| Letton | Mitchell and Gerrard this volume |
| Llandrinio | Lewin in Brown (1983b) and this volume |
| Llandinam | Lewin (1987) this volume |
| Llangorse Lake | Chambers and Jones (1984) (cited: Limbrey 1987) |
| Llanidloes | Lewin (in Brown, 1983b) |
| Llyn Ebre | Oxley, N. C. (1973) (cited: Mitchell, 1980) |
| Llyn Mawr | Rowlands and Shotton (1971) (cited: Brown 1983a) |
| Llyn Mire | Moore and Becket (1971) also work in progress (cited: Limbrey, 1980) |
| Llyn Moelfre | Gerrard (1980) |
| Longdon Marsh | Brown (1983a) |
| Longney | Brown (1983c) |
| Longney Crib | Brown (1983a) |
| Lugwardine | Mitchell (1983) in Barber and Gregory (1983) |
| Maesmawr | Thorne and Lewin (1979) |
| Maismore | Burrin (1980) |
| Marton Pool | Rowlands and Shotton (1971) and work in progress (Limbrey 1980) |
| Moreton Morrell | Shotton (1967) (cited: Barber and Twigger, 1987) |
| Mousenatch | Mitchell (1983) in Barber and Gregory (1983) |
| New Pool | work in prep. Twigger |
| Penstrowed | Lewin (1987), this volume |
| Pilgrim Lock | Burrin (1980), Shotton (1978) |
| Plynlimon | Newson (1979), Moore (1968) (cited: Brown, 1983c) |
| Powick | Brown (1983a) Brown (1985) |
| Preston Montford | Mitchell and Gerrard 1987, this volume |
| Queenhill Bridge | Burrin (1980) |
| Redbrook | Mitchell (1983) in Barber and Gregory (1983) |
| Ripple Brook | Brown (1983a) and Brown and Barber (1985) |
| Rhosgoch Common | Bartley (1960) (Barber and Twigger, 1987) |

TABLE 16.1—*continued*

| | |
|---|---|
| Rush Pool—Hartlebry | Brown (1984) |
| Salwarpe | Coope and Shotton (1981) |
| Severnbank House | Brown (1983a) |
| Shawardine Pool | work in progress (Barber and Twigger, 1987) |
| Shrawley Pool | Brown (1983a) |
| Shrewsbury Cutoff | Pannett and Morey (1976) (cited: Barber and Twigger 1987) |
| Smithmoor Common | Brown (1983a) and Brown and Barber (1985) |
| Stourport | (Roger Constant's Pit) Coope and Shotton (1981) Dawson and Gardiner 1987, this volume, Dawson (1986) |
| Tanllwyth | Mitchell and Gerrard (1987) this volume |
| Tewkesbury | Brown (1983a) |
| Towy Valley | Brown (1983a) |
| Upton-on-Severn | Brown (1983b) |
| Upton Warren | Coope and Shotton (1981) |
| Walmore Common | Brown (1983a) |
| Wasperton | Shotton (1977), Dawson (1986) |
| Welshpool | Lewin (1987) this volume |
| Whattall Moss | Brown (1983c) |
| Whixall Moss | Turner (1964) and (1965) and work in progress (Limbrey, 1980) and Brown (1983) |
| Wigmore | Eyles (1973) (cited: Brown, 1987, this volume) |
| Wilden | Coope and Shotton (1981) |
| Wilden Marsh | Brown (1983b) |
| Wooferton | Brown (1987) this volume |
| Worcester | Burrin (1980) and Shotton (1978) |
| Wroxeter | Brown (1983b) |
| Yarkhill | Mitchell (1983) in Barber and Gregory (1983) |

representative monitoring or central compilation of sediment yield data for Britain, and the possible variety of yields within the whole of the Severn–Wye Basin remains uncertain.

Channel change patterns and rates have been monitored mainly in the Upper Severn and Wye (Chapter 8 and Figure 16.1) where they are known to be active. It is clear that here change rates can be considerable, and probably have eliminated the earlier Holocene floodplain sediments from mid-course floodplain sites. Study of erosion and sedimentation in lowland catchments has perforce to adopt a longer timescale, examining the sediments stored on floodplains in lowland basins (Chapter 4). Here a combination of erodible soils and arable cultivation has been associated with enhanced deposition rates, while artificial control of lowland channel courses may have prevented river reworking of such deposits that might have occurred. It is unfortunate that river change (Chapter 8), sediment yield (Chapter 9) and sediment storage (Chapter 14) data necessarily have to pertain to different timescales.

The data base for prehistoric hydrology is necessarily less detailed than for historic timescales. However, as shown in Chapter 10 by Coope and Barber, it is now possible to identify in considerable detail climatic variations for the Lateglacial and also for the period since 10,000 BP. However, it is necessary to be able to derive transfer functions to link details of climate to other environmental parameters but such linkages are still not easy to establish in view of the considerable spatial diversity that exists. Although the data available on the vegetation history have increased in recent years, Barber and Twigger show in Chapter 11 that when set against the somewhat diverse background of the Severn Basin, there is still a significant gap between the information derived from the Welsh upland sites and that material based upon the Crose Mere–Shrewsbury–Church Stretton area. Despite this spatial gap in the data base, they are able to distinguish the Lateglacial dominated by grass and sedge with associated instability in the fluvial system, from the early Holocene from 10,000 to 5000 which was forest-dominated and which had low sedimentary input to the fluvial system. The early Holocene could then be differentiated from the late Holocene which experienced forest clearance and particularly was characterized by linkages between fluvial processes and the way in which vegetation clearance took place from specific sites.

In Chapter 12, Limbrey shows how it is of continuing significance to be able to elucidate changes in environment as instigated by human activity. However, such human activity is not always unambiguously recorded and it is particularly important to know the significance of domestic animals and of crop plants. Such knowledge then has to be associated with the way in which farmers extended into the uplands and then subsequently retreated. This underlines the need to visualize the spatial data base together with temporal change superimposed like a palimpset upon it. A further dimension is emphasized in Chapter 13 by Gardiner and Dawson where they argue that it is necessary to see the problem of river terraces against the background of the intellectual attitude to a particular problem. They are able to review the context of terrace development in the Severn Basin against developments in terrace understanding and proceed to derive estimates of hydrologic and hydraulic conditions at times of terrace sedimentation. This clearly indicates how effective interpretation of sedimentary sequences can be; where calculations of palaeodischarges have been made these are suggested to approximate to former mean annual flood flows.

Innovation is also a feature of the analysis reported by Brown in Chapter 14 where he scrutinizes long-term sediment budgets. This should really be one of the central objectives of palaeohydrology because it is critical to our understanding of the detail of past processes. It is important to remember that it is not simply a matter of allochthonous development. Indeed it is possible to have storage in the middle and lower reaches of basins which are subject to different residence times. This therefore leads Brown to advocate 'punctuated equilibrium' because as much sediment went into storage as left the Severn and Wye Basins in the Holocene.

Although the above commentary tends to emphasize limitations, the summary of its extent in Figure 16.1 indicates that the data base is really quite substantial. Although the types of data have been summarized for convenience under 5 categories, the majority of information is classified as palaeoecological or stratigraphic with less information pertaining to terrace and floodplain form sequences, to sedimentological sequences and to river channel change. However, some sites have several categories present and occasionally all types of information are derived from one site. Although it was originally hoped (Starkel and Thornes, 1981) that it would be possible to dictate the spatial distribution of data collection when attempting palaeohydrological analysis, what has happened is that the availability of sites very often decrees exactly how the data base has evolved. One dimension which has not been explored fully as yet in relation to the Severn Basin is to establish the way in which the spatially varied terrestrial evidence relates to the estuarine record which has been the subject of recent analysis from the Severn Estuary and from the Bristol Channel.

## THE PATTERN OF PALAEOHYDROLOGY

It is desirable to look at ways in which the data base collected from systematic sources in the preceding chapters relates to provide a chronological picture. In Chapter 15 Starkel summarized the changes that have occurred in continental Europe and proposed that from 13,000 to 8000 ka BP the cold stage was replaced by a warm one. He identified four major types of river system and the first two are pertinent to the Severn Basin. Because the area north of the Ironbridge gorge was originally covered by glacial ice and because the area to the south was periglacial in character so the basin includes the first two of the four river systems recognized by Starkel.

However, it is usual to think of the basin as a whole and so a summary diagram has been compiled to collate results from preceding chapters. This diagram (Figure 16.2) attempts to juxtapose data on climate, land cover and the fluvial system in relation to the timescale of the last 15,000 years. Because of the variations which have occurred throughout the basin, the vegetation pattern has been separated into upland and lowland areas. Wherever possible each of the trends shown in the summary diagram has been drawn to reflect the sequence reconstructed from the Severn Basin. The extent to which correlation exists between the patterns is real rather than a manifestation of forced correlation. Furthermore there is an attempt in the case of precipitation, temperature and hydrology to demonstrate past conditions compared

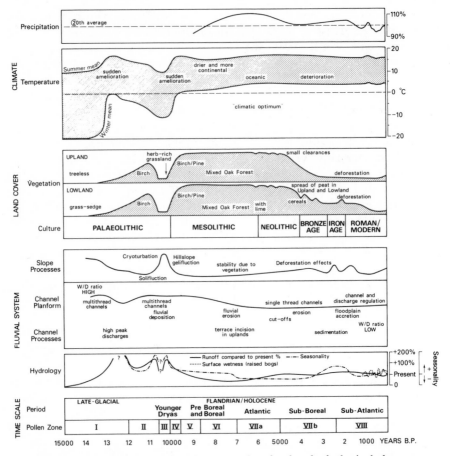

FIGURE 16.2   Sequence of environmental and palaeohydrological change

with present. Particularly in the case of hydrology, runoff in the past is contrasted in percentage terms with present runoff. Surface wetness as indicated by raised bogs and the seasonality of the hydrological regime are also shown.

Although it is not easy to separate phases of change in the patterns that are shown, there may be some justification for identifying five major phases and these are tentatively suggested in Table 16.2. The sequence of events summarized in the table demonstrates that the last 15,000 years may be more appropriate as a more meaningful unit of time in continental Europe than it is in the Severn Basin.

The table of generalized events (Table 16.2) summarizes the change from glacial conditions to late glacial when, with ameliorating temperatures despite cold winters, grass sedge vegetation was gradually succeeded by birch wood-

TABLE 16.2  Sequence of palaeohydrological changes in the Severn Basin

| Approximate years | Phase | Precipitation | Temperature | Vegetation | Hydrology | Channel processes |
|---|---|---|---|---|---|---|
| Pre-15,000 | Glacial | Snow | Low | None | Summer floods | Multithread channels and fluvioglacial systems |
| 15,000–11,000 | Earlier Late Glacial | | Ameliorating cold winters | Grass sedge succeeded by birch woodland | High peak discharges, decreasing with forest spread | Multithread channels |
| 11–10,000 | Lateglacial Zone III | | Extreme cold | Herb-rich grassland | Runoff reduced but sediment supply greatly increased | Fluvial deposition, instability in fluvial system |
| 10,000–4000 | Flandrian/Holocene | Drier then Oceanic | Rising | Mixed oak forest some clearances | Runoff reduced, sometimes lower than present | Fluvial erosion, single-thread meandering channels |
| 4000–Present | Flandrian/Holocene | | Deterioration then fluctuating | Deforestation | Seasonality fluctuating discharge regulated | Lowlands cutoffs Floodplain accretion Sedimentation |

land and high peak discharges were a feature which gradually decreased with the increasing spread of forest. Under these conditions multithread river channels were dominant. Subsequently in Zone 3 with extreme cold and a decrease in forest cover and a return to herb-rich grassland, runoff was reduced but sediment supply was greatly enhanced because of increased hillslope gelifluction. This led to fluvial deposition in some areas with considerable instability in the fluvial system elsewhere. The final two stages in the Flandrian/Holocene could be separated at either 4000 or 5000 years BP. In the first phase precipitation was above the present average conditions, at first drier but then more oceanic and there was complete spread of mixed oak forest although some clearances featured towards the end of the phase. Runoff was reduced, reflecting the extensive vegetation cover, and was sometimes lower than that of the present time. Fluvial erosion was dominant and single-thread meandering channels were instated. Subsequently in the final phase when climatic conditions were fluctuating about the twentieth-century average, deforestation gave rise to a fluctuation in discharge and to considerable spatial variations in fluvial processes. It has been estimated that in the early part of the Flandrian/Holocene erosion estimates might be of the order of 20t km$^{-2}$ yr$^{-1}$ under a high forest cover, whereas yields could be 6 or 7 times greater than this once agriculture begun to have a significance impact on floodplain areas (Brown and Barber, 1985). Also featuring in lowland fluvial systems during this final phase were a number of cutoffs which were associated with high peak discharges, flood plain accretion and significant amounts of sedimentation.

It is not easy to extend the summary diagram (Figure 16.2) to provide quantitative estimates of palaeodischarges. The dimensions of palaeochannels and sediment size data have been used elsewhere (Church, 1978; Costa, 1983; Maizels, 1983; Rotnicki, 1983) to provide palaeodischarge estimates but it has not been possible to utilize this approach extensively in this volume. This is partly because extensive systems of palaeomeanders or of well-defined former channels, as found for example in Poland, do not occur in the Severn Basin, possibly because the Severn Basin effectively represents the upper portion of a European basin and does not include the middle and lower reaches. Furthermore the range of sediment sizes available is constrained by the thinly bedded nature of the Palaeozoic bedrock in many basins, and rivers have proved capable of transporting this bed material over a whole range of probable discharges down to the present day. However, some suggestions are possible, using available methods. There are at least three ways in which palaeohydrologic estimates have been obtained. First has been the use of generalized basin equations using catchment parameters such as the drainage network. Thus Jones (1982) employed measures of drainage density in the Severn Basin and estimated that in the Lateglacial (Table 16.2) discharges could have been approaching 4 times mean annual flood discharges of today. A second way in which palaeohydrologic estimates can be derived is based upon abandoned river channel sections. This approach was used by Brown (1983c) when he estimated from the dimensions of a palaeomeander at Callow End that bankfull discharge would have been some 30 times greater than present average discharge. A similar approach has been employed for a number of years in research pioneered by G. H. Dury and in his analysis of the channels of the Severn Basin he has utilized techniques for the treatment of osage underfitness (Dury, 1985), which allowed him to indicate discharges of the order of 5 times present ones.

A third way in which discharge can be estimated for palaeohydrologic situations is by interpretation of fossil sediments. This approach is well exemplified by the work of Dawson (Chapter 13 this volume and Dawson, 1986a) where he has very carefully derived estimates of discharges possible at the time of the Worcester Terrace and has compared these with present mean annual flood values. He concluded that at the time of the Worcester terrace deposition, based upon hydraulic reconstruction discharges were between 1.4 and 5.5 and possibly up to 11.5 times the present mean annual flood. After very careful scrutiny of the palaeohydraulic estimates Dawson (1986a) concluded that the discharges along the Severn and along the Avon at the time of the Worcester Terrace could have been up to 11 or possibly 20 times the present but that subsequently after 9000 BP in the Flandrian they were nearer 3 times present discharges. Such magnitudes are comparatively slight when it is borne in mind that these comparisons are being made with mean annual flood discharges, themselves values which are exceeded quite frequently. It also has to be borne in mind that palaeohydrologic estimates should have an error margin on them to compare with the range of flows that characterizes the present discharge record.

## APPLICATIONS

This volume has provided long-term perspectives on the Severn–Wye Basin over long timescales that do nevertheless have some practical implications. In the 'historical' past, for 100 years or more, hydrological change has prevailed in two senses. The first is that river courses have altered and discharges have varied as part of the normal course of events. Though the magnitude of rare events and the fact that river channels erode across their flood plains may exceed human expectations, they are by no means without precedent and historical hydrology would indeed give us a better idea of what to anticipate.

Changes also occur, secondly, as a response to environmental change itself. This may be brought about through climatic fluctuations, through land use change and through such phenomena as river channelization and river impoundment. On the whole, changes within the last 100 years as a result of such environmental modifications have not been dramatic but it is possible to detect them in the hydrological record and to compare the degrees of change that arise from different causes. This could be useful for the evalu-ation of engineering works.

In the longer term hydrological changes resulting from wholesale land use change or from climatic change are shown to have been much greater. At the present time, when there is a re-evaluation in the UK of land resources, in particular involving considerable potential shifts in the amount of land use for agriculture, amenity and forestry, these more dramatic changes for which there is no recorded precedent may require information from palaeohy-drology to outline the possible dimensions of the effects which might be anticipated. Again, climatic changes which are now expected in the early years of the twenty-first century as a result of the greenhouse effect and increasing carbon dioxide concentrations could mean that we seek an analogue which is beyond historical hydrology. It really is desirable to evaluate the potential effects of, say, a sustained increase in storm precipi-tation as far as hydrological management is concerned.

We need also to appreciate that floodplains and wetlands in particular are sensitive environments which contain a record of hydrological change and fluctuation which is beyond our present experience but not beyond possible expectations in the not too distant future. Just as Butzer (1980) used the world distribution of climate in the Atlantic to indicate a possible analogue for a future warmer $CO_2$ world, similarly it is possible to utilize data from Figure 16.2 and from the chapters in this book to indicate the possible ways in which environmental responses may occur as further environmental changes take place.

To optimize the use of such palaeohydrological information in relation to interpretations for the future it is necessary to establish clearer transfer functions between the environmental components in the palaeohydrological system. Once models have been further developed it is possible that such models, perhaps utilizing energy transfer, may be the basis for interpretation of the future in the light of the greater knowledge of the past. Perhaps an apt way of summarizing the situation are the words used by A. G. J. MacFar-lane (1986) in the final discussion of a joint symposium of the Royal Society and the British Academy *Predictability in Science and Society*, when he said:

> we live in a small bubble of order, separated from both past and future by entropy barriers. Our local order is largely imposed by using observations which create useful information. The reason we can't reverse time is essentially the same as the reason we can't predict arbitrarily far into the future; both require unbounded amounts of information.

## REFERENCES

Andrews, E. D. (1983). Entrainment of gravel from naturally sorted river bed material, *Geological Society of America Bulletin*, **95**, 371–8.

Baker. V. R. (1973). *Palaeohydrology and sedimentology of Lake Missoula flooding in eastern Washington*, Geological Society of America Special Paper 89.

Baker, V. R. (1979). Erosional processes in channelized water flows on Mars, *Journal of Geophysical Research*, **84**, 7985–93.

Baker, V. R. (1983). Large-scale fluvial palaeohydrology, in Gregory, K. J. (ed.), *Background to Palaeohydrology*, Wiley, Chichester, pp. 453–78.

Bartley, D. D. (1960). Rhosgoch Common, Radnorshire: Stratigraphy and pollen analysis, *New Phytol.*, **59**, 238–62.

Beales, P. W. (1980). Late Devensian and Flandrian vegetational history of Crose Mere, Shropshire, *New Phytol.*, **85**, 133–61.

Beckinsale, R. P. and Richardson, L. (1964). Recent findings on the physical development of the Lower Severn Valley, *Geographical Journal*, **130**, 87–105.

Bridge, J. S. and Jarvis, J. (1976). Flow and sedimentary processes in the meandering River South Esk, Glen Cova, Scotland, *Earth Surface Processes*, **1**, 303–36.

Brown, A. G. (1981). Palaeohydrology of the temperate zone: Subproject A, Fluvial Environments Severn Basin. Summary reports and maps, May 1981, IGCP, Project 158.

Brown, A. G. (1982). Human impact on former floodplain woodlands of the Severn, in Bell, M. and Limbrey, S. (eds), *Archaeological Aspects of Woodland Ecology*, British Archaeological Reports, International Series 146, pp. 93–105.

Brown, A. G. (1983a). Late Quaternary palaeohydrology, palaeoecology and floodplain development of the Lower River Severn, unpublished Ph.D. thesis, University of Southampton.

Brown, A. G. (ed.) (1983b). INQUA Eurosiberian Subcommission for the study of the Holocene, IGCP Project 158, The Severn 1983 Excursion Guide.

Brown, A. G. (1983c). Floodplain deposits and accelerated sedimentation in the lower Severn basin, in Gregory, K. J. (ed.), *Background to Palaeohydrology*, Wiley, Chichester.

Brown, A. G. (1984). The Flandrian vegetational history of Hartlebury Common, Worcestershire, *Proc. Birmingham Natural History Soc.*, **25**, 89–98.

Brown, A. G. (1985). Traditional and multivariate techniques in the interpretation of floodplain sediment grain size variations, *Earth Surface Processes and Landforms*, **10**, 281–91.

Brown, A. G. and Barber, K. E. (1985). Holocene palaeoecology and sedimentary history of a small lowland catchment in Central England, *Quaternary Research*, **24**, 87–102.

Burrin, P. J. (1980). A review of valley fill sediments and floor morphology in the Severn valley, in *Palaeohydrology of the temperate zone*: Previous research and research objectives, May 1981, IGCP Project 158.

Butzer, K. W. (1980). Adapation to global environmental change, *Professional Geographer*, **32**, 269–78.

Church, M. (1978). Palaeohydrological reconstruction from a Holocene valley fill, in Miall, A. D. (ed.), *Fluvial Sedimentology*, Canadian Society of Petroleum Geologists Memoir, 5, 743–72.

Church, M. and Ryder, J. M. (1972). Paraglacial sedimentation: a consideration of fluvial processes conditioned by glaciation, *Bulletin of the Geological Society of America*, **83**, 3059–3072.

Collinson, J. D. and Lewin, J. (eds) (1983) *Modern and Ancient Fluvial Systems*, International Association of Sedimentologists Special Publ. 6, Blackwell, Oxford, 575 pp.

Coope, G. R. and Shotton, F. W. (1981). *Palaeohydrology of the temperate zone*: Subproject A, Fluvial Environments Severn Basin. Summary diagrams and maps, May 1981, International Geological Correlation Programme, Project 158.

Costa, J. E. (1983). Palaeohydraulic reconstruction of flash flood peaks from boulder deposits in the Colorado Front Range, *Bulletin of the Geological Society of America*, **94**, 986–10004.

Cullingford, R. A., Davidson, D. A. and Lewin, J. (eds) (1980). *Timescales in Geomorphology*, Wiley, Chichester, 360 pp.

Dawson, M. R. (1985). Environmental reconstructions of a Late Devensian terrace sequence: some preliminary findings, *Earth Surface Processes and Landforms*, **10**, 237–46.

Dawson, M. R. (1986a). Late Devensian fluvial environments of the Lower Severn basin, unpublished Ph.D. thesis, University of Leicester.

Dawson, M. R. (1986b). Sedimentological aspects of periglacial terrace aggradations, in Boardman, J. (ed.), *Periglacial Environments and Processes in Britain and Ireland*, Cambridge University Press.

Dawson, M. R. and Bryant, I. D. (1987). The three-dimensional geometry of litho-facies associations in Pleistocene terrace gravels, Holt Heath, Worcestershire, U.K., in Ethridge, F. G. (ed.), *Advances in Fluvial Sedimentology*, Procs. Third Int. Conf. on Fluvial Sedimentology, Fort Collins, 1985 SEPM, Special Pub. 39.

Dawson, M. R. and Gardiner, V. (1987). River terraces: a general model and a palaeohydrological and sedimentological examination of the Late Devensian terraces of the Lower Severn Basin, in Gregory, K. J., Lewin, J. and Thornes, J. B, *Palaeohydrology in Practice*, Wiley, Chichester.

Dietrich, W. E., Smith, J. D. and Dunne, T. R. (1979). Flow and sediment transport in a sand-bedded meander, *Journal of Geology*, **87**, 305–15.

Dury, G. H. (1985). Attainable standards of accuracy in the retrodiction of palaeodis-charge from channel dimensions, *Earth Surface Processes and Landforms*, **10**, 205–14.

Gerrard, J. (1980). Large-scale forms of hillslope failure in the Severn and Wye Basins, in *Palaeohydrology of the temperate zone*: Previous research and research objectives, IGCP 158, May 1980.

Gorley, R. S. (1970). Geomorphology of the Country around Ironbridge with special reference to the Origin of the Ironbridge Gorge, unpublished Ph.D. thesis, University of Birmingham.

Graf, W. L. (1979). Mining and channel response, *Annals Association of American Geographers*, **69**, 262–75.

Graf, W. L. (1983). The arroyo problem—palaeohydrology and palaeohydraulics in the short term, in Gregory K. J. (ed), *Background to Palaeohydrology*, Wiley, Chichester, pp. 279–302.

Gregory, K. J. (1987). *The Nature of Physical Geography*, Arnold, London.

Gregory, K. J. (ed.) (1985). *Energetics of Physical Environment*, Wiley, Chichester.

Hughes, D. J. (1977). Rates of erosion on meander arcs, in Gregory, K. J. (ed.), *River Channel Changes*, Wiley, Chichester, pp. 193–205.

Jones, M. D. (1982). The palaeogeography and palaeohydrology of the River Severn. Shropshire, during the late Devensian Glacial Stage and the early Holocene, unpublished M. Phil thesis, University of Reading.

Limbrey, S. (1980). Prehistoric land-use changes, in *Palaeohydrology of the temperate zone*, Fluvial Environments Severn Basin. Previous research and research objectives, IGCP 158, May 1980.

Limbrey. S. (1983). Archaeology and palaeohydrology, in Gregory, K. J. (ed.), *Background to Palaeohydrology*, Wiley, Chichester.

MacFarlane, A. G. J. (1986). In: Predictability in science and society, *Proceedings Royal Society London*, **A407**, 143–5.

Maizels, J. K. (1983). Palaeovelocity and palaeodischarge determination for coarse gravel deposits, in Gregory, K. J. (ed.), *Background to Palaeohydrology*, Wiley, Chichester, pp. 101–39.

Mitchell, D. J. (1979). Aspects of the hydrology and geomorphology of the Farlow Basin, Shropshire, unpublished M.Sc. thesis, University of Birmingham.

Mitchell, D. J. (1980). Valley sediments and changing sediment yields, in *Palaeohydrology of the temperate zone*, Fluvial Environments Severn Basin. Previous research and research objectives, IGCP 158, May 1980.

Mitchell, D. J. (1983). The use of contemporary suspended sediment yields to estimate the erosion of the Wye catchment in the last 8000 years, in Barber, K. E. and Gregory, K. J., INQUA Eurosiberian Subcommission for the Study of the Holocene, IGCP Project 158. *Palaeohydrology of the temperate zone in the last 15,000 years*, Severn 1983.

Moore, P. D. and Chater, E. H. (1969). The changing vegetation of west-central Wales in the light of human history, *J. Ecol.*, **57**, 361–79.

Newson, M. D. (1986). River basin engineering—fluvial geomorphology, *Journal of the Institution of Water Engineers and Scientists*, **40**, 307–24.

Osborne, P. J. (1971). An insect fauna from the Roman site at Alcester, Warwickshire, *Britannia*, **2**, 156–65.

Osborne, P. J. (1972). Insect faunas of late Devensian and Flandrian age from Church Stretton, Shropshire, *Phil. Trans. R. Soc. Lond.*, **263**, 327–67.

Oxley, N. C. (1974). Suspended sediment delivery rates and the solute concentration of stream discharge in two small catchments, in Gregory, K. J. and Walling, D. E. (eds), *Fluvial Processes in Instrumented Watersheds*, Inst. Brit. Geog. Special Pub. 6, 73–85.

Pannet, D. J. and Morey, C. (1976). The origin of the old river bed at Shrewsbury, *Bulletin of the Shropshire Conservation Trust*, **35**, 7–12.

Reid, I., Frostick, L. E. and Layman, J. T. (1985). The incidence and nature of bedload transport during flood flows in coarse-grained alluvial channels, *Earth Surface Processes and Landforms*, **10**, 33–44.

Rotnicki, K. (1983). Modelling past discharges of meandering rivers, in Gregory, K. J. (ed.), *Background to Palaeohydrology*, Wiley, Chichester, pp. 321–354.

Rowlands, P. H. and Shotton, F. W. (1971). Environmental change and palaeobotany: Pleistocene deposits of Church Stretton, Shropshire and its neighbourhood, *J. Geol. Soc.*, **127**, 599–622.

Ryder, J. M. and Church, M. (1986). The Lillooet terraces of Fraser River: a palaeoenvironmental enquiry, *Canadian Journal of Earth Sciences*, **23**, 869–84.

Schumm, S. A. (1968). Speculations concerning palaeohydrologic controls of terrestrial sedimentation, *Bulletin of the Geological Society of America*, **79**, 1573–88.

Shaw, J. (1969). Aspects of glacigenic sedimentation, with special reference to the area around Shrewsbury, unpublished Ph.D. thesis, University of Reading.

Shaw, J. (1972). The Irish Sea Glaciation of North Shropshire, some environmental reconstructions, *Field Studies*, **3**, 603–31.

Shotton, F. W. (1965). Borings along the line of the Sowe Valley Sewer and their relation to the alluvium and terrace gravels of river, *Proc. Cov. Nat. Hist. & Sci. Soc.*, **3**, 228–35.

Shotton, F. W. (1967). Investigation of an old peat moor at Moreton Morrell, Warwickshire, *Proceedings of the Coventry and District Natural History and Scientific Society*, **4**, 13–16.

Shotton, F. W. (1968). The Pleistocene succession around Brandon, Warwickshire. *Phil. Trans. Roy. Soc. Lond., Ser B*, **254**, 387–400.

Shotton, F. W. (1977). The English Midlands, INQUA Guidebook A2, Tenth Congress, Birmingham.

Shotton, F. W. (1978). Archaeological inferences from the study of alluvium in the Lower Severn–Avon valleys, in Limbrey, S. and Evans, I. G. (eds), *Man's Effect on The Landscape: The Lowland Zone*, Council for British Archaeology Research Report 21, 27–32.

Starkel, L. and Thornes, J. B. (1981). *Palaeohydrology of river basins*, British Geomorphological Research Group Technical Bulletin No. 28, 107 pp.

Thorne, C. R. and Lewin, J. (1979). Bank processes, bed material movement and planform development in a meandering river, in Rhodes, D. D. and Williams, G. P. (eds), *Adjustments of the Fluvial System*, Kendall-Hunt, Dubuque, Iowa, pp. 117–37.

Trimble, S. W. (1983). A sediment budget for Coon Creek basin in the driftless area. Wisconsin, 1853–1977, *American Journal of Science*, **283**, 454–74.

Walling, D. E. and Webb, B. W. (1983). Patterns of sediment yield, in Gregory K. J. (ed.), *Background to Palaeohydrology*, Wiley Chichester, pp. 69–100.

Williams, G. P. (1983). Palaeohydrological methods and some examples from Swedish fluvial environments. I. Cobble and boulder deposits, *Geografika Annaler*, **65A**, **227–43**.

# Location Index

References to Severn and Wye Basins are not given as these occur throughout the text. Material in illustrations is indicated by page references in italics.

Abercannon Farm 154, *154*
Abergwnyu *103*
Abermule *40*, 79, *85*, *135*, 137, 139
Afon Cyff 181, 188, 192, *193*
Alberbury 180
Alcester *346*, *347*
Alkmund Park Pool 222, *346*, *347*
Andernach 213
Apley Park 278
Arrow River 325
Ashleworth 82, 92
Ashmoor Common 28, 229, 232, 242, *326*, *328*, *346*, *347*
Aston Mill *346*, *347*
Avon River 11, 39, *40*, 41, 57, 63, 67, *315*
Axe Valley 6

Baggy Moor 229
Bagmere 224
Balcombe *111*
Bank Cynon Isaf *114*, *135*, 147, 148, 149, *149*, *151*, *152*, *153*, 154, *154*, 155
Baschurch Pools 217, 223, 232, *233*, 234, 235, 238, 241, *346*, *347*
Batheaston *103*
Beacons Reservoir *110*
Beckford 6, *346*, *347*
Belmont *183*, *184*, *185*, *346*, *347*
Berth Pool 228, 234, 235, *236*, 239, 240, 241, 242
Bewdley *40*, 79, 81, 83, *86*, 88, 134, 137, *139*, *140*, *141*, *142*, 153, *154*, 155, 277, 278, 297, 345, *346*, *347*

Bidford on Avon *346*, *347*
Birchgrove Pool 234, 235, *237*, 238, 239, 240, 241
Birmingham 51, 108, *109*
Black Mountains 257
Blelham Bog 224, 225
Bobbitshole *45*
Bodmin Moor 204
Bogs *346*, *347*
Bolton Fell Moss *213*
Boreatton Moss 235, 238, *346*, *347*
Borth Bog 232
Brandon *346*, *347*
Brecon Beacons 263
Bredon 325, *346*, *347*
Bredons Hardwick *346*, *347*
Breiddin 258, *346*, *347*
Bridgenorth 39, 278, 279
Brithdir *135*, 147, 149, 153, *154*
Bromfield 258, 262, 264
Brook 5
Bubbenhall *346*, *347*
Buildwas 44, 180, *346*, *347*
Bushley Green 277
Bushley Brook Valley 320
Butts Bridge *183*, *184*, *185*, *346*, *347*

Cadora 37
Caenedd *346*, *347*
Caersws *135*, 137, *138*, 142, *142*, *143*, 144, 146, 156, 179, 181, *346*, *347*
Callow End 222, 228, 229, *326*, *346*, *347*
Camp Century 208
Canca River, Columbia 276

361

# Subject Index

Major items are indexed with illustrations in italics.

365

r